Springer
Proceedings in Physics 25

Springer Proceedings in Physics

Managing Editor: H. K. V. Lotsch

Volume 1 *Fluctuations and Sensitivity in Nonequilibrium Systems* Editors: W. Horsthemke and D. K. Kondepudi

Volume 2 *EXAFS and Near Edge Structure III* Editors: K. O. Hodgson, B. Hedman, and J. E. Penner-Hahn

Volume 3 *Nonlinear Phenomena in Physics* Editor: F. Claro

Volume 4 *Time-Resolved Vibrational Spectroscopy* Editors: A. Laubereau and M. Stockburger

Volume 5 *Physics of Finely Divided Matter* Editors: N. Boccara and M. Daoud

Volume 6 *Aerogels* Editor: J. Fricke

Volume 7 *Nonlinear Optics: Materials and Devices* Editors: C. Flytzanis and J. L. Oudar

Volume 8 *Optical Bistability III* Editors: H. M. Gibbs, P. Mandel, N. Peyghambarian, and S. D. Smith

Volume 9 *Ion Formation from Organic Solids (IFOS III)* Editor: A. Benninghoven

Volume 10 *Atomic Transport and Defects in Metals by Neutron Scattering* Editors: C. Janot, W. Petry, D. Richter, and T. Springer

Volume 11 *Biophysical Effects of Steady Magnetic Fields* Editors: G. Maret, J. Kiepenheuer, and N. Boccara

Volume 12 *Quantum Optics IV* Editors: J. D. Harvey and D. F. Walls

Volume 13 *The Physics and Fabrication of Microstructures and Microdevices* Editors: M. J. Kelly and C. Weisbuch

Volume 14 *Magnetic Properties of Low-Dimensional Systems* Editors: L. M. Falicov and J. L. Morán-López

Volume 15 *Gas Flow and Chemical Lasers* Editor: S. Rosenwaks

Volume 16 *Photons and Continuum States of Atoms and Molecules* Editors: N. K. Rahman, C. Guidotti, and M. Allegrini

Volume 17 *Quantum Aspects of Molecular Motions in Solids* Editors: A. Heidemann, A. Magerl, M. Prager, D. Richter, and T. Springer

Volume 18 *Electro-optic and Photorefractive Materials* Editor: P. Günter

Volume 19 *Lasers and Synergetics* Editors: R. Graham and A. Wunderlin

Volume 20 *Primary Processes in Photobiology* Editor: T. Kobayashi

Volume 21 *Physics of Amphiphilic Layers* Editors: J. Meunier, D. Langevin, and N. Boccara

Volume 22 *Semiconductor Interfaces: Formation and Properties* Editors: G. Le Lay, J. Derrien, and N. Boccara

Volume 23 *Magnetic Excitations and Fluctuations II* Editors: U. Balucani, S. W. Lovesey, M. G. Rasetti, and V. Tognetti

Volume 24 *Recent Topics in Theoretical Physics* Editor: H. Takayama

Volume 25 *Excitons in Confined Systems* Editors: R. Del Sole, A. D'Andrea, and A. Lapiccirella

Excitons in Confined Systems

Proceedings of the International Meeting,
Rome, Italy, April 13–16, 1987

Editors: R. Del Sole,
A. D'Andrea, and A. Lapiccirella

With 152 Figures

Springer-Verlag Berlin Heidelberg New York
London Paris Tokyo

Professor Rodolfo Del Sole

Dipartimento di Fisica, II Università di Roma "Tor Vergata", Via Orazio Raimondo,
I-00173 Roma, Italy

Dr. Andrea D'Andrea

Ist. Metod. Avanzate Inorg., Area della Ricerca di Roma, CNR, CP 10,
I-00016 Monterotondo Staz., Italy

Dr. Andrea Lapiccirella

Ist. Teoria e Strutt.-Elettron., Area della Ricerca di Roma, CNR, CP 10,
I-00016 Monterotondo Staz., Italy

ISBN 3-540-18707-3 Springer-Verlag Berlin Heidelberg New York
ISBN 0-387-18707-3 Springer-Verlag New York Berlin Heidelberg

Library of Congress Cataloging-in-Publication Data. Excitons in confined systems. (Springer proceedings in Physics; v. 25) Includes index. 1. Exciton theory – Congresses. I. Del Sole, R. (Rodolfo), 1944–. II. D'Andrea, A. (Andrea), 1943–. III. Lapiccirella, A. (Andrea), 1947–. IV. Series.
QC176.8E9E96 1987 530.4'1 87-36925

Printing: Weihert-Druck GmbH, D-6100 Darmstadt
Binding: J. Schäffer GmbH & Co. KG., D-6718 Grünstadt
2153/3150-543210

Preface

This volume contains the proceedings of the International Meeting on Excitons in Confined Systems, organized by the National Research Council of Italy (CNR) and by the Second University of Rome "Tor Vergata". The meeting was held at the CNR Research Area of Rome in Montelibretti, 13–16 April, 1987. About 50 participants came together for this period to exchange their ideas in intense scientific discussions.

The aim of the meeting was to discuss the theoretical and experimental aspects of excitons in many different confined systems, from semi-infinite solids to quantum wells. The main idea was to bring together people with different cultural backgrounds to discuss and compare the concepts involved in the different systems.

This volume is divided into three parts, according to the type of confinement. The first part is concerned with excitons in semi-infinite solids. Its main purpose is to summarize and contribute to the 30-year-old discussion, opened by Pekar in 1957, on exciton reflectance, involving the concepts of additional boundary conditions and of the dead layer. The complexity of the problem originates from the so-called spatial dispersion (i.e., the exciton motion) and from the exciton confinement in the semi-infinite crystal. The recent findings about exciton masses and lifetimes and the new developments of the quantum theory of exciton polaritons are also discussed.

The second part, on excitons in thin films, is intended to fill the gap between the popular research fields of excitons in semi-infinite semiconductors and excitons in quantum wells. Interesting experimental and theoretical results on exciton quantization are presented.

The third part is about excitons in quantum wells. Many interesting results of this recently developed research field are discussed. For instance, studies on excitons in II–VI heterostructures and on the effects of electric field on optical properties are presented. This part also includes some contributions on the growth and electronic structure of quantum wells, which are important ingredients in determining exciton states. The very new topic of excitons in quantum wires is also treated.

We are especially grateful for the cooperation of the Department of Physics of the Second University of Rome, the Institute of Advanced Inorganic Meth-
their efforts, which were crucial in making this meeting possible.

We are especially grateful for the cooperation of the Department of Physics of the Second University of Rome, the Institute of Advanced Inorganic Meth odologies of the CNR, and the Institute of Electronic Theory and Structure and Spectrochemical Behaviour of Coordination Compounds of the CNR, which enabled us to meet optimistic deadlines.

Rome, August 1987

R. Del Sole
A. D'Andrea
A. Lapiccirella

Contents

Part III Excitons in Superlattices and in Quantum Wells

Part I

Excitons in Semi-infinite Solids

Exciton Reflectivity and Additional Boundary Conditions

P. Halevi

Departamento de Física del Instituto de Ciencias,
Universidad Autónoma de Puebla, Apdo. Post. J-48,
Puebla, Pue. 72570, Mexico

1. Introduction

Thirty years have passed by since Pekar came to the conclusion that in semiconductor crystals "there can exist several plane monochromatic waves of the same frequency, direction of propagation, and polarization but with different refractive indices. This phenomenon differs from ordinary birefringence, where the waves of different refractive indices must be polarized at right angles to each other". Several pioneering works by PEKAR [1] and by HOPFIELD and THOMAS [2] have started a new branch of Solid State Optics. In this brief review it is impossible to do justice to the hundreds of papers on the subject published to date. The book by AGRANOVICH and GINZBURG [3] gives a listing until 1983.

Our interest lies in undoped, direct-gap semiconductors in the spectral region of exciton transitions. Typical examples are CdS of the II-VI and GaAs of the III-V group. In order to prevent rapid recombination of the electrons and holes, the experiments are conducted at liquid helium or lower temperatures. The Mott-Wannier excitons in these semiconductors have diameters of the order of tens of Å. I will limit the discussion to an isotropic configuration and to an isolated excitonic transition (of energy $\hbar\omega_T$) from the top of the valence band to a certain quantum level n just below the bottom of the conduction band. The contributions from any other transitions are lumped together in the background dielectric constant ε_0, assumed to be a real constant. The bulk response in phase space is well described by a modified single-oscillator model for the frequency (ω) and wavevector-dependent dielectric function:

$$\varepsilon(\omega,\vec{q}) = \varepsilon_0 + \omega_P^2/[\omega_T^2 + (\hbar\omega_T/m)q^2 - \omega^2 - i\nu\omega] . \tag{1a}$$

The frequency ω_P is a measure of the oscillator strength, and the phenomenological frequency ν is assumed to incorporate all the effects of interactions with phonons, impurities, etc. Any possible wavevector-dependence of ω_P and ν is neglected, thus the nonlocal effects arise only from the term $(\hbar\omega_T/m)q^2$, m being the sum of the electron and hole masses. This term comes from the kinetic energy $\hbar^2q^2/2m$ of the exciton. Choosing the z-coordinate perpendicular to the surface and the y-coordinate perpendicular to the plane of incidence of the light (1a) is conveniently rewritten in the form

$$\varepsilon(q_z) = \varepsilon_0 + (\varepsilon_0 m/\hbar\omega_T)(\omega_L^2 - \omega_T^2)/[q_z^2 - \Gamma^2(\omega)], \tag{1b}$$

where $\Gamma^2(\omega) = (\omega^2 - \omega_T^2 - \hbar\omega_T q_x^2/m + i\nu\omega)m/\hbar\omega_T$, $\omega_L^2 = \omega_T^2 + \omega_P^2/\varepsilon_0$, $q_x = \omega\sin\theta/c$ is the parallel component of the wavevector, and θ is the angle of incidence.

In order to solve Maxwell's equations for a dielectric half-space we need a linear response relating the displacement vector $\vec{D}(z)$ and the electric field $\vec{E}(z')$:

$$\vec{D}(z) = \int_0^\infty \varepsilon(z,z')\, E\,(z')\, dz', \tag{2}$$

where z and z' are coordinates perpendicular to the surface; the x and y coordinates have been ignored on account of the invariance of our geometry in these directions. The kernel $\varepsilon(z,z')$ is the dielectric function in real space.

2. Surface Dielectric Response and Additional Boundary Conditions

What is the relation between $\varepsilon(z,z')$ in (2) and $\varepsilon(q_z)$ in (1b)? In the case of bulk response the lower limit of integration in (2) is extended to $-\infty$ and $\varepsilon(z,z') = \varepsilon(z - z')$ because in this case we have translational invariance in the z-direction, as well as in the x- and y- directions. Then the answer is straightforward, namely $\varepsilon(z - z')$ is the Fourier transform, with respect to the component q_z, of $\varepsilon(\omega,\vec{q})$. In the case of a surface, however, there is no obvious prescription that relates $\varepsilon(z,z')$ to $\varepsilon(\omega,\vec{q})$. That is, the "bulk response" (1b) must be supplemented by an explicit "surface response" $\varepsilon(z,z')$. There is no way to proceed without additional information.

To restate this difficulty more categorically we consider the transverse-wave solutions gotten by substituting (1b) in the dispersion relation

$$\varepsilon(q_z) = (q_x^2 + q_z^2)c^2/\omega^2. \tag{3}$$

There are two solutions for q_z, say $q_1(\omega)$ and $q_2(\omega)$. For normal incidence of the light, or s-polarized oblique incidence, the electric field in the semiconductor must be a superposition of two plane waves:

$$E(\vec{r},t) = (E_1 e^{iq_1 z} + E_2 e^{iq_2 z})e^{i(q_x x-\omega t)} . \tag{4}$$

Maxwell's boundary conditions (continuity of E and H) are insufficient to determine both amplitudes E_1 and E_2, as well as that of the reflected wave. Hence the need for Additional Boundary Conditions (ABC's). A number of ABC's has been proposed, usually stated in terms of some condition to be satisfied at the surface by the excitonic polarization vector,

$$P(z) = (4\pi)^{-1} \int_0^\infty \varepsilon(z,z')E(z')dz' - (4\pi)^{-1}\varepsilon_0 E(z). \tag{5}$$

This definition excludes the background contribution. The properties of the three most popular ABC's are listed in Table I for normal incidence or s-polarized light. Presumably, the "correct ABC" for a given semiconductor should be justified by a comparison with realistic microscopic theories and/or by comparison with experiments on high-quality samples.

Table 1. Characteristics of three special cases of additional boundary conditions for s-polarized light. The prime denotes the derivative with respect to z.

ABC	U	Excitonic polarization	References
Pekar&Hopfield-Thomas	-1	$P(\ell) = 0$	1,2
Ting et al.	1	$P'(\ell) = 0$	4
Agarwal et al.	0	$i\Gamma P(\ell) + P'(\ell) = 0$	5-7

The problem of determining the "correct ABC" is intimately related to the surface-response of the semiconductor: if $\varepsilon(z,z')$ is explicitly known then it should be possible to determine the value of $P(z)$ at the surface, which in turn would yield the amplitude of the additional wave (absent if $m \to \infty$). I will adopt a simple phenomenological model, Fig. 1a.

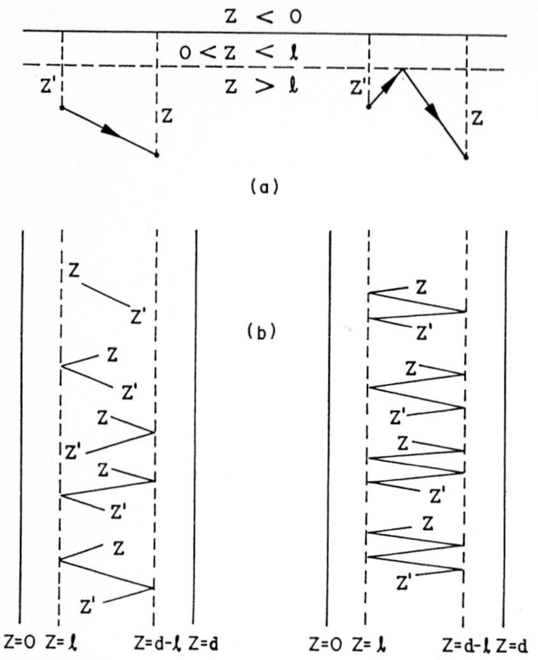

Fig. 1. An exciton is "created" at the point $\vec{r}'$, and "observed" at the point $\vec{r}$ in the non-local medium. It may proceed from $\vec{r}'$ to $\vec{r}$ directly, giving rise to a bulk process, or else it may interact with one or both surface layers any number of times. The possible processes are schematically indicated. The normal components of $\vec{r}$ and $\vec{r}'$ are z and z' and they are measured from the left-hand surface. An exciton-free layer is included in the model. (a) Surface; (b) Thin film.

Next to the (nominal) surface there is an exciton-free layer of thickness ℓ; it is often called the "dead layer". Such a layer describes, in a crude manner, the presence of a repulsive potential which acts on an exciton as it approaches the surface [2]. Even for an ideally perfect surface there must exist an "intrinsic" exciton-free layer, a consequence of the image-charge force of the distortion of the exciton wave functions near the surface. There is also an "extrinsic" contribution to ℓ, produced by charged impurities (Schottky-barrier effects). It is difficult to separate these two contributions, and I simply assume that in the dead layer $\omega_p = 0$, i.e. $\varepsilon(\omega,\vec{q}) = \varepsilon_0$. Beyond this layer $(z,z' \geq \ell)$ one has to distinguish between bulk and surface response. If an exciton is created at a distance z' from the surface and "observed" at a distance z then a bulk process, by definition, should be independent of the distances z and z' to the surface; it may only depend on the difference $|z - z'|$. Thus the bulk contribution to $\varepsilon(z,z')$ is still described by $\varepsilon(z - z')$, the Fourier transform of (1b). The dominant surface process is assumed to be a reflection of the exciton as a whole, as it approaches the boundary of the exciton-free layer. The total vertical distance traversed between an excitation at z' and observation at z is now $z + z'$ $(z,z' > 0)$. The contribution to $\varepsilon(z,z')$ from these processes is then assumed to be $\varepsilon(z + z')$, also the Fourier transform of (1b), but with a different argument. The interaction with the surface is not necessarily elastic, so the surface contribution is weighted by means of a reflection coefficient U. Thus our model dielectric function is [8-10]

$$\varepsilon(z,z') = \varepsilon(z - z') + U\,\varepsilon(z + z'), \quad (z,z' > \ell) \tag{6a}$$

$$= \epsilon_0 \delta(z - z') + \frac{i\omega_p^2 m}{2h\omega_T \Gamma(\omega)} \left[e^{i\Gamma|z-z'|} + Ue^{i\Gamma(z+z')} \right] . \tag{6b}$$

Next I substitute (6b) in (5), to give

$$P(z) = \frac{i\omega_p^2 m}{8\pi\omega_T \Gamma(\omega)} \int_{\ell}^{\infty} \left[e^{i\Gamma|z-z'|} + Ue^{i\Gamma(z+z')} \right] E(z')dz'.$$

The lower limit of the integral has been changed from 0 to ℓ in order to allow for the exciton-free layer. From here an integral expression for $P'(z) = dP/dz$ is readily found. Then evaluating $P(\ell)$ and $P'(\ell)$ it follows that [8-10]

$$i\Gamma(\omega)\ (1 - U)P(\ell^+) + (1 + U)\ P'(\ell^+) = 0. \tag{7}$$

This is the generalized ABC, a relation satisfied by the excitonic polarization at the inner interface between the dead layer and the bulk. The important point to notice is that (7) has been derived from the response-function (6) - corresponding to the model of Fig. 1a - without any additional assumptions. This ABC now makes it possible to solve for the extra field amplitude in (4), and from there one may proceed and calculate the optical properties of the semiconductor, such as reflectivity.

In the special cases $U = \pm 1$ or 0 the generalized ABC (7) reduces to some well-known ABC's, as given in Table I. However, now one may vary U continuously, and even select complex values. Thus the introduction of the reflection coefficient U in (6a) corresponds to a parameterization of the ABC (7). Other forms of linear relations connecting P and dP/dz have been considered early on [2, 11, 12].

The question of conservation of energy has been also studied [12,13], with the conclusion that in general the energy of an electromagnetic wave incident at the crystal surface is not transmitted continuously into the medium. The reason for this behavior is that, in addition to an electromagnetic flux the exciton also has a mechanical flux. This latter flux vanishes only in special cases. Applying a criterion given in [12] and assuming normal incidence and $\nu = 0$ the total energy is conserved in the region $\omega > \omega_T$ provided that $|U| = 1$. However, for $\omega < \omega_T$ the condition is different, namely $\mathrm{Im}U = 0$. If U is assumed to be independent of the frequency then this implies that $U = \pm 1$. (However, models with a frequency-dependent U have been also considered [11]) Otherwise the surface acts as a sink or as a source of energy. Or else, for $\nu \neq 0$, U must be complex, with its phase depending on $|U|$ and on $\Gamma(\omega)$ [13]. If a given, constant value of U ($\neq \pm 1$) is chosen - supposedly describing the specific nature of the interaction between the exciton and the surface - then in order to preserve the continuity of energy flow across the surface $\vec{E}(\vec{r})$ must be given by an expression that includes a term proportional to $\delta(z)$, rather than (4); the ABC also takes a form that is much more complicated than (7) [13b]. Still another way to restore energy conservation is the introduction of an inhomogeneous term, proportional to $\vec{E}$, on the right side of (7) [14]; such an ABC was proposed in [3].

Real surfaces of semiconductors are far from perfection. In fact, samples prepared according to the same specifications and conditions yield quite different optical spectra! Thus it should not be surprising that a surface may contribute to the energy balance. Supplementing this argument with that of "simplicity of description" I will limit the parameter U to real values between -1 and +1. The next step is comparison with experiments, in the hope of determining the numerical value of U.

3. Critical Comparison with an Experimental Reflectivity Spectrum (normal incidence).

The description of a semiconductor surface in the exciton regime requires numerous parameters: ϵ_0, ω , ω_T, m, ν, ℓ, and U. There is no difficulty in fitting theoreti-

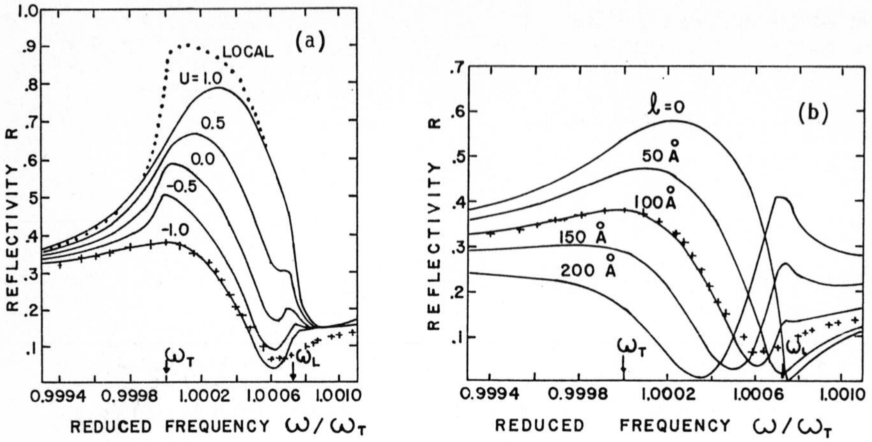

Fig. 2. Normal-incidence reflectivity for the A(n=1) exciton of CdS. The parameter U corresponds to a particular ABC, e.g., U = -1 describes the Pekar ABC. The experimental spectrum is due to Patella, Evangelisti, and Capizzi [15]. (a) Thickness of exciton-free layer ℓ = 100 Å; (b) U = -1; for various values of ℓ. From ref. [9].

cal results to a given experimental spectrum. For instance, in [15] the same normal-incidence reflectivity R(ω) was fitted for three values of U - each time resulting in different values of ν and ℓ! Clearly, an independent measurement of the parameters is necessary. Moreover these parameters should be determined directly, that is independently of any assumptions on ABC's. Raman or Brillouin scattering are just such methods. Fortunately the values of all the parameters except ℓ and U have been measured by Brillouin scattering [16] for the same sample of CdS (much-quoted 505/1) for which, previously, the reflectivity spectrum was taken [15]. The spectrum for the A(n = 1) exciton (crosses) is shown in Figs. 2a, b. The theoretical curves are based on the generalized ABC (7) [9], with the parameters from [16]. In Fig. 2a, ℓ = 100 Å is chosen and U is varied in steps of 0.5 from -1 to +1. In Fig. 2b, U is given the value -1, while ℓ is varied in steps of 50 Å from 0 to 200 Å. It is evident that the values ℓ = 100 Å and U = -1 not only give the best fit, but an excellent fit as well [there is, though, a small discrepancy near $\omega_L^2 = (\omega_T^2 + \omega_P^2/\varepsilon_0)^{\frac{1}{2}}$]. This suggests, at least for one high-quality sample of CdS, that the excitonic polarization $P(\ell)$ vanishes at the interface between the bulk and the dead layer (Pekar-Hopfield ABC). It is gratifying that this ABC does conserve energy, as mentioned.

The accuracy of normalization of the R (ω) spectrum in [15] (also used in [9]) was questioned in [17]. Now, according to [15], "the absolute scale is established by normalization of the experimental reflectivity outside the excitonic region to the value computed using ε_0 = 8.1". This is a perfectly reasonable procedure. It also leads to a dead layer thickness ($\ell \simeq$ 100 Å) that is consistent with values given in the literature for numerous samples of CdS. References [2, 15, 16, 18-22] all give 70 Å $\leq \ell \leq$ 112 Å. Thus a value $\ell \simeq 3a_B$ (where a_B is the excitonic Bohr radius) seems to be consistent with a wide range of measurements. It was claimed [17] that a normalization process based on refraction-index measurements [23] is more accurate than that of [15, 9]. However, in the spectral region (4,800 Å $\lesssim \lambda \lesssim$ 6,500 Å) where the measurement [17] was performed, the refraction index is far from being constant, indicating the importance of excitons. According to the author [23], "the effect of excitons on the index of refraction probably will also have to be included as they are known to play an important if not dominant role in the absorption, reflectivity and emission spectra of CdS". Therefore it seems that the normalization [15, 9] is actually more reliable than that of [23, 17].

4. Oblique Incidence Reflectivity

For oblique incidence, if the light is s-polarized, the physics is much the same as for normal incidence. The angle of incidence θ enters into the problem only through q_x in $\Gamma(\omega)$. A convenient method for calculating optical properties is based on the surface impedance, defined as

$$Z_s \equiv -E_y(0^+)/B_x(0^+). \tag{8}$$

The reflectivity is given by

$$R_s = |Z_s \cos\theta - 1|^2 / |Z_s \cos\theta + 1|^2. \tag{9}$$

Allowing for an exciton-free layer one has to match the fields $E_y(z)$ and $B_x(z)$ at both sides of the surface $z = 0$. The same boundary conditions also apply at the interface $z = \ell$, however at $z = \ell^+$ use must be also made of the generalized ABC (7). The electrodynamics of the problem is readily solved and Z_s is found to be

$$Z_s = \frac{\omega}{cq_\ell} \frac{q_\ell(a_1 - a_2) - i(q_2a_1 - q_1a_2)\tan q_\ell \ell}{(q_2a_1 - q_1a_2) - iq_\ell(a_1 - a_2)\tan q_\ell \ell} \tag{10}$$

where $q_\ell = (\omega/c)(\varepsilon_0 - \sin^2\theta)^{\frac{1}{2}}$ is the value of q_z inside the dead layer; q_1 and q_2 are the two permitted values of q_z in the spatially dispersive bulk - solutions of (3) - and

$$a_k = \frac{1}{q_k - \Gamma(\omega)} + \frac{U}{q_k + \Gamma(\omega)} . \tag{11}$$

Equivalent formulas for the special case $\ell = 0$ have been derived in [10, 24, 25].

In the case of p-polarized incidence the electric field also has a normal component E_z, and then longitudinal oscillations (polarization waves) are also possible. Their dispersion is governed by the equation

$$\varepsilon(q_z) = 0. \tag{12}$$

With $\varepsilon(q_z)$ still given by (1b) there is then a third solution, $q_z \equiv q_3(\omega)$, in addition to the two (transverse) solutions of (3). Thus the electric field in the spatially dispersive bulk has the form

$$\vec{E}(\vec{r},t) = \sum_{k=1}^{3} \vec{E}^{(k)} e^{iq_k z} e^{i(q_x x - \omega t)} . \tag{13}$$

Because of the anisotropy associated with the semiconductor surface the responses $D_x(z) = D_x[E_x(z')]$ and $D_z(z) = D_z[E_z(x')]$ are not necessarily equivalent. In order to model, in a simple way, the possibility that the parallel and perpendicular components of the exciton's dipole moment interact with the surface in different manners I assume that $D_x(z)$ and $D_z(z)$ are given by (2) (with the lower limit of the integral replaced by ℓ) and (6), however with different values of the parameters U, say U_x and U_z. This leads to two ABC's (for the x- and z- components of the excitonic polarization):

$$i\Gamma(\omega)(1 - U_j)\mathcal{P}_j(\ell^+) + (1 + U_j)\mathcal{P}'_j(\ell) = 0, \quad j = x,z. \tag{14}$$

These two equations are sufficient in order to determine the amplitudes of the two "additional waves" (either x- or z- component in (13)).

Now we have to specify two scattering parameters, U_x and U_z, in order to obtain numerical results for some optical property of the semiconductor. Various less general ABC's have been proposed, including "anisotropic" ones ($U_z = -U_x$), as summarized in Table 2.

Table 2. Characteristics of five special cases of additional boundary conditions for p-polarized light. The prime denotes the derivative with respect to z.

ABC	U_x	U_z	Behaviour of excitonic polarization: Parallel component	Normal component	Refs.
Pekar	-1	-1	$P_x(\ell)=0$	$P_z(\ell)=0$	1,2
Rimbey-Mahan	-1	1	$P_x(\ell)=0$	$P'_z(\ell)=0$	26
Fuchs-Kliewer	1	-1	$P'_x(\ell)=0$	$P_z(\ell)=0$	27
Ting et al.	1	1	$P'_x(\ell)=0$	$P'_z(\ell)=0$	4
Agarwal et al.	0	0	$i\Gamma P_x(\ell)+P_x(\ell)=0$	$i\Gamma P_z(\ell)+P_z(\ell)=0$	5-7

Again we express the reflectivity in terms of a surface impedance Z_p

$$R_p = |Z_p - \cos\theta|^2/|Z_p + \cos\theta|^2, \tag{15}$$

$$Z_p = E_x(0^+)/B_y(0^+). \tag{16}$$

In order to calculate Z_p one has to use Maxwell's boundary conditions at $z = 0$ and $z = \ell$ and, in addition, the two ABC's (14). The result is [28]

$$Z_p = \frac{cq_\ell}{\omega\varepsilon_0}\,\frac{(\varepsilon_0/q_\ell)[(1,2) + (2,3) + (3,1)] - i[(\varepsilon_1/q_1)(2,3) + (\varepsilon_2/q_2)(3,1)]\tan q_\ell\ell}{[(\varepsilon_1/q_1)(2,3) + (\varepsilon_2/q_2)(3,1)] - i(\varepsilon_0/q_\ell)[(1,2) + (2,3) + (3,1)]\tan q_\ell\ell} \tag{17}$$

$$(i,j) = a_i b_j - b_i a_j , \tag{18}$$

$$a_k = \frac{1}{q_k - \Gamma(\omega)} + \frac{U_x}{q_k + \Gamma(\omega)}, \quad b_k = \left[\frac{1}{q_k - \Gamma(\omega)} + \frac{U_z}{q_k + \Gamma(\omega)}\right]\gamma_k , \tag{19}$$

$$\gamma_{1,2} = -q_x/q_{1,2}, \qquad \gamma_3 = q_3/q_x . \tag{20}$$

In the special case that $\ell = 0$ these expressions reduce to those given in [8, 10]. The case $\ell = 0$ and $U_x = U_z$ has been solved in [24]. The ABC's (14) have been recently generalized for several resonances [29]. The Brewster effect for the Pekar-Hopfield model has been also studied [30].

The semiconductors in which our interest lies are usually uniaxial (optical axis $\hat{c}$). Therefore, in general we would have to allow for an anisotropy of the exciton mass in our theoretical description. However, this complication may be avoided if the crystal is cleaved in a plane that is parallel to $\hat{c}$ and the electric fields are constrained to a plane that is perpendicular to $\hat{c}$. Clearly this "isotropic configuration" corresponds to p-polarized light, the plane of incidence being perpendicular to $\hat{c}$ ($\parallel \hat{y}$). (As for s-polarized light, there is no isotropic configuration; for anisotropic exciton-polaritons see [31]). Reflectivity spectra for p-polarized light have been taken mainly for the A(n = 1) exciton of CdS [20, 21, 32, 33]. In [33] spectra for phase shifts, in addition to $R(\omega)$ were obtained.

For oblique incidence the comparison between experimental and theoretical spectra is plagued by the fact that ABC - independent measurements (by Brillouin or Raman scattering) of the exciton parameters are not available. Once such measurements are performed, a fit similar to that in Fig. 2 [9] should be attempted. Hopefully, for a certain pair of scattering parameters U_x and U_z, a good fit to experimental $R_p(\omega)$ spectra is obtained for a series of angles of incidence θ - with the same set of exciton parameters. It should be mentioned, however, that optical spectra are in general very insensitive to the parameter U_z. This indicates that the optical properties are determined, mainly, by the parallel component $P_x(\ell)$ of the excitonic polarization. In Fig. 3 I show a comparison between theoretical and experimental reflectance and thermoreflectance spectra for a series of eight angles between $\theta = 13°$ and $\theta = 82°$ [20b]. The fit is quite reasonable, however it should be born in mind that the parameters have not been measured independently. In particular, it is found that $\ell = 109$ Å, and the calculations are based on the Pekar ABC. This gives further support to the ABC $\vec{P}(\ell) = 0$ ($U_x = U_z = -1$).

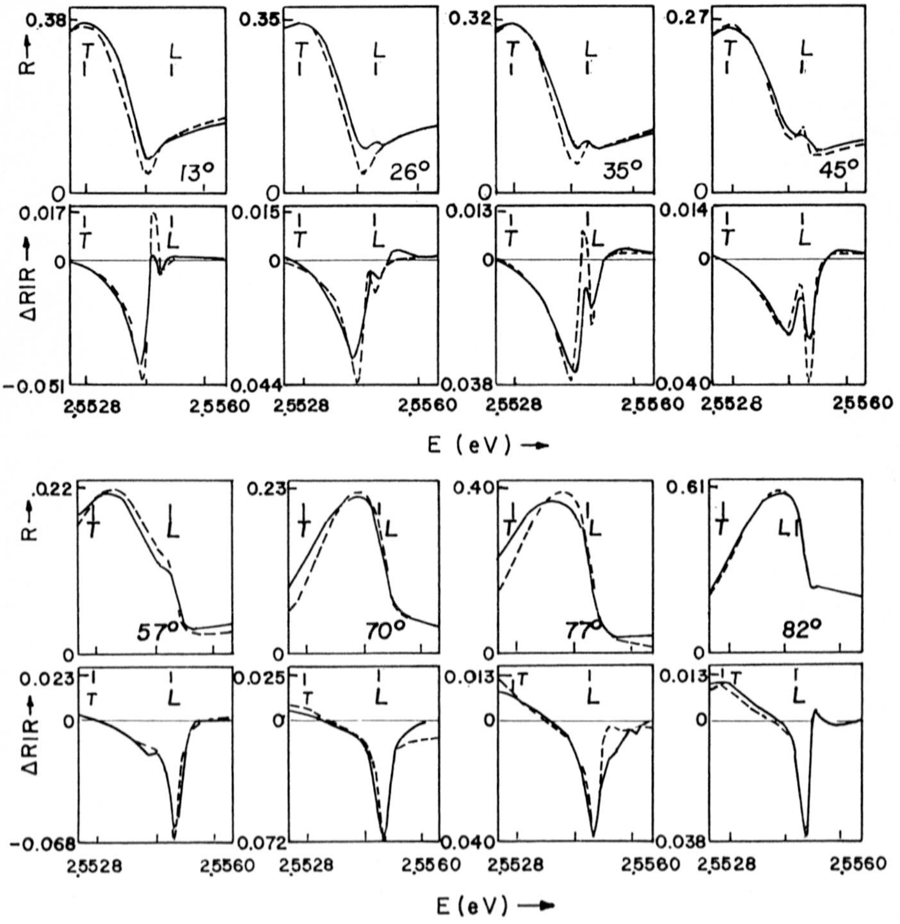

Fig. 3. Reflectance and thermoreflectance spectra of the A(n = 1) exciton of CdS in the symmetric configuration at different angles of incidence (p-polarized light). Continuous lines: experiments on sample No.12/9. Dashed lines: fitting calculation with Pekar's ABC and $\ell = 109$ Å. The letters T and L stand for ω_T and ω_L. From STÖSSEL and WAGNER [20b].

5. Thin Films

The first studies of transmissivity and reflectivity of thin films are due to PEKAR [1]. In the case of normal incidence calculations have been done for the ABC's listed in Table 1 [34], and for p-polarization of the light the anisotropic ABC's ($U_z = -U_x$) have been also explored [35]. In [36] the main interest lies in damping effects.

A reasonably realistic (macroscopic) model of the interaction of an exciton with the two surfaces of a thin film should include exciton-free layers and allow for scattering at the interfaces between these layers and the spatially dispersive bulk. The generalization of the previously introduced model of a surface (Fig. 1a) is shown in Fig. 1b. Now in addition to bulk processes and single scattering events at either surface, multiple scatterings are also possible. The contribution to the dielectric function from bulk processes is still the same as in (6). The two surfaces contribute a term

$$\frac{i\omega_p^2 m}{2\hbar\omega_T\Gamma(\omega)} \sum_{n=1}^{\infty} U^n (e^{i\Gamma\zeta_n} + e^{i\Gamma\zeta'_n}). \tag{21}$$

Here n is the number of scatterings that take place between the "excitation" at z' and the "observation" at z. Each one of these events contributes a "reflection coefficient U; hence the factor U^n. The ζ_n and ζ'_n are the projections on the z-axis of the trajectories followed by the exciton after n interactions with the surface. Here $\zeta_n(\zeta'_n)$ refers to the case that that the first interaction, as we proceed from z' to z takes place at the surface $z = \ell$ $(z = d - \ell)$. These quantities describe the phase changes, in the direction perpendicular to the film, of the exciton after undergoing a n'th order scattering. Upon performing the summation in (21) for all the processes shown in Fig. 1b the dielectric function in the region $\ell \leq z, z' \leq d - \ell$ assumes the form [37]

$$\varepsilon(z,z') = \varepsilon_0(z - z') + [i\omega_p^2 m/2\hbar\omega_T\Gamma(\omega)]e^{i\Gamma|z-z'|}$$

$$+ \frac{i\omega_p^2 mUe^{i\Gamma(d-2\ell)}}{\hbar\omega_T\Gamma(\omega)[1 - U^2e^{2i\Gamma(d-2\ell)}]} \left\{ \cos[\Gamma(z+z' - d)] + Ue^{i\Gamma(d-2\ell)}\cos[\Gamma(z - z')] \right\}. \tag{22}$$

From here the excitonic polarization is found and, after some algebra, the ABC's are derived. For normal incidence (or oblique incidence with s-polarized light) one ABC is identical with (7), and the other is

$$i\Gamma(\omega)(1 - U)\mathcal{P}(d - \ell) - (1 + U)\mathcal{P}'(d - \ell) = 0. \tag{23}$$

It is interesting that these are just the (decoupled) ABC's corresponding to two independent interfaces at $z = \ell$ and $z = d - \ell$.

The reflectivity is calculated by means of a scheme that involves a sequence of surface impedances Z(z). The reflectivity R (ω) is still given by (9) in terms of $Z(0^+)$. We express $Z(0^+)$ in terms of $Z(\ell^+)$, then express $Z(\ell^+)$ in terms of $Z(d - \ell^+)$, and finally express $Z(d - \ell^+)$ in terms of $Z(d^+)$ which, of course, is the surface impedance of vacuum. The calculation is based on the usual Maxwell's boundary conditions plus (7) and (23). The scheme is easy to program, and it yields a very general result for the reflectivity, transmissivity, and absorptivity as a function of ω, θ, d, ℓ, and U [37]. In particular, one may follow the changes in a spectrum as U is changed in a continuous fashion [38].

In Fig. 4 I show $R(\omega)$ spectra for a thin film of CdS in the neighborhood of the A(n = 1) exciton. The parameter U is given three values: 0, -0.5, and -1 [37]. The structure is rather complicated, a consequence of interference of four waves with values of q_z given by $\pm q_1$ and $\pm q_2$. An attempt to interpret the minima in terms of simple Fabry - Perot resonances for standing waves has met only with

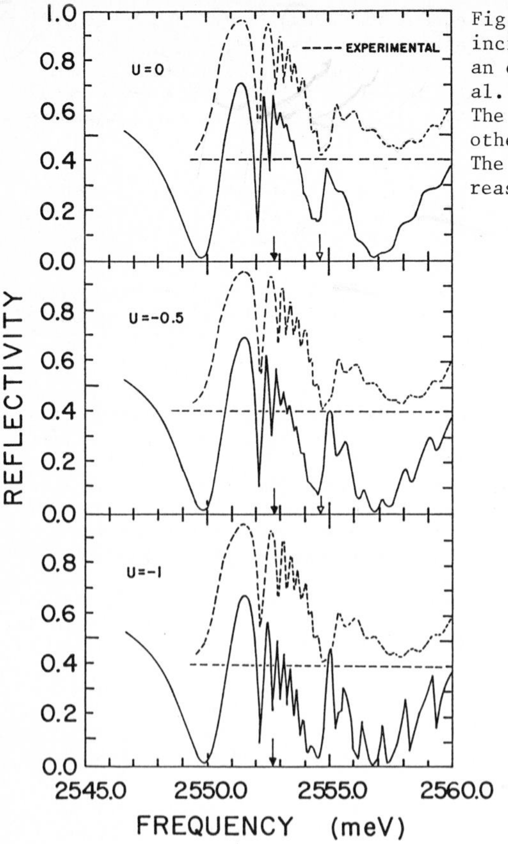

Fig. 4. Comparison of calculated, normal-incidence reflectivity spectra R (ω) with an experimental spectrum by MAKARENKO et al. [40] for the A(n = 1) exciton of CdS. The film thickness is 0.2 μm; all the other parameters are the same as in Fig. 2. The ABC corresponding to U = -0.5 gives a reasonable fit to the experimental curve.

partial success. Of course such a criterion is only valid if spatial dispersion, damping and dead layers are all absent. The important point is that the spectra change in a qualitative way as U is gradually changed from +1 to -1. Also there is considerable sensitivity to the thickness of the dead layer. For example, the positions of the minima above ω_L shift to higher frequencies as we increase ℓ from 0 to 100 Å. The shift increases with $(\omega - \omega_T)$ and, for sufficiently high frequencies, is well accounted for by the formula [37]

$$\Delta\omega_n \simeq 4(\omega_n - \omega_T)\Delta\ell/(d - 2\ell). \tag{24}$$

Experimental work on interference in thin films has been conducted in [39, 40], and one of the spectra [40], for d = 0.2 μm, is also shown in Fig. 4 (dashed lines). The exciton parameters in our calculation are the same as used for Fig. 2 [16]. While these were measured by Brillouin scattering, unfortunately this was not done on the CdS sample used in [40]. Thus one should not expect too good a fit between the experimental curve and any one of the theoretical curves. Nevertheless, granted the complexity of the structure, the agreement is quite satisfying for $U \simeq -0.5$. In [40] the Pekar ABC (U = -1) was used, however in order to achieve a better fit the damping frequency ν was assumed to change abruptly to a higher value for $\omega > \omega_L$. The precise values of $\nu_<$ and $\nu_>$ depend strongly on the film thickness d. Therefore this behavior does not seem to be related to the ν (ω) dependence found in [41].

Interestingly, a frequency-dependence of ν has been also judged important in [42]. From a simultaneous measurement of reflectivity and transmissivity spectra it was found that $\nu(\omega) \propto d\omega/dq$. This dependence, with Pekar's ABC and $\ell = 0$ gave a very good fit to an experimental transmissivity spectrum

It is also worthwhile noting that a thin semiconductor film on a metallic substrate exhibits an R (ω) spectrum that depends very strongly on the ABC employed [43].

Transmission of pulses through spatially dispersive thin films has been also studied [44 - 46] and was reviewed elsewhere [47].

6. Conclusion

For limitations of space I have not dealed with the important topic of Attenuated Total Reflection (ATR) spectroscopy and the corresponding surface - exciton polaritons that are excited by this method. The reader is referred to several papers [7, 9, 38, 48, 49] and reviews [3,50] on the subject.

In general, the Pekar ABC, supplemented by an exciton-free layer, provides a good fit to experimental reflectivity spectra. Unfortunately, judgement is hindered by a lack of independently measured (by Brillouin scattering) exciton parameters needed in order to fit these spectra. Also, there exists evidence that a more sophisticated description of surface response - in the form of a repulsive potential V(z), rather than a simple dead layer - is necessary for a realistic description of exciton-dynamics. This theme will be explored in detail in other chapters of these proceedings.

References

1. S.I. Pekar, J. Expt. Theoret. Phys. 33, 1022 (1957) [Sov. Phys. JETP 6, 785 (1958)]; Sov. Phys. JETP 7, 813 (1958); J. Phys. Chem. Solids 5, 11 (1958); Sov. Phys. JETP 9, 314 (1959)
2. J.J. Hopfield: Phys. Rev. 112, 1555 (1958); J.J. Hopfield and D.G. Thomas: Phys. Rev. 132, 563 (1963); J.J. Hopfield: J. Phys. Soc. Japan, Suppl. 21, 77 (1966)
3. V.M. Agranovich and V.L. Ginzburg: Crystal Optics with Spatial Dispersion, and Excitons, 2nd. ed., Springer Ser. Solid-State Sci., Vol. 42 (Springer, Berlin, Heidelberg 1984)
4. C.S. Ting, M.J. Frankel, J.L. Birman: Solid State Commun. 17, 1285 (1975)
5. G.S. Agarwal, D.N. Pattanayak, E. Wolf: Phys. Rev. Lett. 27, 1022 (1971); Opt. Commun. 4, 255 (1971); G.S. Agarwal: Opt. Commun. 6, 221 (1972); Phys. Rev. B 8, 4768 (1973); 10, 1447 (1974)
6. J.L. Birman and J.J. Sein: Phys. Rev. B 6, 2482 (1972); J.L. Birman and R. Zeyher: in Polaritons, ed. by E. Burstein and F. De Martini (Pergamon, New York 1974) p. 161
7. A.A. Maradudin and D.L. Mills: Phys. Rev. B 7, 2787 (1973); D.L. Mills: in Polaritons, ed. by E. Burstein and F. De Martini (Pergamon, New York 1974) p. 147
8. P. Halevi and R. Fuchs: in Physics of Semiconductors 1978, ed. by B.L.H. Wilson, Conf. Ser. No. 43, 863 (The Institute of Physics, Bristol and London 1979)
9. P. Halevi and G. Hernández-Cocoletzi: Phys. Rev. Lett. 48, 1500 (1982)
10. P.Halevi and R. Fuchs: J. Phys. C: Solid State Phys. 17, 3869 (1984)
11. T. Skettrup and I. Balslev: Phys. Rev. B 3, 1457 (1971)
R. Zeyher, J.L. Birman, W. Brenig: Phys. Rev. B 6, 4613 (1972);
V.M. Agranovich and V.I. Yudson: Opt. Commun. 7, 121 (1973);
V.A. Kiselev: Sov. Phys. Solid State 15, 2338 (1974); V.A. Kiselev, B.S. Razbirin; I.N. Uraltsev: Phys. Stat. Solidi b 72, 161 (1975);
D.L. Johnson and P.R. Rimbey: Phys. Rev. B 15, 5087 (1976);
O.V. Konstantinov and Sh. R. Saifullaev: Sov. Phys. Solid State 18, 1998 (1976); F. Flores, F. García-Moliner, and R. Monreal: Phys. Rev. B 15, 5087 (1977)
12. M.F. Bishop and A.A. Maradudin: Phys. Rev. B 14, 3384 (1976); 21, 884 (1980) (erratum)

13. (a) R. Monreal, F. García-Moliner, F. Flores: Solid State Commun. 32, 613 (1979); (b) Phys. Scripta 22, 155 (1980)
14. N.N. Akhmediev and V.V. Yatsishen: Solid State Commun. 27, 357 (1978)
15. F. Patella, F. Evangelisti, M. Capizzi: Solid State Commun. 20, 23 (1976)
16. P. Y. Yu and F. Evangelisti: Phys. Rev. Lett. 42, 1642 (1979)
17. A. D'Andrea and R. Del Sole: Phys. Rev. B 29, 4782 (1984)
18. F. Evangelisti, A. Frova, F. Patella: Phys. Rev. B 10, 4253 (1974)
19. G.V. Benemanskaya, B.V. Novikov, A.E. Cherednichenko: Sov. Phys. Solid State 19, 806 (1977); A.V. Komarov, S.M. Ryabchenko, M.I. Strashnikova: Sov. Phys. JETP 47, 128 (1978); J. Wicksted, M. Matsushita, H.Z. Cummins, T. Shigenari, X.Z. Lu: Phys. Rev. B 29, 3350 (1984)
20. W. Stössel and H.J. Wagner: Phys. Stat. Solidi (b) 89, 403 (1978); 96, 369 (1979)
21. I. Broser, M. Rosenzweig, R. Broser, M. Richard, E. Birkicht: Phys. Stat. Solidi (b) 90, 77 (1978)
22. A.B. Pevtsov and A.V. Sel'kin: Sov. Phys. JETP 56, 282 (1982)
23. W. Langer: J. Appl. Phys. 37, 3530 (1966)
24. N.N. Akhmediev and V.V. Yatsyshen: Sov. Phys. Solid State 18, 975 (1976)
25. D.R. Tilley: J. Phys. C: Solid State Phys. 13, 781 (1980); J.S. Nkoma: J. Phys. C: Solid State Phys. 16, 3713 (1983)
26. P.R. Rimbey and G.D. Mahan: Solid State Commun. 15, 35 (1974); P.R. Rimbey: Phys. Stat. Solidi (b) 68, 617 (1975), Phys. Rev. B 15, 1215 (1977); 18, 977 (1978); D.L. Johnson and P.R. Rimbey: Phys. Rev. B 15, 5087 (1976)
27. K.L. Kliewer and R. Fuchs: Phys. Rev. 172, 607 (1968); R. Fuchs and K.L. Kliewer: Phys. Rev. B 3, 2270 (1971)
28. G. Hernández-Cocoletzi: M.Sc. Thesis, Universidad Nacional Autónoma de México, 1980
29. P. Dub. Czech. J. Phys. B 36, 1041 (1986)
30. A.B. Pevtsov: Sov. Phys. JETP 56, 282 (1983)
31. S.A. Permogorov, A.V. Sel'kin, V.V. Travnikov: Sov. Phys. Solid State 15, 1215 (1973); E. Tosatti and G. Harbeke: Nuovo Cim. B 22, 87 (1974); K. Hümmer and P. Gerbhardt: Phys. Stat. Solidi (b) 85, 271 (1978)
32. S.A. Permogorov, V.V. Travnikov, A.V. Sel'kin: Sov. Phys. Solid State 14, 3051 (1973)
33. A.B. Pevtsov, S.A. Permogorov, Sh. R. Saifullaev, A.V. Sel'kin: Sov. Phys. Solid State 22, 1396 (1980); 23, 1644 (1982)
34. M.F. Bishop: Solid State Commun. 20, 779 (1976)
35. D.L. Johnson: Phys. Rev. B 18, 1942 (1978)
36. B. Dietrich and J. Voigt: Phys. Stat. Solidi (b) 93, 669 (1979)
37. J.A. Gaspar-Armenta and P. Halevi: Rev. Mex. Fís. 33 (1987), in press
38. P. Halevi, G. Hernández-Cocoletzi, J.A. Gaspar-Armenta: Thin Solid Films 89, 271 (1982)
39. V.A. Kiselev, B.S. Razbirin, I.N. Uraltsev: ZhETF Pis. Red. 18, 504 (1973); Phys. Stat. Solidi (b) 72, 161 (1975)
40. I.V. Makarenko, I.N. Uraltsev, V.A. Kiselev: Phys. Stat. Solidi (b) 98, 773 (1980)
41. J. Wicksted, M. Matsushita, H.Z. Cummins, T. Shigenari, X.Z. Lu: Phys. Rev. B 29, 3350 (1984)
42. I. Broser, K.H. Pantke, M. Rosenzweig: Solid State Commun. 58, 441 (1986)
43. V.M. Agranovich, V.E. Kravtsov, T.A. Leskova, A.G. Malshukov, G. Hernández-Cocoletzi, A.A. Maradudin: Phys. Rev. B 29, 976 (1984)
44. D.L. Johnson: Phys. Rev. Lett. 41, 417 (1978)
45. A. Puri and J.L. Birman: Phys. Rev. A 27, 1044 (1983)
46. J.A. Gaspar-Armenta and P. Halevi: preprint
47. P. Halevi and J.A. Gaspar-Armenta: in Proc. 2nd. Intern. Conf. on Surface Waves in Plasmas and Solids, Ohrid, 1985, ed. by S. Vukovic, World Scientific Pub. Co., Singapore 1987)
48. P. Halevi and R. Fuchs: J. Phys. C: Solid State Phys. 17, 3899 (1984)
49. J. Lagois: Phys. Rev. B 23, 5511 (1981)
50. J. Lagois and B. Fischer: in Surface Polaritons, ed. by D.L. Mills and V.M. Agranovich (North-Holland, Amsterdam 1982).

Microscopic Theory of Exciton Polaritons in Semi-infinite Solids

R. Del Sole[1] *and A. D'Andrea*[2]

[1]Dipartimento di Fisica, II Università di Roma "Tor Vergata", I-Roma, Italy
[2]Istituto di Metodologie Avanzate Inorganiche del CNR, I-Roma, Italy

1. INTRODUCTION

Thirty years after the pioneering work of Pekar /1/, the fundamental problem of exciton reflectance does not yet have a generally accepted solution.
Two polariton branches propagate in the presence of spatial dispersion, so that an additional boundary condition (ABC) is needed to determine their relative amplitudes, in addition to Maxwell's boundary conditions.
Hopfield and Thomas /2/ showed that the ABC consists of the vanishing of the exciton polarization at the surface, as already hypothesized by Pekar /1/, at least in the case of a tight-binding model of Frenkel excitons with nearest-neighbour interactions.
They also introduced the concept of dead layer in order to account for the repulsive image potential that extended excitons feel near the surface. Since then, the correct ABC and the existence of the dead layer have been the object of a not-yet-settled controversy.

It is clear that the ABC and an eventual dead layer must be embodied in the nonlocal nonhomogeneous dielectric susceptibility of the vacuum-crystal system. This was shown first in the framework of the so-called dielectric approximation /3-5/, which assumes that the bulk, translationally invariant, dielectric susceptibility $\varepsilon(\omega,\vec{r}-\vec{r}')$ holds up to the surface. This approximation leads to an ABC different from that of Pekar /1/ and Hopfield and Thomas /2/ and to no dead layer, but it completely neglects surface effects.

Further insight was provided by some papers /6,7/ dealing with the relation between the kind of exciton reflection at the surface and ABC. In general, exciton reflection at the surface leads to a nonhomogeneous dielectric susceptibility. The dielectric approximation results to be a particular case, corresponding to diffuse reflection. These papers, however, neglect the effect of the surface on exciton internal motion, limiting themselves to the case of Frenkel excitons. These have been studied in great detail in a number of papers /8-11/ where microscopic calculations of optical properties have been carried out for quite sophisticated models. Pekar's ABC has been confirmed in the case of semi-infinite isotropic crystals /8/.

On the other hand, microscopic calculations for extended (Wannier-Mott) excitons have also been performed /12-19/. The existence of a transition region near the surface, where excitons are less probably found than in the bulk (dead layer), is confirmed, but there is no agreement on its depth.

A number of experiments have been performed /20-28/ and interpreted in terms of a homogeneous dead layer and various ABC's. It has been shown /21/ that any ABC gives a good fit of normal-incidence reflectivity if the dead layer depth is a fitting parameter, while oblique incidence reflectivity seems to favor Pekar's ABC /24/. A further complication arises, since extrinsic dead layers often exist in semiconductors, due, for instance, to built-in electric fields which may ionize excitons. For these reasons, dead layer depths determined by fitting experimental line shapes are not conclusive, and the theoretical work is essential in order to have a full understanding of the behavior of extended excitons near the crystal boundary.

The purpose of this article is to review the work carried out by the authors /16-19,29/ on a realistic model of the crystal-vacuum interface, suitable to describe extended (Wannier-Mott) excitons in semi-infinite semiconductors. The model will be described in Section 2. It is possible to find very accurate exciton wave functions within the model /19/, and to compute from them the dielectric susceptibility and the reflection coefficient /17,18,29/. Essentially no approximations are introduced in carrying out the calculations.

The comparison with experiments is carried out in Section 3. This is done using the lowest possible number of adjustable parameters, seeking a reasonable -yet not perfect- agreement between theory and experiment. This procedure avoids the imprecise determination of some parameters - as, for istance, the dead layer depth - that only weakly affectsthe reflectance line shapes. We find that the microscopic moael considered here gives a good description of exciton reflectance in a number of semiconductors, whereas some important discrepancies occur in other materials, as in InP and GaAs.

The conclusion of this work and comparison with another method of attacking the exciton reflectance problem, namely the coherent wave approach /30/, are discussed in Section 4.

2. MICROSCOPIC THEORY

2.1 The Model

We consider excitons at the fundamental edge of a semiconductor with parabolic nondegenerate valence and conduction bands, occupying the half-space $z>0$. The effective-mass Schroedinger equation is

$$[-\frac{\hbar^2}{2M}\frac{d^2}{dZ^2} - \frac{\hbar^2}{2\mu}\Delta_{\vec{r}} - \frac{e^2}{\epsilon r} - E]\psi(\vec{r},Z) = 0, \qquad (1)$$

where Z is the center-of-mass position, $\vec{r} = \vec{r}_e - \vec{r}_h$, $\vec{r}_e$ and $\vec{r}_h$ are the electron and hole position respectively, the exciton energy E is measured from the conduction-band bottom, and ϵ is the static dielectric constant. We assume the boundary condition (BC) that the envelope wave function $\psi(\vec{r},Z)$ vanishes when either the electron or the hole are at the surface:

$$\psi(z_e = 0) = \psi(z_h = 0) = 0. \qquad (2)$$

This BC has been shown to be correct in absence of surface states when a large surface barrier prevents electron and hole escape from the surface /31/. It has been derived /32/ from the image potential infinite barrier that electrons and holes experience approaching the surface, and was assumed in previous works on this topic /12-19,28/. The image potential has been neglected here, to make the problem tractable. Indeed its effect should be small, since the BC (2) repels electrons and holes far from the surface, where the image potential vanishes. In a sense, we can say that the repulsive image potential has been approximately accounted for by means of (2).

This model also neglects built-in **electric fields** that can be present in the space charge region below the surface.

2.2 Wave Functions

Equation (1) is separable in the relative $\vec{r}$ and center-of-mass Z coordinate. However, they are mixed by the boundary conditions (2). The solution of this problem can be written as a linear combination of products of center-of-mass (plane waves) and relative motion (hydrogenic) wave functions, with coefficients chosen in order to fulfil the BC's (2) /17,19/:

$$\Psi(\vec{r},Z) = [\pi(1+|A|^2)]^{-1/2} \, [(e^{-iK_zZ}+Ae^{iK_zZ})\phi_1(r) + \sum_n \phi_n(\vec{r})e^{-P_nZ}c_n] \,, \qquad (3)$$

where $\phi_1(r)$ and $\phi_n(\vec{r})$ are, respectively, the ground state and excited hydrogenic orbitals with energy E_n, $\mathcal{E}=E-\hbar^2K_\parallel^2/2M$, $\vec{K}=(\vec{K}_\parallel,K_z)$ is the center-of-mass wave vector, such that $\mathcal{E} = -R^*+\hbar^2K_z^2/2M$, $R^*=\mu e^4/(2\hbar^2\varepsilon^2)$ is the effective Rydberg, and

$$P_n = [2M(E_n - \mathcal{E})/\hbar^2]^{1/2} \,. \qquad (4)$$

This wave function is very good as can be seen from its amplitude at any point $z_e=0$ or $z_h=0$, where it should vanish. We always find very small values, less than 10^{-2} times the peak value of the 1s hydrogenic orbital.

The square modulus of the exciton wave function in CdS at r=0, that is relevant for calculation of the dielectric function, is plotted in Figs.1(a) and 1(b) (solid lines) for different values of the center-of-mass energy $E_{c.m.}=\hbar^2K_z^2/2M$ as a function of the center-of-mass depth Z. The accuracy of the wave function is also clear from its vanishing at Z=0, in agreement with the boundary conditions (2). Other details of the wave functions are shown in Fig.2, where constant-$|\psi|^2$ curves are drawn in the (ρ,z) plane for different center-of-mass positions Z (namely, $2a_B$, a_B, and $0.75a_B$, where a_B is the Bohr radius). Reduction and distortion of the wave function, as the surface is approached, is clearly apparent from the figures. The distortion is a consequence of the different electron and hole masses, while the reduction originates from the "no escape" boundary conditions. The latter is the main effect for values of $M/\mu \leq 7$. This is at variance with the result for $m_h=\infty$ /33/, where a very large distortion of the 1s ground state occurs, changing it into $2p_z$.

The wave function given by (3) is too complicated to allow the calculation of the dielectric susceptibility, for which $\psi(0,z)$ is relevant. This can be done however using a simple analytical approximation of the form

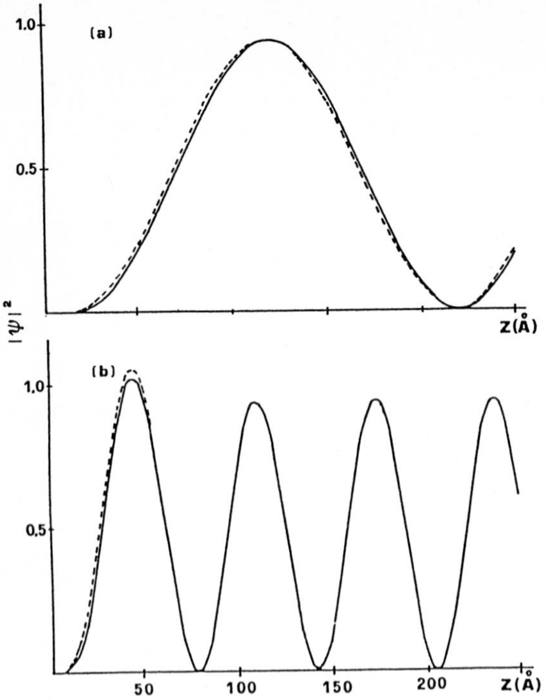

Fig.1 Square modulus of exciton wave functions of CdS, $|\psi(0,Z)|^2 \pi^2 a_B^3$, for center-of-mass energies: (a) 1 meV and (b) 10 meV. Solid lines are the nonadiabatic wave functions computed according to Eq.(3) and dashed lines are the analytical wave functions Eq.(5). The physical parameters are m_e=0.22, m_h=0.72, ϵ=9, M=0.94, and R*=28.7 meV.

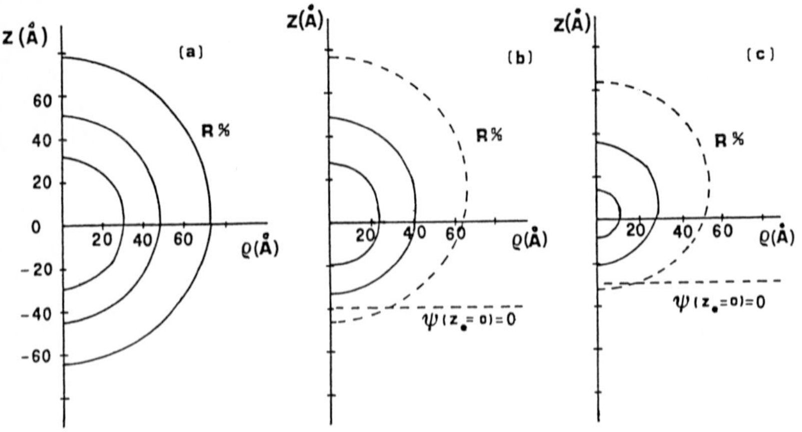

Fig.2 Level curves of exciton wave functions of CdS, $R=|\psi(z,\rho,Z)/\psi_{max}|^2$ for three different values of Z: (a) $Z=2a_B$, (b) $Z=a_B$, (c) $Z=3a_B/4$. The values of R for the level curves, starting from the origin r=0, are R=8.5%, 2.7%, and 0.5%. The center-of-mass energy is 10 meV and the values of physical parameters are the same as in Fig.1.

$$\psi(0,Z)=(2\pi)^{-1/2}[e^{-iK_zZ}+Ae^{iK_zZ}-(1+A)e^{-PZ}]\,\phi_1(r), \tag{5}$$

which has been developed in Ref.16 with

$$A = -(P - iK_z)/(P + iK_z). \tag{6}$$

In Ref.16, this form was derived from the Schroedinger equation, and P was taken as some average of the P_n's defined in Eq.(4).
The decay length 1/P of the evanescent wave in (5) is the transition layer depth for $K_z \to 0$.
The comparison with the numerical wave function (3) shows only a qualitative agreement /17/. The agreement however can be improved by treating P as a fitting parameter /19/, to be adjusted to reproduce the numerical wave function. A very good agreement is found between the numerical and the analytical wave functions in this way (see Fig.1), by using the same value of P in the relevant range of center-of-mass energies.

The parameter P ranges from $1.8a_B^{-1}$ for $m_e=m_h$ to a_B^{-1} for $m_h/m_e=5$, while the dead-layer depth 1/P correspondingly changes from $0.55a_B$ to a_B. It is remarkable that the analytical approximation describes not only the large-Z oscillations but also the $Z\to 0$ behavior. This has been shown in Figs.1(a) and 1(b), where the analytical approximation (dashed lines) is nearly indistinguishable from the numerical wave function. More generally, similar results are obtained in the range $m_e/m_h \gtrsim 0.2$, shown in Fig.3.

As m_e/m_h becomes smaller than 0.2, the fitted value of P increases. However, also the mismatch of boundary conditions begins to increase, and the wave function with c_n and A determined variationally becomes unreliable. Increasing the number of the bound states in the sum of Eq.(3) does not reduce the mismatch. This is due to the large electron-hole mass anisotropy which, in order to satisfy the boundary conditions for both electrons and holes at the surface, requires the

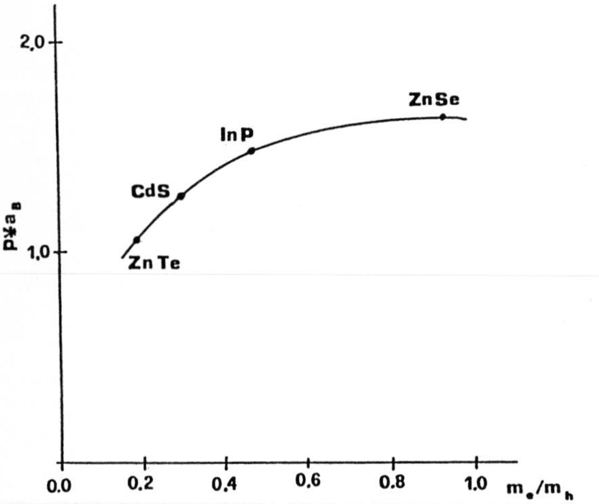

Fig.3 The fit parameter P of analytical wave function (Eq.5), i.e., the inverse dead-layer depth for vanishing center-of-mass energy, plotted as a function of the mass ratio m_e/m_h.

presence of a large number of states in Eq.(3), including those with positive energy. Since the inclusion of such states is extremely difficult, our approach is limited to the values $m_h/m_e \leq 5$, or $M/\mu \leq 7$, which are those of interest in semiconductors. In this range, the analytical approximation is very close to the numerical solution (3), and allows a quantitative determination of nonlocal dielectric tensor, embodying the ABC.

A different approach to treat exciton wave functions in semi-infinite semiconductors is that based on the adiabatic separation of the fast relative e-h motion (characterized by the reduced mass μ) from the slow center-of-mass motion (characterized by the total mass M) /13-15,19/. It can be shown /19/ that this approach also leads to good wave functions, even for $\frac{M}{\mu} \lesssim 7$, provided that a very accurate solution of the relative motion is carried out. Of course the range $M/\mu > 7$ is also well described, but the analytical approximation (5) fails to reproduce the wave functions /19/. It is possible that a suitable approximation to the adiabatic wave functions in the range $M/\mu > 7$ might be obtained considering a homogeneous dead-layer together with the transition layer 1/P.

2.3 Dielectric Susceptibility

The dielectric susceptibility can be computed from linear response theory /17/:

$$\varepsilon(\omega,q_{\|},z,z') = \varepsilon_0\delta(z-z') + 8\pi e^2|p_{vc}|^2/(m^2\omega^2)$$
$$\cdot \int_0^\infty dk_z \left[\frac{\psi_{k_z}(0,Z)\,\psi^*_{k_z}(0,Z')}{\hbar\omega_0 + \hbar^2(q_{\|}^2+k_z^2)/(2M) - \hbar\omega - i\gamma} + \frac{\psi_{k_z}(0,Z')\,\psi^*_{k_z}(0,Z)}{\hbar\omega_0 + \hbar^2(q_{\|}^2+k_z^2)/(2M) + \hbar\omega + i\gamma}\right], \qquad (7)$$

where m is the electron mass and p_{vc} is the momentum matrix element between k=0 valence and conduction band Bloch wave functions. The wave function (5) allows one to compute it in closed form /17/:

$$\varepsilon(\omega,q_{\|},z,z') = \varepsilon_0\delta(z-z') + i4\pi\alpha M\omega_0\{e^{iq|z-z'|} + Ue^{iq(z+z')}$$
$$-2q[\exp(iqz-Pz') + \exp(iqz'-Pz) - \exp(-Pz-Pz')]\,(q-iP)^{-1}\}\,(2\hbar q)^{-1} \qquad (8)$$

with

$$q = [2M(\hbar\omega - \hbar\omega_0 + i\Gamma)/\hbar^2 - q_{\|}^2]^{1/2} \qquad (9)$$

and

$$U = (q + iP)\,/\,(q - iP)\,. \qquad (10)$$

We emphasize that no other approximation, in addition to those involved in the very formulation of the problem, has been introduced to derive this expression.

The exciton contribution to $\varepsilon(\omega,q_{\|},z,z')$ is contained in the curly brackets of (8). The first term is the bulk contribution, the second term derives from exciton reflection at the surface, and the other terms vanish after a transition layer of depth 1/P. The exciton contribution vanishes at z or z'=0, as a consequence of BC's (2). Since the first two terms are partially cancelled by the other ones, the dielectric susceptibility is strongly reduced in the transition layer (z or z'< 1/P), which corresponds to the homogeneous dead layer of the Hopfield-Thomas model /2/. The transition-layer depth 1/P results vary, as a function of m_e/m_h, from 0.7 a_B to a_B, in contrast with a popular choice of $2a_B$ for the dead-layer depth /20/.

After the transition layer, the dielectric susceptibility assumes the parametrized form of Halevi and Hernandez-Cocoletzi /34/ with $U_x=U_z=U$, given by Eq.(10). In the model employing a homogeneous dead layer, U is related to the ABC after the dead layer; here the ABC is the vanishing of the exciton polarizability at the surface and is related in a complicated way to exciton parameters /17/. However, it is worthwhile to discuss U because it is related to the type of exciton scattering at the surface and it describes surface effects on the dielectric properties at long range, well after the transition-layer depth.

Note that U calculated according to Eq.(10) depends on the frequency and the angle of incidence ($\vec{K}_{||}$) through q, and on the transition-layer depth. At normal incidence, and at $\omega \cong \omega_0$, U is -1, corresponding to Pekar's ABC. However, as soon as $\omega > \omega_0$, the phase of U drastically changes and, moreover, when $\omega < \omega_0$, or in attenuated-total-reflectance (ATR), we have $|U|>1$, which is different from any values given by the ABC's quoted in the literature. Spatial dispersion and dead layer are not mutually independent. Their connection is evidenced by the dependence of U on the inverse transition-layer depth, P. In the extreme case of no dead layer ($P \to \infty$, for Frenkel excitons) the dependence of U on ω and $\vec{K}_{||}$ disappears and Pekar's ABC is recovered.

2.4 Optical Properties

The reflection coefficient of s- and p-polarized light is obtained by solving Maxwell's equations involving the nonlocal dielectric susceptibility (8), in Ref.17 e 29 respectively. ATR spectra, suitable to detect surface polaritons, can also be calculated according to the formulas given in /29/.

The transmission coefficient for a slab has been derived in Refs.35 and 36. The ABC has also been derived (see equation (43) of /17/) as a complicated expression involving the exciton parameters. Its physical meaning is however simple, being related to the vanishing of the exciton polarization at the surface, embodied in (8). As 1/P tends to zero, Pekar's ABC /1/ is recovered.

3. COMPARISON WITH EXPERIMENTS

We know that the present theory is able to describe exciton optical properties in a number of different experimental situations. We consider CdS reflectivity, as measured at various angles of incidence on high-purity samples, where extrinsic effects were minimized /20,26/. The same sample used to measure normal-incidence reflectivity /20/ was later used for Brillouin scattering /37/ from which reliable exciton parameters are extracted. A problem in interpreting experimental data is that of the proper normalization of the measured reflectivity, on which depends the height of the major peak above ω_0. In Ref.20 the normal-incidence spectrum was arbitrarily normalized to 0.31 at the frequency of 4890 cm^{-1}. We have normalized the same spectrum to get $R=(\epsilon_0^{1/2}-1)^2/(\epsilon_0^{1/2}+1)=0.24$ at this frequency (where the exciton contribution vanishes) and the value $\epsilon_0=8.1$ was extracted from the refraction-index measurements of Langer /38/.

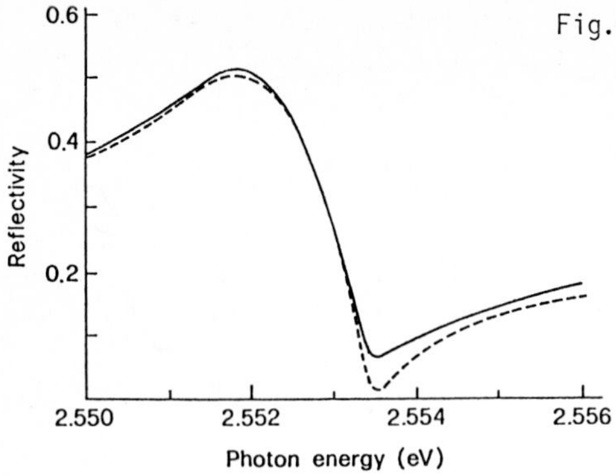

Fig.4 Normal incidence reflectivity for the $A_{n=1}$ exciton of CdS. The experimental spectrum (solid line) is of the 505/1 sample of Ref.20. Theoretical reflectivity (dashed line) is computed using $\hbar\omega_0$=2.5515 eV, ε_0=8.1, M=0.94m, $4\pi\alpha$= 0.013, and Γ=0.1 meV.

Figure 4 shows the good agreement with our calculations performed using a set of parameters which are in perfect agreement with those derived from Brillouin scattering experiments by Yu and Evangelisti /37/.

A less accurate normalization of these data was previously used /16,17,21,34/, obtained by normalizing the reflectivity to 0.24 inside the region where the exciton contribution is important. The resulting peak reflectivity is 0.38, instead of the present value 0.51. As a consequence, the calculations based on macroscopic models adjusted the dead-layer depth to 100 Å, a much bigger value than that given by the present microscopic approach of 20 Å, in order to describe the low (0.38) reflectivity peak (the reflectivity peak decreases as the dead-layer depth increases; see Fig.2 of Ref.34).

The proper normalization of experimental data adopted here nicely reconciles the present microscopic theory, yielding a small not adjustable transition layer, with reflectivity and Brillouin scattering results. A further confirmation of the small dead layer is the absence of the spike at the longitudinal-exciton frequency in experimental data, which instead appears in the best fit of Ref.34.
A crucial test for theory is that of angle-dependent reflectivity, which cannot be reproduced by the homogeneous dead-layer model without assuming angle-dependent exciton parameters /26/. Instead, the microscopic theory gives a very good account of p-wave reflectivity (Fig.5) as measured (and correctly normalized) at a number of incidence angles by Broser et al. /26/, using the same values of parameters as for the normal-incidence spectrum.

Good agreement between theory and experiment is also obtained for ZnO and ZnSe /29/. Here we show the ATR spectra of ZnSe. Our calculations (Fig.6) compare well with experimental results /39,40/, using parameters which well agree with those used to described normal-incidence reflectivity /17/, and also with those quoted in Tokura, Koda, Hirabayashi, and Nakada /40/. The high-energy side peaks, until now unexplained, might be due to the light-exciton branch.

There are however other materials where the agreement between theory and experiment is not so good. One among them is ZnTe: we did not succeed in obtaining a good description of normal- and oblique-incidence reflectivity using the same set of parameters /29/. This however does not mean that our model is not suitable for ZnTe, because there is no agreement in the literature on exciton parameters, and we did not carry out a full fitting procedure.

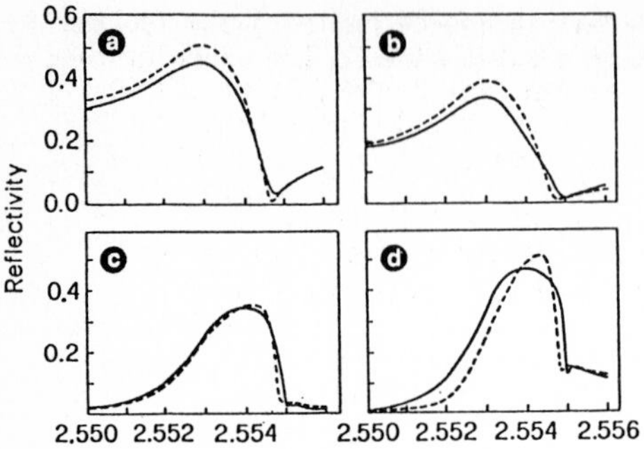

Fig.5

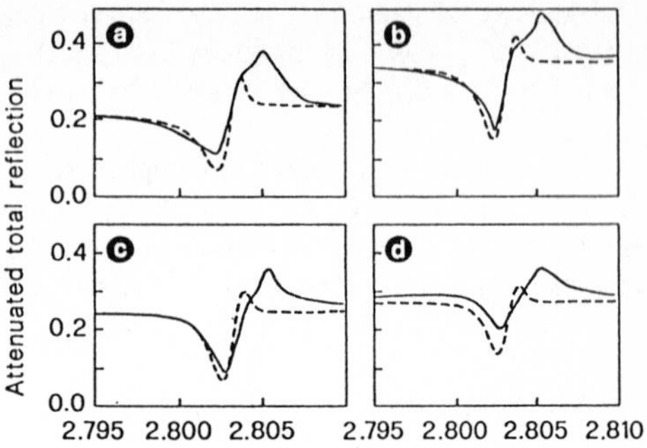

Fig.6

Fig.5 p-polarized angle-dependent reflectivity for the $A_{n=1}$ exciton of CdS. The incidence angles are $\theta_a=11°$, $\theta_b=48°$, $\theta_c=71°$, and $\theta_d=78°$. The experimental spectra (solid lines) are of the B-109 sample of Ref.26. Theoretical reflectivities (dashed lines) are computed using $\hbar\omega_0=2.5525$ eV, $\varepsilon_0=8.1$, M=0.94m, $4\pi\alpha=0.013$, and $\Gamma=0.1$ meV.

Fig.6 ATR in ZnSe. The incidence angles are $\theta_a=57.2°$, $\theta_b=62.8°$, $\theta_c=68.5°$, and $\theta_d=73.5°$. The experimental spectra (solid lines) are the type-A sample of Ref.40. Theoretical spectra (dashed lines) are computed with $\hbar\omega_0=2.802$ eV, $\varepsilon_0=8.1$, M=0.6m, $4\pi\alpha=0.88\times10^{-2}$, P=0.045 $\mathring{A}^{-1}$, and $\Gamma=0.4$ meV.

The cases of GaAs (shown in Fig.7) and of InP /29/ are clearer, since reliable values of exciton parameters can be taken from the literature. The experimental curve shown in Fig.7 was measured on a sample cleaved in helium atmosphere, in order to minimize the effects of extrinsic dead-layer /27/. We did not succeed, by varying P, to reproduce the step-like rise of reflectance on the high-energy side of Fig.7. In InP the discrepancy is smaller, but also evident. A better agreement is obtained within the homogeneous dead-layer approach /2/, using however a large dead-layer depth, of $2a_B=240$ Å /29/.

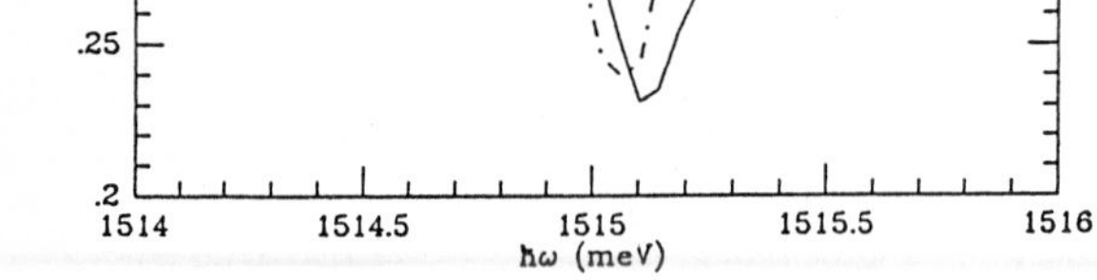

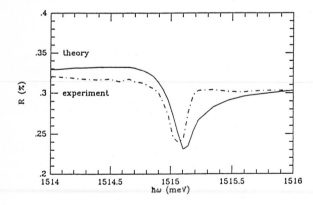

Fig.7 Exciton reflectivity of GaAs. Experiment from /27/.

In view of the similarity between GaAs and InP excitons, we believe that the origin of these discrepancies must be intrinsic. It might be due to the image potential or band degeneracy, both neglected in our model.

Finally we would like to emphasize that our calculations never yield the spike at the longitudinal exciton frequency. The same phenomenon occurs in experimental spectra of high-quality samples, so that we may conclude that the spike is a characteristic of low-quality samples, where large extrinsic dead layers occur.

4. CONCLUSION

We have shown that it is possible to compute the dielectric susceptibility of the vacuum-semiconductor system within a simple - yet realistic - model, essentially without introducing further approximations in addition to those embodied in the model itself. In this way the correct ABC and the transition-layer depth are derived from wave functions. The results obtained using Peker's ABC, without dead layer, are qualitatively similar to ours, and also to experiments /29/ carried out on good-quality samples. We may conclude that the first idea about the problem of exciton optical properties was essentially correct.

The present model neglects band degeneracy, the image potential, surface states and the electric field of the space charge region. In spite of these approximations, it gives a good description of exciton reflectance in several materials. We believe that it can be implemented to yield a good description also for other materials, where the agreement between experiment and theory is at present unsatisfactory.

A different method to treat excitons is the coherent wave approach /30/. In this case one deals with equations involving directly the exciton polarization, driven by the electromagnetic field, rather than the wave function. As far as this approach is restricted to linear optics, it is in principle completely equivalent to the traditional approach, dealing with wave functions and linear response. However the coherent wave formalism seems to be more complicated from the mathematical point of view, so that up to now it has been applied only to one-dimensional model semiconductors. It is therefore not meaningful to compare its results with those obtained within the realistic model described here.

5. REFERENCES

1. S.I. Pekar, Zh. Eksp. Teor. Fiz. 33, 1022 (1957) [Sov. Phys.-JEPT 6, 785 (1958)]; Fiz. Tverd. Tela (Leningrad) 4, 1301 (1962) [Sov. Phys.-Solid State 4, 953 (1962)]
2. J.J. Hopfield and D.G. Thomas, Phys. Rev. 132, 563 (1963)
3. A.A. Maradudin and D.L. Mills, Phys. Rev. B 6, 2787 (1973)
4. G.S. Agarwal, D.N. Pattanyak and E. Wolf, Phys. Rev. Lett. 27, 1022 (1971); Opt. Commun. 4, 255 (1971); 4, 260 (1971)
5. J.L. Birman and J.J. Sein, Phys. Rev. B 6, 2482 (1972)
6. R. Zeyher, J.L. Birman and W. Brening, Phys. Rev. B 6, 4613 (1972)
7. D.L. Johnson and P.R. Rimbey, Phys. Rev. B 14, 2398 (1976)
8. M.R. Philpott, Phys. Rev. B 14, 3471 (1976)
9. C.A. Mead, Phys. Rev. B 15, 519 (1977)

10. C.A. Mead and M.R. Philpott, Phys. Rev. B 17, 914 (1978)
11. C.A. Mead, Phys. Rev. B 17, 4644 (1978)
12. C.S. Ting, M.J.Frankel and J.L.Birman, Solid State Commun. 17, 1285 (1975)
13. S. Sakoda, J. Phys. Soc. Jpn. 40, 152 (1976)
14. I. Balslev, Phys. Status Solidi B 88, 155 (1978)
15. I. Balslev, Solid State Commun. 39, 359 (1981)
16. A. D'Andrea and R. Del Sole, Solid State Commun. 30, 145 (1979)
17. A. D'Andrea and R. Del Sole, Phys. Rev. B 25, 3714 (1982)
18. A. D'Andrea and R. Del Sole, Phys. Rev. B 29, 4782 (1984)
19. A. D'Andrea and R. Del Sole, Phys. Rev. B 32, 2337 (1985)
20. F. Evangelisti, A. Frova and F. Patella, Phys. Rev. B 10, 4253 (1974)
21. F. Patella, F. Evangelisti and M. Capizzi, Solid State Commun. 20, 23 (1976)
22. D.D. Sell, S.E. Stokowski, R. Dingle and J.V. Di Lorenzo, Phys. Rev. B 7, 4568 (1973)
23. S. Feierabend and H.G. Weber, Solid State Commun. 26, 191 (1978)
24. W. Stossel and H.J. Wagner, Phys. Status Solidi (b) 89, 403 (1978)
25. W. Stossel and H.J. Wagner, Phys. Status Solidi (b) 96, 369 (1979)
26. I. Broser, M. Rosenzweig, R. Broser, M. Richard and E. Birkicht, Phys. Status Solidi (b) 90, 77 (1978)
27. L. Schultheis and I. Balslev, Phys. Rev. B 28, 2292 (1984)
28. J. Lagois, Phys. Rev. B 10, 5511 (1981)
29. A. D'Andrea and R. Del Sole, to be published
30. I. Balslev, these proceedings, next paper; and references therein
31. R. Del Sole, J. Phys. C 8, 2971 (1975)
32. M.F. Deigen and M.D. Glinchuck, Fiz. Tverd. Tela (Leningrad) 5, 3250 (1964) [Sov. Phys.-Solid State 5, 2377 (1964)]
33. S. Satpathy, Phys. Rev. B 28, 4585 (1983)
34. P. Halevi and G. Hernandez-Cocoletzi, Phys. Rev. Lett. 48, 1500 (1982)
35. K. Cho, J. Phys. Soc. Jpn. 54, 4431 (1985)
36. A. D'Andrea and R. Del Sole, these proceedings.
37. P. Yu and F. Evangelisti, Phys. Rev. Lett. 42, 1642 (1975)
38. D.W. Langer, J. Appl. Phys. 37, 3530 (1966)
39. Y. Tokura, I. Hirabayashi and T. Koda, J. Phys. Soc. Jpn. 42, 1071 (1977)
40. Y. Tokura, T. Koda, I. Hirabayashi and S.Nakada, J. Phys. Soc. Jpn. 50, 145 (1981)

Exciton Polaritons near Surfaces: The Coherent Wave Approach

I. Balslev

Fysisk Institut, Odense Universitet, DK-5230 Odense M, Denmark

It is shown that the old problem of describing the coupling through a semiconductor surface between external light and excitons can only be handled rigorously by taking full account of the coherent coupling between electromagnetic and quantum-mechanical waves. The coherent wave approach developed in the last 5 years is applied and discussed.

1. Introduction

A persistent problem in semiconductor optics has been the understanding of the coupling through a surface between external light and excitons in direct gap materials. Since the first works by PEKAR [1] and by HOPFIELD and THOMAS [2] a large number of experimental and theoretical studies have been devoted to exciton polaritons near semiconductor surfaces [3-17].

The purpose of the present paper is to give a supplementary clarification of the problem and to demonstrate that the necessary theoretical approach must include a coherent coupling between quantum and electromagnetic wave phenomena. The mathematically simplest theory with this property is the coherent wave theory developed by STAHL [15]. In order to demonstrate the power of this approach the present work includes a thorough comparison with standard theories of the optical response of excitons such as those of ELLIOTT [18] and HOPFIELD-THOMAS [2]. In this comparison it is particularly helpful to use a one-dimensional exciton polariton model [15].

The structure of bulk exciton polaritons will be discussed briefly in Sect. 2. In the simple case of infinite hole mass discussed in Sect. 3 the origin of the excitonfree layer below a semiconductor surface becomes clear. This case is also suitable for comparing the optical response of a one- and a three-dimensional exciton model. The full complexity of the surface effects of exciton polaritons is present when the electron and hole masses are comparable and at the same time the optical retardation over one exciton Bohr radius is significant. This case is discussed in Sect. 4.

2. Infinite Crystal

Let us consider a direct gap semiconductor with a gap frequency ω_g, electron mass m_e and hole mass m_h. In a conventional treatment the bulk optical response of Wannier excitons is described by a susceptibility χ_{nonloc} dependent on frequency ω and wave vector k as follows:

$$\chi_{nonloc}(\omega,k) = \sum \frac{f_n}{\omega_n^2+\beta_n k^2+i\Gamma_n\omega-\omega^2} + \int \frac{g(\omega_o)d\omega_o}{\omega_o^2 - \omega^2} . \tag{1}$$

The first term includes the excitonic resonances while the second term includes the absorption continuum above the gap frequency (for simplicity we neglect the nonlocal response of this continuum). f_n, ω_n, β_n, and $g(\omega)$ can be found by solving for the electron-hole motion for which the Hamiltonian is given by

$$H_o = \hbar\omega_g - \frac{\hbar^2}{2m_e}\nabla_e^2 - \frac{\hbar^2}{2m_h}\nabla_h^2 + V_{eh} , \tag{2}$$

where V_{eh} is the electron hole interaction. In particular, the ground state (n=1) has $\omega_1 = \omega_g-\omega_x$ where $\hbar\omega_x$ is the exciton Rydberg and the wave function in relative space has the effective Bohr radius a_B. Bulk polariton modes are then derived by inserting the above susceptibility into Maxwell's equations [2].

In the coherent wave approach the quantum-mechanical and electromagnetic dynamics is combined in a set of directly coupled equations relating the electric field $\underline{E}$ and the macroscopic electron-hole amplitude Y [19]:

$$i\dot{Y}(\underline{R},\underline{r},t) - \Omega_{eh}Y(\underline{R},\underline{r},t) = -(\underline{M}_o/\hbar)\delta(\underline{r})\underline{E}(\underline{R},t) , \tag{3a}$$

$$\varepsilon_r\ddot{\underline{E}}(\underline{R},t) - c^2\nabla_{\underline{R}}^2\underline{E}(\underline{R},t) = -(2\underline{M}_o/\varepsilon_o)Re(\ddot{Y}(\underline{R},0,t)) . \tag{3b}$$

Here $Y(\underline{R},\underline{r},t)$ is the expectation value of the operator which at time t annihilates a pair with center of mass at $\underline{R}$ and relative coordinate $\underline{r}$, ε_r is a background dielectric constant and $\underline{M}_o$ is the transition dipole moment. (In principle the delta function in (3a) should have a finite width. There is a well defined relation between this width and the parameter ε_r [19]). The spatial operator $\hbar\Omega_{eh}$ is identical with the effective mass Hamiltonian (2) operating on the coordinates $\underline{R},\underline{r}$. The bulk modes of Eq.(3) are identical with those derived from an Elliott-Hopfield-Thomas treatment including all the essentially undamped polaritons below ω_g and the damped mode above ω_g creating real electron-hole pairs.

The nonlocal character of the excitonic response is a result of the translational degree of freedom. Thus β_n in (1) is given by $\beta_n \approx \hbar\omega_g/2m$ where $m = m_e+m_h$. The typical spatial range of the nonlocal response is the reciprocal wave vector of the translational exciton motion, and this length becomes large near resonances. Such a length associated with the n=1 resonance is given by [20]

$$\lambda_{transl} = \sqrt{\frac{\hbar}{2m|\omega_g-\omega_x-\omega|}} . \tag{4}$$

If any of the quantal lengths a_B and λ_{transl} become comparable to the polariton wavelength, then a surface introduces a complicated interference between quantal and electromagnetical waves.

3. Surface Effects in Case of Infinite Hole Mass

Let us now consider the simple case of $m_h \to \infty$ and electron motion in the z direction only. A suitable electron-hole interaction is then of short range [15,17]:

$$V_{eh} = -2\hbar\omega_{x1}a_{B1}\delta(z_e-z_h), \tag{5}$$

where a_{B1} and $\hbar\omega_{x1}$ are Bohr radius and binding energy of the bulk one-dimensional exciton. Let the system consist of a crystal in the halfspace z>0, in which case the Hamiltonian is given by

$$H(Z,z) = H_o + B\Theta(-Z-z), \tag{6}$$

where the center of mass Z is z_h and the relative coordinate z is z_e-z_h. Θ is the unit step function. The barrier height B can be considered to be much larger than $\hbar\omega_{x1}$ and so we may apply the limit $B \to \infty$.

Treating now Z as a parameter in the quantum-mechanical problem, an Elliott-type approach leads to a single bound state with the wave function

$$\psi(z) = \psi(0) \begin{cases} \exp(-\lambda z); & z>0, \\ \sinh(\lambda(Z+z))/\sin(\lambda Z); & -Z<z<0 \end{cases} \tag{7}$$

and the eigen energy

$$\hbar\omega_o(Z) = \hbar\omega_g-\hbar^2\lambda^2/(2\mu), \tag{8}$$

μ is the reduced mass and λ is a Z dependent parameter. The ionized excitons have the wave functions

$$\psi(z) = \psi(0) \begin{cases} \sin(kz+\varphi)/\sin\varphi; & z>0, \\ \sin(k(Z+z))/\sin(kZ); & -Z<z<0, \end{cases} \tag{9}$$

where the energy is $\hbar\omega_g+\hbar^2k^2/(2\mu)$. While the relative-space wave number k is a free parameter, λ, φ and $\psi(0)$ are determined by normalization and the jump condition invoked by the interaction (5). The inverse function $Z(\omega_o)$ of $\omega_o(Z)$ can be expressed as

$$Z(\omega_o) = - \frac{a_{B1} \log(1 - \sqrt{(\omega_g - \omega_o)/\omega_{x1}})}{2\sqrt{(\omega_g - \omega_o)/\omega_{x1}}} \quad . \tag{10}$$

This function is shown in Fig. 1. Note that there is no bound state for $Z<a_{B1}/2$. The oscillator strength proportional to $\psi^2(0)$ (not shown) goes monotonically to zero when Z approaches $a_{B1}/2$. The fact that the surface region is detuned justifies the use of an excitonfree layer ℓ as first suggested by HOPFIELD and THOMAS [2]. This approximation corresponds to considering $\omega_o(Z)$ equal to $\omega_o(\infty)$ for $Z>\ell$ and infinite for $Z<\ell$.

The optical response is that of an inhomogeneous medium with an imaginary part of the susceptibility. In the absence of damping, χ'' is given by

$$\chi''(\omega,Z) = C_1\delta(\omega_o(Z)-\omega) + C_2\theta(\omega-\omega_g)/k(\omega), \tag{11}$$

where C_1, C_2 are Z-dependent constants and $k(\omega)$ is the relative-space wave number of ionized exciton with energy $\hbar\omega$. The real part of $\chi(\omega,Z)$ can be found by a Kramers-Kronig (K-K) transformation of (11).

The result (11) and its K-K transform is derived much more elegantly when the coherent wave theory is applied to the one-dimensional model with $\hbar\Omega_{eh} = H(Z,z)$ of Eqs.(2,5,6). First we note that the right hand side of Eq.(3b) is essentially the excitonic susceptibility times $-\ddot{\underline{E}}(\underline{R})$. Next we insert in (3) a harmonic time variation which gives rise to a rotation and a counterrotating component of Y. Each of the components Y_- and Y_+ can be written as a product of functions of the variables Z, z, and t. The resulting complex susceptibility found in this way can be written as

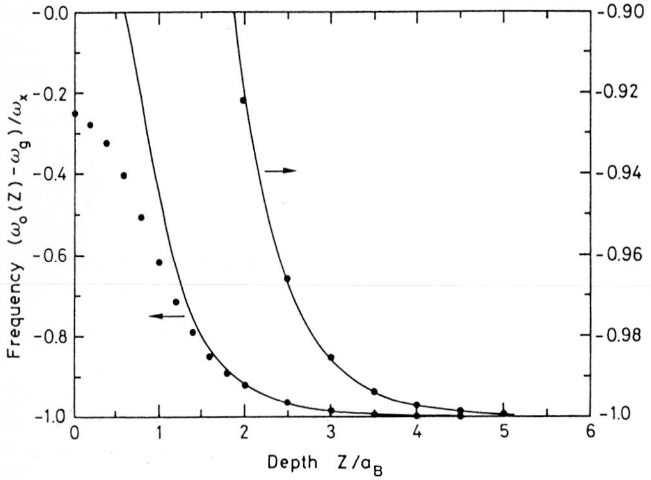

Fig.1.- Local resonance frequency ω_o as function of the hole coordinate Z. Full curves are calculated from the one-dimensional model (Eq. 10) with $a_{B1}=1.2a_B$. The dots are calculated for 3d Wannier excitons [22].

$$\chi(\omega,Z) = \chi_-(\omega,Z) + \chi_+(\omega,Z), \tag{12}$$

$$\chi_\pm(\omega,Z) = \frac{2M_o^2}{\varepsilon_o \hbar\omega_x a_{B1}} \frac{1}{\kappa_\pm a_{B1}(\coth(\kappa_\pm Z)+1) - 2}, \tag{13}$$

$$\kappa_\pm = a_{B1}\sqrt{(\omega_g \pm \omega - i\gamma)/\omega_x}, \tag{14}$$

where γ is a dephasing rate (or equal to 0_+ in the absence of irreversible processes). Both above and below the ω_g the z dependence of $Y_\pm$ is the same as that given for $\psi(z)$ in Eq.(7) with λ replaced by $\kappa_\pm$. Near resonance the anti-resonant part $\chi_+(\omega,Z)$ is real and a smooth function of ω, and it is easily verified that $\chi_-(\omega,Z)$ has poles at $\omega = \omega_0(Z)$ given by the inverse function of (10). Other comparisons between the result in (12-14) and the conventionally derived response in (11) show that the two approaches lead to the same inhomogeneous response. The spectrum of the bulk susceptibility $\chi_-(\omega,\infty)$ is shown in Fig. 2.

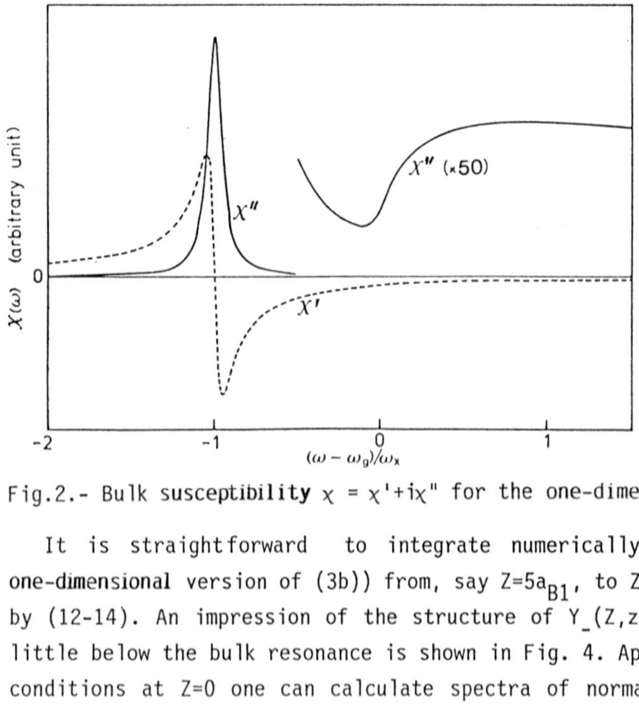

Fig.2.- Bulk susceptibility $\chi = \chi' + i\chi''$ for the one-dimensional model.

It is straightforward to integrate numerically Maxwell's equations (a one-dimensional version of (3b)) from, say $Z=5a_{B1}$, to Z=0 with the reponse given by (12-14). An impression of the structure of $Y_-(Z,z)$ near a surface for ω a little below the bulk resonance is shown in Fig. 4. Applying Maxwell's boundary conditions at Z=0 one can calculate spectra of normal incidence reflectance. Typical results are shown in Fig. 3. The spectral shape of the reflectance is sensitively dependent on the parameter $k_b a_{B1}$ ($k_b^2 = \varepsilon_r(\omega_g/c)^2$) determining the ratio between the Bohr radius and the typical electromagnetic length. The classical response is obtained for $k_b a_{B1} \approx 0$ while strong modifications occur for $k_b a_{B1} > 0.1$. It is important to note that these calculated spectra for $k_b a_{B1} = 0.5$ and 0.1 contain several essential features found in experimental spectra of GaAs [8,21] and CdS [6].

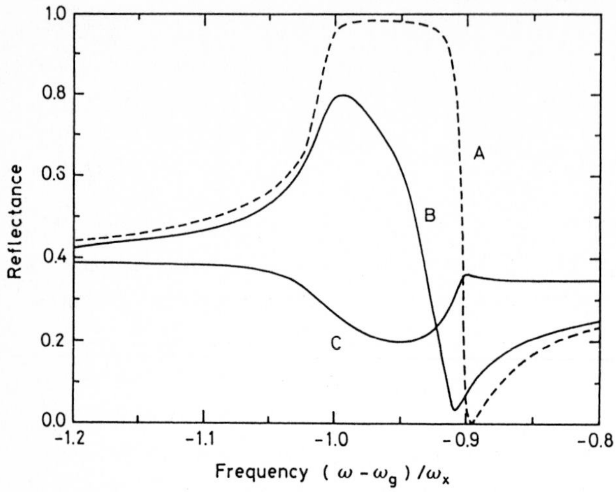

Fig. 3 - Reflectance spectra calculated from the one-dimensional model with infinite hole mass, $\gamma = 0.005\ E_X$, $k_b = 3\omega/c$, and three values of $k_b a_B$: 0.0001 (curve A), 0.1 (curve B), 0.5 (curve C)

In order to relate the parameters of the one-dimensional model to those of a three-dimensional (3d) one it is relevant to compare the result (10) with the corresponding detuning of 3d Wannier excitons near a surface. SATPATHY [22] has calculated $\omega_0(Z)$ in this case and the result is shown in Fig. 1. Except in regions with a very large detuning ($0<Z<2a_{B1}$), $\omega_0(Z)$ is the same in the one- and the three-dimensional model if

$$a_{B1} = 1.2\ a_B\ ; \quad \hbar\omega_{X1} = \hbar\omega_X\ . \tag{15}$$

This identification of a_{B1} leads to a good quantitative agreement with the reponse given by (12-14) and experimental spectra. For example curve C in Fig.3 with $k_b a_{B1} = 0.5$ is close to the experimental spectrum of GaAs [8,21] and in this material $k_b a_B$ is known from other considerations to be 0.38.

The surface properties of infinite-hole-mass-exciton polaritons can be studied in three dimensions without severe problems [10], but the susceptibility can not be expressed by standard functions as is the case in Eq. (13). Not only Maxwell's equations, but also the relative motion must be solved by suitable numerical methods, and in this case it is easy to incorporate in an adiabatic approximation the influence of a finite hole mass, as discussed in the next section.

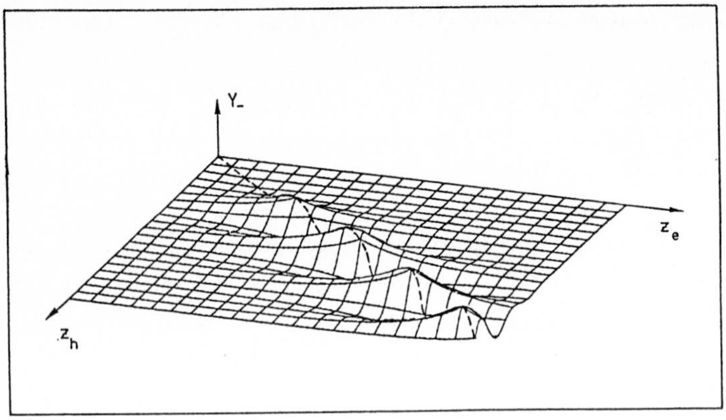

Fig. 4. - Real part of the electron hole amplitude $Y\_$ as function of electron and hole coordinates z_e and z_h in case of a polariton frequency slightly below resonance. Note that the dead layer near a surface is a dead corner in the z_e-z_h space.

4. Finite Translational Exciton Mass

Let us now consider the general situation with finite translational exciton mass. Then there are three typical lengths involved in the polariton propagation, namely a_B, λ_{transl}, and $1/k_b$. We saw in the previous section that the limit $a_B \gg \lambda_{transl}$ can be handled without severe problems. As for the opposite limit we are dealing with a spatially dispersive Frenkel-type resonance. In this case the surface creates the need for an additional boundary condition (ABC), a problem which has been discussed in several works [1-6,16] in the last 30 years. It turns out, however, that the limit $a_B \ll \lambda_{transl}$ is of minor importance for semiconductor excitons because a_B and λ_{transl} are comparable in essential parts of the spectra. The typical width of a resonance, splitting ω_{LT}, is of the order $0.1\omega_x$ in common semiconductors. Therefore one has the typical value

$$\lambda_{transl}/a_B \approx \sqrt{10\mu/m} \tag{16}$$

and this number is of the order unity in most direct gap semiconductors. This fact makes the excitonic response near a semiconductor surface extremely complicated: at resonance the polariton wavelength becomes comparable to both quantal lengths, and in that case one can reduce the surface problems to neither an ABC nor a question of inhomogeneous response. The first to see this were HOPFIELD and THOMAS [2] who introduced a phenomenological excitonfree layer with adjustable thickness and suggested the Pekar ABC at the interface.

If the onedimensional model includes a finite exciton mass, then the Hamiltonian $H(Z,z)$ contains a center-of-mass kinetic energy and two potential barriers, one at $z_e=0$ and one at $z_h=0$. Within the framework of a conventional approach one can apply the adiabatic approximation in which the center-of-mass kinetic energy is neglected when solving for the relative motion between the above two barriers. The relative motion can be studied as function of the center-of-mass depth Z and defines in this way local oscillator parameters (frequency $\omega_n(Z)$ and strength $f_n(Z)$) of each bulk resonance, but the nonlocal coupling between these oscillators cannot be derived from the adiabatic approach. As a crude approximation (not justified by microscopic considerations) one can assume that the nonlocal coefficients β_n of (1) are independent of Z. One can then integrate numerically the equation of motion for the resonant polarization together with Maxwell's equation (3b). There is still a need for an ABC, but as the detuning at the surface is very large (see Fig. 1) the calculated transmission properties of the surface depend little on the ABC.

The coupled equations relevant when the coherent wave theory is applied to the one-dimensional model are as follows:

$$(H(Z,z)-\hbar\omega-i\hbar\Omega)Y_-(Z,z) = \frac{1}{2} M_0\delta(z)E(Z), \tag{17a}$$

$$(\frac{c^2}{\omega^2}\frac{d^2}{dZ^2} + \varepsilon_r)\varepsilon_0 E(Z) = -2M_0 Y_-(Z,0), \tag{17b}$$

where ε_r now includes the bulk antiresonance contribution to the edge response. The halfspace solutions to (17) are valid for any mass ratio, but can only be found by numerical methods. Previous works have employed real-space integration in Z-z-space [17] and Fourier methods [25]. Such studies can be used for checking the validity range of the mass ratio when using the adiabatic approximation.

A more realistic application of the coherent wave theory is to set up equations similar to (17) in three dimensions. The resulting six-dimensional boundary value problem can be reduced to a three-dimensional one by symmetry, but even a 3d boundary value problem is known to be very demanding with respect to computer time and core memory if brute-force numerical methods are used. Further work is needed in the search for fast iterative or variational methods which can bring an 'exact' numerical treatment within reach of ordinary computers.

It is appropriate here to discuss two approximate calculation schemes which can be applied to a threedimensional model without severe calculational problems, namely the adiabatic approximation and the superposition method developed by ZEYHER [16] and D'ANDREA and DEL SOLE [12, 26].

If the mass ratio m_e/m_h is not too close to unity then the adiabatic approximation should be rather accurate. As discussed in Refs. 10, 13 the motion in a threedimensional relative space can be explored by suitable variational methods. The most recent trial function is suggested by SCHULTHEIS and BALSLEV [21]. It is worth noticing that the adiabatic calculation scheme is suitable for including image charge forces, band bending (ref. 21 and the Author's second paper in the present proceeding), and n=2 resonances [24]. Oblique incidence and attenuated total reflection can also be handled, but such calculations have not been reported so far.

D'ANDREA and DEL SOLE [26] also developed a model based on "no escape" boundary conditions in electron-hole configuration space. In contrast to the coherent wave approach this theory is based on expansions in terms of stationary states. A particularly simple approximation is the use of a single evanescent electron-hole wave describing the admixture near the surface of exciton states which for zero wave vector have higher energies [26]. As discussed in other contributions to the present proceedings (papers by CHO, D'ANDREA and DEL SOLE) the superposition method can be extended to include several excited exciton states. So far, the inclusion of the continuum states of ionized excitons in the basis has not been reported. If this is done and all closed orbit excitons are included, then the superposition method should produce spectra identical with those obtained from the coherent wave theory.

5. Concluding Remarks

Although the coherent wave theory provides certain computational problems when applied to a three-dimensional halfspace geometry, it clarifies conceptually the complexity of exciton confinement. Whenever the translational symmetry is broken, the resulting coupling between quantum waves in electron-hole (e-h) space and electromagnetic waves in center-of-mass space must be treated without loss of coherence. The boundary condition on Y is applied in a higher dimensional space than is the case for ordinary electrodynamics. But once applied, the boundary condition in e-h space generates the halfspace response of the edge as a whole, including all excitonic resonances and the Coulomb-enhanced continuum.

Acknowledgement

Valuable and frequent discussions with A. Stahl are gratefully acknowledged.

References:

1. S.I. Pekar, Zh. Exper. Fiz. 33, 1022 (1957) [Eng. transl.: Sov. Phys.-JETP 6, 785].
2. J.J. Hopfield and D. G. Thomas, Phys. Rev. 132 (1963).
3. V.M. Agranovich and V.L. Ginzburg, Spatial Dispersion in Crystal Optics and the Theory of Excitons (Springer, Berlin, 1966), Crystal Optics with Spatial Dispersion and Excitons (Springer, Berlin, 1984).
4. R. Zeyher, J.L. Birman and W. Brenig, Phys. Rev. B6, 4613 (1972).
5. T. Skettrup and I. Balslev, Phys. Rev. B3, 1457 (1971).
6. A.A. Maradudin and D.L. Mills, Phys. Rev. B7, 2787 (1973)
7. F. Evangelisti, A. Frova, and F. Patella, Phys. Rev. B10, 4253 (1974).
8. D.D. Sell, S.E. Stokowski, R. Dingle, and J.V. DiLorenzo, Phys. Rev. B7, 4568 (1973).
9. S. Sakoda, J. Phys. Soc. Japan 40, 152 (1976).
10. I. Balslev, Phys. Stat. Sol. (b) 88, 155 (1978).
11. P. Halevi and R. Fuchs, in Proc. 14th Int. Conf. Phys. Semicond., Edinburgh, 1978, B.H.L. Wilson, ed. (Institute of Physics, Bristol, 1979), p. 863, P. Halevi and G. Hernandez-Cocoletzi, Phys. Rev. Lett. 48, 1500 (1982).
12. A. D'Andrea and R. Del Sole, Solid State Commun. 30, 145 (1979).
13. I. Broser and M. Rosenzweig, Phys. Stat. Sol. (b) 95, 141 (1979).
14. J. Lagois, Phys. Rev. B23, 5511 (1981).
15. A. Stahl, Phys. Stat. Sol. (b) 106, 575 (1981).
16. R. Zeyher, in Condensed Matter Physics Vol. 1, J.T. Devreese, ed. (Plenum, New York,1981).
17. I. Balslev and A. Stahl, Phys. Stat. Sol. (b) 111, 531 (1982).
18. R.J. Elliott, Phys. Rev. 108, 1384 (1957).
19. A. Stahl and I. Balslev, Electrodynamics of the Semiconductor Band Edge, Springer Tracts in Modern Physics 110 (1987).
20. M. F. Bishop and A.A. Maradudin, Phys. Rev. B14, 3384 (1976).
21. L. Schultheis and I. Balslev, Phys. Rev. B28, 2292 (1983).
22. S. Satpathy, Phys. Rev. B28, 4585 (1983).
23. A. D'Andrea and R. Del Sole, Phys. Rev. B32, 3227 (1985).
24. I. Balslev, Solid State Commun. 45, 661 (1983), Phys. Rev. B28, 5665 (1983).
25. L. Gotthard, A. Stahl, and G. Czajkowski, J. Phys. C 17, 4865 (1984)
26. A. D'Andrea and R. Del Sole, Phys. Rev. B29, 4782 (1984).

Surface-Exciton Polaritons: A Few Fresh Ideas

P. Halevi[1], *G. Hernández-Cocoletzi*[1], *and F. Ramos*[2]

[1]Universidad Autónoma de Puebla, Apdo. Post. J-48,
Puebla, Pue. 72570, México

[2]Universidad de Sonora, Apdo. Post. A088,
Hermosillo, Son. 83190, México

We discuss three ideas related to surface and interface exciton polaritons.(1) It is shown, for the A-exciton of CdS, that a large exciton-free layer is no impediment to observing surface polaritons. Our calculation leads to deep Attenuated Total Reflection (ATR) minima. A 100 Å thick exciton-free layer causes the minima to shift ~ 0.5 meV to lower energies, almost independently of the angle of incidence. (2) The dispersion relation for surface-exciton polaritons has been solved in the presence of wavevector-linear splitting of energy bands as characteristic of the B-exciton of CdS. Assuming that the propagation vector q_x is a real quantity and that the frequency ω is complex we find two branches $\mathrm{Re}\omega(q_x)$. The higher branch is similar to the dispersion curve of the surface A-exciton. For the lower branch $\omega \lesssim \omega_T$ (the transverse-exciton frequency), with little dispersion. This branch seems to correspond to additional minima recently found in ATR spectra of the B-exciton (P. Halevi et al., Phys. Rev. B 32, 6986 (1985)), strengthening the case for a new "nonlocal" surface polariton.(3) It is shown that excitons and plasmons may interact across a semiconductor/metal interface, leading to a coupled polariton. For a simple model two propagation regions are found, one below ω_T and the other above ω_L. An ATR calculation shows that this mode can be readily excited.

1. Introduction

Generally speaking, the properties of surface-exciton polaritons are well understood [1,2]. In this paper we briefly describe three simple and new ideas. They deal with the effect of the exciton-free layer on the surface A-exciton; a "nonlocal" surface B-exciton; and exciton-plasmon polaritons at the interface between a semiconductor and a metal. We will deal with these topics in secs. 2, 3, and 4, respectively. All three involve Attenuated Total Reflection (ATR) spectroscopy and we hope to spur experimental interest in these ideas.

2. The surface A-exciton in CdS : is it observable ?

The A(n = 1) exciton of CdS has been, by far, the preferred choice of experimentalists in their studies of bulk excitons by means of ordinary reflection spectroscopy. Thus it is suprising that - to our knowledge - no ATR spectra have been reported for this semiconductor. Perhaps the explanation is given by a (misleading) statement by Lagois and Fischer [3] upon reporting the first observation of surface-exciton polaritons : "The exciton-free surface layer has to be as thin as possible in order to excite excitons by an ATR technique... Only for thin exciton-free surface layers are the exponentially decaying electromagnetic waves still sufficiently intense at the crystal regions where excitons exist. From the free-exciton reflectance spectra it is known that the exciton-free surface layer on ZnO is less than about 30 Å, probably the smallest one of the II - VI semiconducting compounds. For this reason we use ZnO crystals in our experiment." This conveys the impression that it is not possible to excite surface-exciton polaritons in CdS; in most studies of this semiconductor it is concluded that the "dead layer" has a thickness of the order of 100 Å [4].

The fallacy of the above-quoted reasoning is due to the fact that the amplitude of the surface polariton does not decrease exponentially inside the exciton-free layer. In an ATR experiment (frequency-scan) the polariton wavevector is

$q_x = (\omega_{min}/c)\varepsilon_p^{1/2} \sin\theta$, where ε_p is the dielectric constant of the prism and θ is the angle of incidence inside the prism. Therefore the perpendicular component of the wavevector inside the dead layer is

$$q_z = (\varepsilon_o \omega_{min}^2/c^2 - q_x^2)^{1/2} = (\omega_{min}/c)(\varepsilon_o - \varepsilon_p \sin^2\theta)^{1/2}. \quad (1)$$

We have characterised the dead layer by a real and frequency-independent dielectric constant ε_o. For typical values of ε_o and ε_p one always has $\varepsilon_o > \varepsilon_p$. It follows that q_z is a real quantity (neglecting damping). Therefore inside the exciton-free layer the fields have a waveguide character; the exponential decay does not start from the surface, but rather from the interface between the dead layer and the bulk. The conclusion is that, in an ATR experiment, there should be no difficulty in reaching the region of existence of the excitons.

We have performed an ATR calculation for the configuration shown in the inset of Fig. 1. The dielectric function $\varepsilon(\omega,q)$ of the nonlocal bulk is given by the usual Hopfield-Thomas model [4]. The problem of Additional Boundary Conditions (ABC's) has been discussed in detail in [4]. On the basis of the findings reviewed there we may apply the Pekar ABC at the interface between the bulk and the surface layer. That is, we assume that the excitonic polarization vanishes at this interface. Then the surface impedance Z_p (for p-polarised light) of the semiconductor, including dead layer, is given by eq. (17) in [4] with the surface-scattering parameters $U_x = U_z = -1$, corresponding to the Pekar ABC. The calculation of the reflectivity R_p involves straightforward matching of E_x and H_x field-components at the interfaces prism/air, air/dead-layer, and dead-layer/bulk. We use parameters appropriate to the A(n = 1) exciton of CdS, determined by means of Brillouin scattering [5]. In addition, the thickness of the exciton-free layer is chosen as ℓ = 100 Å [4], ε_p = 2.25, and the width of the air-gap is 0.3μm. The ATR spectra, for a series of angles of incidence θ, are shown in Fig. 1.

The first thing to notice is that well-defined, deep minima are obtained. In fact, for the same θ, the ATR minima of the A-exciton of CdS are much deeper than those of the C-exciton of ZnO. Thus it is clear that the presence of a substantial dead layer is no obstacle at all to ATR spectroscopy of surface-exciton polaritons. If we plot ω_{min} versus q_x for the different θ we obtain the lower (continuous) curve in Fig. 2. This, of course, is the "optical" dispersion relation of the

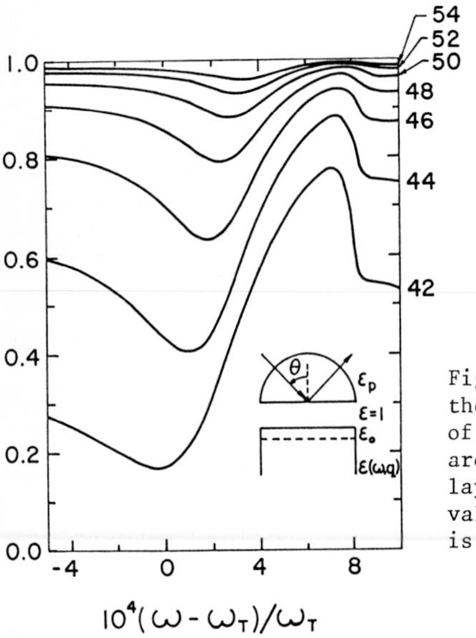

Fig. 1 Attenuated Total Reflection spectra of the A(n = 1) exciton of CdS for several angles of incidence θ(in degrees). The calculations are based on the Pekar ABC and an exciton-free layer 100 Å thick. See text and [5] for the values of the other parameters. The geometry is shown in the inset.

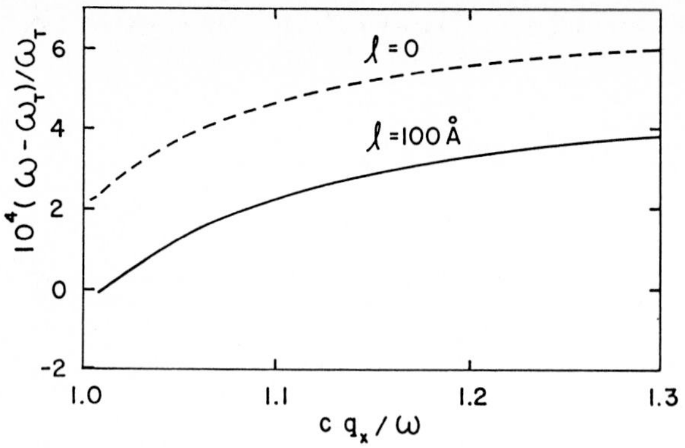

Fig. 2 The "optical" dispersion relation of the A(n = 1) surface-exciton polariton of CdS, determined from the ATR minima of Fig. 1 (lower, continuous curve). For comparison we also show the corresponding dispersion relation with the exciton-free layer absent (upper, dashed curve). The vertical shift is about 0.5 meV. The propagation vector q_x is normalized to the vacuum wavevector ω/c, so the light-line coincides with the ordinate.

surface-exciton polariton. A similar calculation without allowance for the exciton-free layer ($\ell = 0$) gives the upper (dashed) curve. We see that the dead layer causes the surface-exciton frequencies to shift to lower values, as noted before for ZnO [6]. The shift for CdS is about − 0.5meV, almost independent of q_x. This dispersion of the surface-exciton-polariton frequencies may be understood from the following simple model. We neglect the nonlocal and damping effects ($m \to \infty$, $\nu = 0$) and extend the dead layer to become a semiinfinite bounding medium ($\ell \to \infty$). As is well known, the limiting ($q_x \to \infty$) frequency of the surface polariton is given by $\varepsilon(\omega) = -\varepsilon_o$. In the absence of a dead layer the bounding medium is vacuum, so the limiting frequency ω_s satisfies the equation $\varepsilon(\omega_s) = -1$. Using that $\varepsilon(\omega) = \varepsilon_o(\omega^2 - \omega_L^2) / (\omega^2 - \omega_T^2)$ we readily find that the difference in limiting frequencies is

$$\Delta\omega_s \simeq - \frac{1}{2} \frac{\varepsilon_o - 1}{\varepsilon_o + 1} (\omega_L - \omega_T) \simeq - 0.7 \text{ meV} . \tag{2}$$

Comparison with the value −0.5 meV found from Fig. 2 suggests that an exciton-free layer 100 Å thick may be considered as almost "massive" in this respect.

In ref. [4] a discussion is given of a controversy surrounding the thickness of the dead layer in CdS. While most workers in the field quote values in the range $\ell = (90 \pm 20)$ Å it has been also claimed, on the basis of microscopic calculations, that $\ell \simeq 20$ Å [7]. We believe that an ATR experiment would prove very valuable in shedding light on the value of ℓ or, possibly, on a surface structure. The strong shift in Fig. 2 suggests that resolution would not pose a difficulty to the measurement.

3. The additional (nonlocal) surface B-exciton of CdS : does it exist ?

The B-exciton of CdS is known to be more complex - and interesting - than the A-exciton. If the wavevector $\vec{q}$ is perpendicular to the crystalline axis $\hat{c}$ then symmetry considerations permit a $\vec{q}$ - linear splitting of one valence band [8]. The excitons formed by coupling of holes in this band and electrons in the conduction band exhibit a similar splitting, however only for electric fields $\vec{E}$ that are perpendicular to the $\hat{c}$ - axis. The energies of the split excitons are

$$\hbar\omega_\pm(q) = \hbar\omega_T + \hbar^2 q^2/2m \pm \phi q . \tag{3}$$

The oscillator-strength of the degenerate band being equally divided between the "+" and the "-" excitons, the model dielectric function is [8]

$$\varepsilon(\omega,q) = \varepsilon_o + \frac{\frac{1}{2}\omega_p^2}{\omega_+^2(q) - \omega^2 - i\nu\omega} + \frac{\frac{1}{2}\omega_p^2}{\omega_-^2(q) - \omega^2 - i\nu\omega} \quad . \tag{4}$$

The conditions $\vec{q} \perp \vec{c}$ and $\vec{E} \perp \vec{c}$ imply that the B-exciton exists only for p-polarised light with the plane of incidence perpendicular to the crystalline axis. It is of interest to compare the ATR spectra of the B-exciton with those of the A-exciton; this should shed light on the $\vec{q}$- linear term in (3). Such a study has been performed recently [9], and we refer the reader to this reference for details of the calculation. A typical ATR spectrum is shown in Fig. 3. The width of the exciton-free layer is assumed to be 70 Å; the other parameters are quoted in [9]. We see that for θ= 50°, 70°, and 85° (substantially above the nominal critical angle) there are two minima. The broad and deep minimum between ω_T and ω_L behaves in much the same way as for the A-exciton, Fig. 1. If we take the limits $m \to \infty$ and $\phi \to 0$ this minimum persists, so it is clear that it corresponds to the usual surface-exciton polariton. However, in Fig. 3 there is an additional, narrow minimum located at $\omega \simeq \omega_T$. This minimum disappears in the limit $\phi \to 0$, indicating that it is a consequence of the wavevector-linear splitting.

On the basis of Fig. 3 and other ATR spectra (for different values of ϕ, ν, θ, and also another ABC) it was concluded that the minimum at ω_T most likely represents an additional, new surface polariton. However, one also has to allow for the possibility that this minimum corresponds to some energy - loss channel. Therefore it is important to seek additional evidence by deriving, directly from Maxwell's equations, the dispersion relation for surface-exciton polaritons. The calculation is based on a generalization of the Pekar ABC, namely it is assumed that the excitonic polarization vector (i.e. the parallel and the normal components of $\vec{P}(z)$ vanish at the bulk/dead-layer interface for both the "+" and the "-" excitons. This gives four ABC's. The dispersion relation is very complicated and here we give only the solution for real propagation vector q_x and complex frequency ω. In Fig. 4 we plot the real and imaginary parts of ω/ω_T versus $q_x c/\omega_T$. We observe the following features.

(a) Two separate surface polariton branches are obtained.

(b) The higher branch in (a) starts above ω_T and, for high values of q_x keeps increasing to frequencies greater than ω_L; this branch behaves in much the same way as that of the A-exciton [10].

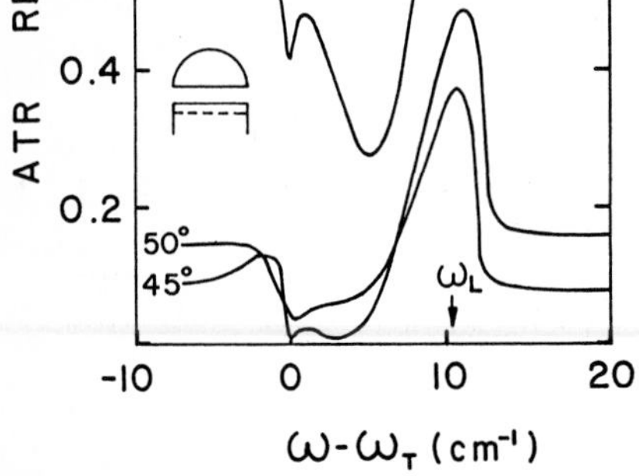

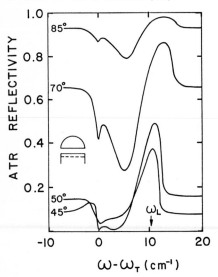

Fig. 3 Attenuated Total Reflection Spectra of the B(n = 1) exciton of CdS in the "isotropic configuration" ($\vec{q}$, $\vec{E} \perp \hat{c}$) as a function of $\omega - \omega_T$ in cm^{-1}. The longitudinal-transverse splitting $\omega_L - \omega_T$ is 10.1 cm^{-1} and the exciton-free layer has a thickness of 70 Å. The other parameters are given in [9a], wherefrom this figure is taken.

(c) The lower branch in (a) starts below ω_T and, for $q_x \to \infty$, reaches the limiting frequency $\omega \approx \omega_T$.

(d) Except for the immediate vicinity of the vacuum light-line the lower dispersion curve is very flat, that is the group velocity $d(\mathrm{Re}\,\omega)/dq_x$ is very small.

(e) The damping (b) is very small: $|\mathrm{Im}\,\omega|/\mathrm{Re}\,\omega \sim 10^{-4}$ for both modes. The damping of the lower branch is actually smaller than of the higher branch (except for very small propagation vectors).

These features must be checked against the "optical" dispersion curves which may be determined from the frequency minima $\omega_{min}(q_x)$ in Fig. 3. Generally speaking, the comparison is quite favorable, however two reservations are in order. Firstly, the maximum propagation vector attainable in ATR is $(\omega_{min}/c)\epsilon_p^{1/2}$ and, therefore, the region $q_x c/\omega_T > 1.5$ in Fig. 4 cannot be reached with a prism of index 1.5. Secondly, in Fig. 3 there are no minima that have $\omega_{min} < \omega_T$, so the corresponding portion of the lower branch in Fig.4(a) does not resemble the "optical" dispersion. The first restriction is a general one for the ATR method and may be relaxed by selecting a higher index prism. As for the second point, we should say that, in order to account for a frequency-scan ATR "experiment" (as in Fig. 3), a better way to solve the dispersion relation is choosing q_x, as well as ω, as a complex quantity [11]. The ratio q_x/ω is a real quantity given by $\epsilon_p^{1/2} \sin\theta/c$ corresponding to the ATR frequency-scan for a given θ. Such a solution, and a detailed comparison between the "optical" and "electromagnetic" dispersion relations will be presented elsewhere.

Fig. 4 and the overall agreement with Fig. 3 strongly suggest that the lower branch in Fig. 4a represents a new surface-exciton polariton. We note that this

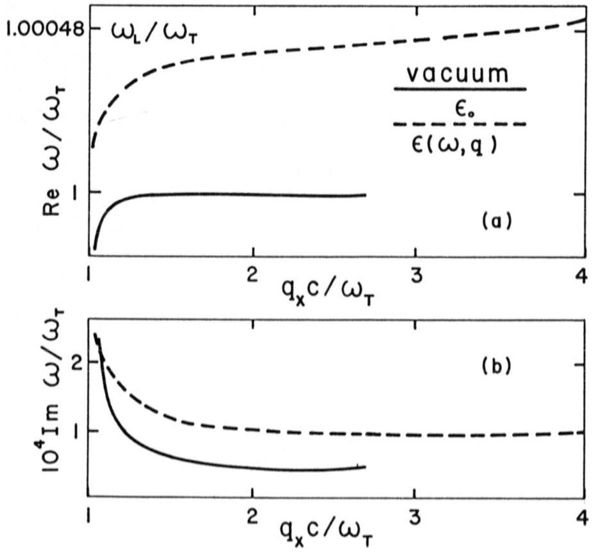

Fig. 4 Surface-exciton polariton dispersion relation for the B-exciton of CdS, with allowance for an exciton-free layer. These curves have been derived directly from Maxwell's equations (rather than ATR spectra) using a suitable generalization of the Pekar ABC. It has been assumed that the propagation vector q_x is a real quantity, and that the frequency ω is complex. (a) The dispersion: Re ω/ω_T versus $q_x c/\omega_T$. The upper mode is a modification of the ordinary surface-exciton polariton. The lower mode is a new,"nonlocal" surface-exciton polariton, a consequence of the wave-vector-linear splitting. (b) The damping: Im ω/ω_T versus $q_x c/\omega_T$. There is a correspondence between the continuous and the dashed curves in (a) and (b).

mode is unique in the sense that it is caused by spatial dispersion - a feature that, to our knowledge, has not been claimed for any other surface mode. For instance, while the finiteness of the exciton mass gives rise to important nonlocal effects, it merely modifies the properties of the surface-exciton polariton that exists even for $m \to \infty$. On the other hand the new surface polariton described in this section disappears in the limit of vanishing wavevector - linear splitting ($\phi \to 0$).

4. Excitons and plasmons : can they interact ?

Consider the following possibilities. In an insulating dielectric, plasma oscillations may occur only due to the collective excitations of the core electrons, in the far ultraviolet - usually far away from the exciton region. In a metal, of course, excitons do not exist. The third possibility is a doped semiconductor; however, even with the highest levels of doping, the plasmon frequencies are in the far infrared while the excitons are in the visible or near ultraviolet. Moreover, an elevated density of charge carriers would lead to screening effects that are too strong for an exciton to survive. Clearly, exciton-plasmon interaction in the bulk of conventional crystals is very weak - probably too weak to be observed.

Now let us bring into contact an undoped semiconductor and a metal. What are the conditions for coupled polariton modes propagating along the semiconductor/metal interface? We will answer this question for a very naive model of the interface. We ignore the possibility of the formation of a Schottky-barrier, assuming a sharp, plane interface between the two media. Effects of spatial dispersion and damping are neglected. Also, the semiconductor is supposed to have a negligibly small dead layer. Then the dielectric function of the semiconductor is simply written as

$$\varepsilon_S(\omega) = \varepsilon_o(\omega^2 - \omega_L^2) / (\omega^2 - \omega_T^2) . \qquad (5)$$

The dielectric function of the metal, $\varepsilon_M(\omega)$, does not need to be specified, however for most metals the plasma frequency is in the ultraviolet, so we may assume that $\varepsilon_M(\omega) < 0$. Then it is easy to show [12] that the existence of interface polaritons requires that

$$0 < \varepsilon_S(\omega) \leq |\varepsilon_M(\omega)| . \qquad (6)$$

This condition has been applied to the dielectric functions $\varepsilon_S(\omega)$ and $\varepsilon_M(\omega)$, plotted schematically in Fig. 5. We see that "propagation-windows" exist only in the region marked by wavy lines, and the limiting frequencies ω_1 and ω_2 are given by (6) with the equality sign applying. The conclusion is that there is one propagation window extending from "low" frequencies to ω_1 and a second one extending from ω_L to ω_2. Clearly, ω_1 and ω_2 depend on the plasmon, as well as on the exciton parameters. So there are two frequency-windows in which excitons and plasmons may interact across a semiconductor/metal interface, propagate along this interface, and decay exponentially away from it into both media.

The dispersion relation of the interface polaritons is given by

$$q_x^2 = \frac{\omega^2}{c^2} \frac{\varepsilon_S(\omega)\varepsilon_M(\omega)}{\varepsilon_S(\omega) + \varepsilon_M(\omega)} . \qquad (7)$$

From this equation and Fig. 5 we may deduce the schematic dispersion curves $\omega(q_x)$ of the low-frequency and the high-frequency polaritons, Fig. 6. There is no vacuum "light-line", because we are dealing with an interface, rather than with a surface, so vacuum is just not involved in the problem. The vacuum light-line is replaced by the "semiconductor light-line", which is simply the bulk polariton dispersion curve $q_x^2 = (\omega/c)^2\varepsilon_S(\omega)$. We note that both modes are located to the right of the low-frequency and high-frequency "light lines" (dashed), thus ensuring proper exponential decay into the semiconductor.

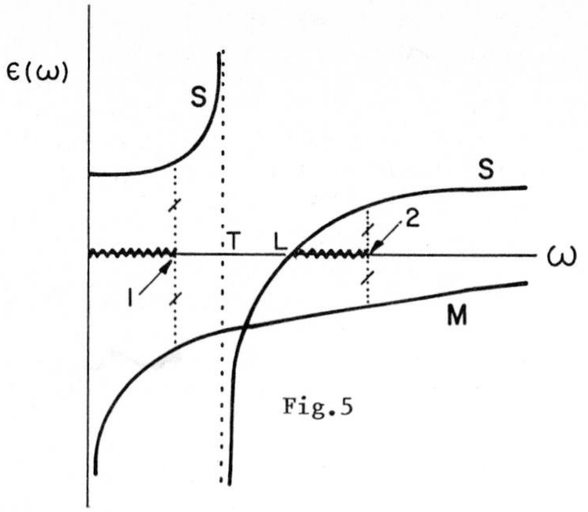

Fig.5

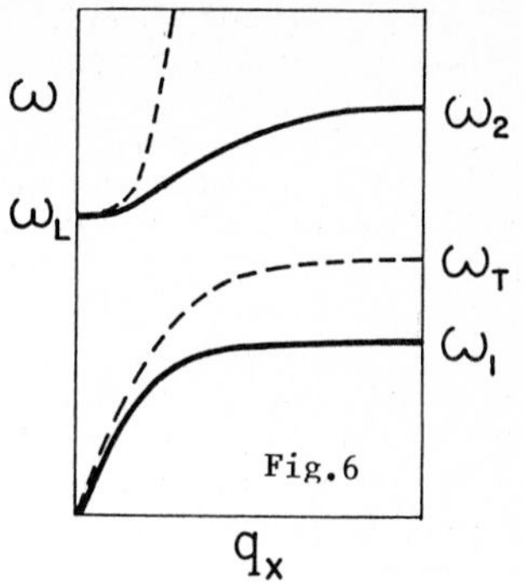

Fig.6

Fig. 5 Schematic plots of the dielectric functions of a semiconductor (S) and of a metal (M) in the vicinity of an excitonic transition. The semiconductor/metal interface supports coupled exciton-plasmon polaritons provided that $\varepsilon_s(\omega) \leq |\varepsilon_M(\omega)|$. This condition is satisfied in the regions marked with wavy lines. These regions terminate at frequencies such that $\varepsilon_s(\omega_i) = |\varepsilon_M(\omega_i)|$. Note that T, L, 1, and 2 stand for ω_T, ω_L, ω_1, and ω_2.

Fig. 6 Dispersion relation $\omega(q_x)$ of coupled exciton-plasmon polaritons at the interface between an undoped semiconductor and a metal (schematic). The frequency ranges of the low - and high - frequency modes are $(0,\omega_1)$ and (ω_L, ω_2) respectively, where ω_1 and ω_2 may be determined graphically, as in Fig.5. The dashed curves are given by $q_x^2 = (\omega/c)^2\varepsilon_s(\omega)$, corresponding to the bulk polariton in the semiconductor.

The lower mode extends over a wide frequency range, while the upper mode exists, typically, in a range of the order of a few meV. Focusing our attention on the latter mode, it is a good approximation to replace $\varepsilon_M(\omega)$ by a constant, average value $\bar{\varepsilon}_M(<0)$. Then, using (5), with some algebra (7) may be rewritten as

$$q_x^2 = \frac{\omega^2}{c^2} \; \frac{\varepsilon_o(-\bar{\varepsilon}_M)}{\varepsilon_o + \bar{\varepsilon}_M} \; \frac{\omega^2 - \omega_L^2}{\omega_2^2 - \omega^2}, \tag{8}$$

$$\omega_2^2 = \frac{\varepsilon_o\omega_L^2 + \bar{\varepsilon}_M\omega_T^2}{\varepsilon_o + \bar{\varepsilon}_M} . \tag{9}$$

Eq. (8) shows the simple structure of the high-frequency mode, and (9) gives its limiting frequency ω_2.

The next issue to discuss is the possibility of excitation and detection of these interface polaritons by means of ATR spectroscopy. Suppose that a thin metallic film is sandwiched between a prism and an undoped semiconductor. This is a generalization of the Kretchmann-Raether configuration [13] with the air on the free side of the film being replaced by the dispersive semiconductor. The dielectric constant of the prism must be greater than $\varepsilon_s(\omega)$. Again considering the high-frequency mode we note that, just above ω_L we may have $\varepsilon_s(\omega)$ smaller than one, that is smaller than the dielectric constant of vacuum (see Fig. 5). If we are only interested in the corresponding frequency range then we may actually "use an air prism"! [14] This certainly simplifies the experimental geometry, which is now reduced to a thin metallic film evaporated on a semiconductor substrate.

Attractive choices, from the experimental point of view, are ZnO for the semiconductor and silver for the metal. The plasma frequency of silver is quite near to the C(n = 1) exciton of ZnO, thus increasing the plasmon-exciton coupling. Also, the thickness of the dead layer in ZnO is sufficiently small to be neglected, as we have assumed in our calculation above. The results of an ATR calculation (with the "air prism") are shown in Fig. 7. The parameters for ZnO are taken from [15] and $\bar{\varepsilon}_M = -2 + 0.4i$. Thus here we allow for damping, which was neglected before. The film thickness is 420 Å and the angle of incidence is 45°.

Fig. 7 indicates a strong resonance, the intensity of the reflected light being lowered by tens of percents. This should be compared with a loss of ~ 6% in the intensity for surface-exciton polaritons [3]. Thus the presence of the metallic film actually enhances the resonance! It is interesting to compare the position of the minimum with the corresponding prediction of (8). Then $q_x^2 = (\omega/c)^2 \sin^2\theta = 0.5\omega^2/c^2$. With $\bar{\varepsilon}_M = -2$ the solution of (8) gives $\hbar\omega = 3.4330$ eV, precisely the value found in Fig. 7. This is 0.7 meV above $\hbar\omega_L$. The width of the resonance is ~ 0.5 meV. Also, from (9), the limiting frequency is $\hbar\omega_2 = 3.4375$ eV. Therefore the available range for the high-frequency mode is $\hbar\omega_2 - \hbar\omega_L \simeq 5$ meV.

The deep minimum in Fig. 7 and the agreement between the position of this minimum and the prediction of the dispersion relation (8) point to the feasibility of experimental excitation of exciton-plasmon polaritons propagating at a semiconductor/metal interface. Future theoretical work on this subject should take into account the Schottky barrier at the interface, an exciton-free layer in the semicondutor as well as spatial dispersion.

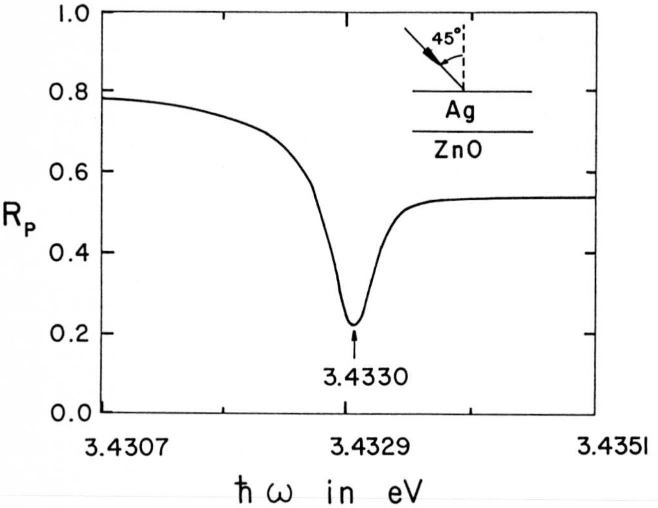

Fig. 7 Attenuated Total Reflection Spectrum for the configuration in the inset. The resonance corresponds to the excitation of coupled exciton-plasmon polaritons at the interface between the ZnO substrate and the silver film (420 Å thick). No prism is needed because, in the frequency-range of interest, $0 < \varepsilon_s(\omega) < 1$.

Acknowledgement

A part of the research leading to this paper was performed at the University of California, Irvine, and the first two authors (PH and GHC) wish to thank Professors A. A. Maradudin and R. F. Wallis for useful talks.

References

1. J. Lagois and B. Fischer: in Surface Polaritons, ed. by D. L. Mills and V. M. Agranovich (North-Holland, Amsterdam 1982) p. 69.
2. V. M. Agranovich and V. L. Ginzburg: Crystal Optics with Spatial Dispersion, and Excitons, 2nd. ed., Springer Ser. Solid-State Sci., Vol. 42 (Springer,Berlin 1984); V. M. Agranovich: in Surface Excitations, ed. by V.M. Agranovich and R. Loudon (North-Holland, Amsterdam 1984) p. 513.
3. J. Lagois and B. Fischer : Phys. Rev. Lett. 36, 680 (1976).
4. P. Halevi : in Excitons in Confined Systems, ed. by R. Del Sole, A. D'Andrea and A. Lapiccirella, Springer Proc. Phys. (Springer, Berlin, Heidelberg 1987) (first paper in this book).
5. P. Y. Yu and F. Evangelisti : Phys. Rev. Lett. 42, 1642 (1979).
6. J. Lagois and B. Fischer : in Physics of Semiconductors, ed. by F.G. Fumi (Tipografía Marves, Rome 1976) p. 788.
7. A. D'Andrea and R. Del Sole : Phys. Rev. B 29, 4782 (1984).
8. G. D. Mahan and J. J. Hopfield, Phys. Rev. 135, A 428 (1964).
9. P. Halevi, O. B. M. Hardouin Duparc, A. A. Maradudin, and R. F. Wallis, (a) Phys. Rev. B 32, 6986 (1985) and (b) ibid, in press.
10. J. Lagois and B. Fischer, Phys. Rev. B 17, 3814 (1978).
11. A. Otto.: in Polaritons, ed. by E. Burstein and F. De Martini (Pergamon, New York 1974), p. 117.
12. P. Halevi, in Electromagnetic Surface Modes, ed. by A. D. Boardman (Wiley 1982), p. 249.
13. E. Kretschmann and H. Raether, Z. Naturforsch. 23a, 2135 (1968).
14. P. Halevi and G. Hernández-Cocoletzi, Phys. Rev. B 18, 590 (1978).
15. J. Lagois and B. Fischer, Solid State Commun. 18, 1519 (1976).

Screening Effect on Exciton Polaritons in Semi-infinite Semiconductors

S. Jaziri[2] and R. Bennaceur[1]

[1]Laboratoire de Physique de la Matière Condensée, Faculté des Sciences de Tunis, Campus Universitaire, 1060 Tunis, Tunisia
[2]Ecole Normale Supérieure de Bizerte, 7021 Zarzouna-Bizerte, Tunisia

1. ABSTRACT

We study in this work the behavior of the statically plasma-screened exciton gas near the surface of a semiconductor. We observe a complete absence of bound states at lengths smaller than 0.84 a_B corresponding to a Mott concentration $n_M = 4.10^{14}$ cm^{-3} at 2K for GaAs. We have analysed the extension of the inhomogeneous layer near the surface. The thickness of this layer increases with increasing screening concentration. The effect of the screening on reflectivity curves for GaAs and CdS has been analysed.

2. INTRODUCTION

The screening of the exciton can be classified in two groups: plasma screening mediated through free e-h and excitonic screening caused by the very existence of the excitons themselves (dielectric screening). The plasma screening can be observed in the following cases: injection or photogeneration at photon energy greater than E_g. In the case of exciton screening the photon energy is equal to the bound exciton energies /1-3/ (high excitation region).

In this work we will consider the static Debye-Huckel screening appropriate in the case of weak nondegenerate plasma screening where the binding energy E_b is higher than the plasma energy $\hbar\omega p$. When E_b is comparable to $\hbar\omega_p$ we need to take into account the retarded interaction. This will correspond to the dynamical screening regime. In this region the Debye-Huckel screening model overestimates the reduction of the binding energy due to the screening free carriers. H.Haug, D.B.Tran Thoai /4/ and W.D.Kraeft et al. /5/ have studied this problem by solving the Bethe-Salpeter equation for the corresponding polarisation function. Their studies show that the dynamical screening tends to reduce the effect of the statical one and to increase the value of the Mott concentration n_M (Exciton-free e-h plasma transition).

The objective of this work is the study of the behavior of the statically screened exciton gas near the surface of a semiconductor. In recent works D'Andrea and Del Sole have developed a microscopic model of exciton polaritons in semi-infinite solids /6,7/. In order to study the screening effect of the free carriers on the exciton gas near the surface we propose to extend their method to the screened coulombic potential of Yukawa type with a Debye screening length

$$D = \left(\frac{\varepsilon kT}{8\pi Ne^2}\right)^{1/2} .$$

3. EXCITON WAVE FUNCTION

The exciton wave function in a semi-infinite semiconductor is the solution of the Schrödinger equation (the image potential V_{im} is neglected as in /6,7/; the screening tends to decrease the effect of V_{im})

$$\left(- \frac{h^2}{2M} \Delta_R - \frac{h^2}{2\mu} \Delta_r - \frac{e^2}{\varepsilon_o r} \exp\left(-\frac{r}{D}\right) \right) \Phi(r,R) = \bar{E}\, \Phi(r,R) \qquad (1)$$

with a boundary condition: $\Phi(z_e=0)=\Phi(z_h=0)=0$, where z_e and z_h are the z components of electron and hole positions. We seek the solution of equation (1) in the form

$$\Phi(r,R) = \frac{\exp(ik_{//}R_{//})}{2\pi^2 (1+/A/)^{0.5}} \{ (\exp(-i_z k Z) + A\exp(i K_z Z))\Phi_1(r)$$

$$+ \sum_{n,1} \Phi_{n,1}(r) \exp(-P_{n,1} Z)\, c_{n,1} \}, \qquad (2)$$

where $\Phi_{n,1}(r)$ are the screened hydrogenic orbitals. The wave functions are

$$\Phi_{n,1}(r) = \frac{1}{r} \exp(-\gamma_{n,1} r) \sum_k a_{k,n,1} (2\gamma_{n,1} r)^k \qquad (3)$$

with $\gamma_{n,1} = \frac{1}{\hbar^2} \sqrt{-2\mu E_n}$, and $a_{k,n,1}$ satisfy the following recurrence relations:

$$a_{k,n,1} = \frac{1+k-n}{k(k+21+1)} a_{k-1,n,1} + \frac{n}{k(k+21+1)} \sum_{s=1}^{k-1} (-1)^{s+1} \left(\frac{1}{2\gamma_{n,1} D}\right)^s a_{k-s-1,n,1} . \qquad (4)$$

In the first iteration we have calculated the wave functions $\Phi_{n,1}(r)$ using the hydrogenic energies E_n. Then we have recalculated the new energies $E_{n,1}$ and reinjected them in the expression of $\Phi_{n,1}(r)$. We have continued the iteration until convergence, with an error on $E_{n,1}$ of the order of 10^{-3} meV. We have calculated the wave function with one hundred terms $a_{k,n,1}$ in the recurrence relation and normalized them (for great values of r the wave functions have been extrapolated by a decaying exponential). We have analysed the dependance of the eigenvalues on the screening length in Fig.1 where we represent $n^2 E_{n,1}$ versus $V_o/N = 16\pi a_B^3 (R^*)/KTN$. For high screening we observe that the degeneracy is slightly removed and for some critical length the bound states start to disappear for the states n=5 until n=1. This result is in good agreement with results deduced numerically for plasma of hydrogen /8,9/. We observe a complete absence of bound states at lengths smaller than 0.846 a_B corresponding to a Mott concentration of $n_M=4.10^{14}$ cm^{-3} at temperature T=2K for GaAs and $n_M=2.10^{16}$ cm^{-3} at temperature T=4K for CdS.

4. ANALYTICAL-VARIATIONAL APPROACH. DEAD LAYER.

The coefficients A and $C_{n,1}$ have been calculated by the variational procedure introduced by D'Andrea and Del Sole /6/; then we have calculated the wave function $\Phi(0,Z)$. Fig.2 represents the plot of the square modulus of the exciton wave function for GaAs for three different screening lengths 500 a_B, 100 a_B and 30 a_B and two different center of mass energies E_k=0.2 meV and E_k=2 meV. We observe a lowering of the maxima and a small shift of their positions to higher values of Z with the decreasing screening length. In order to analyse the optical properties of the medium it is more appropriate to work with the analytical wave function given by

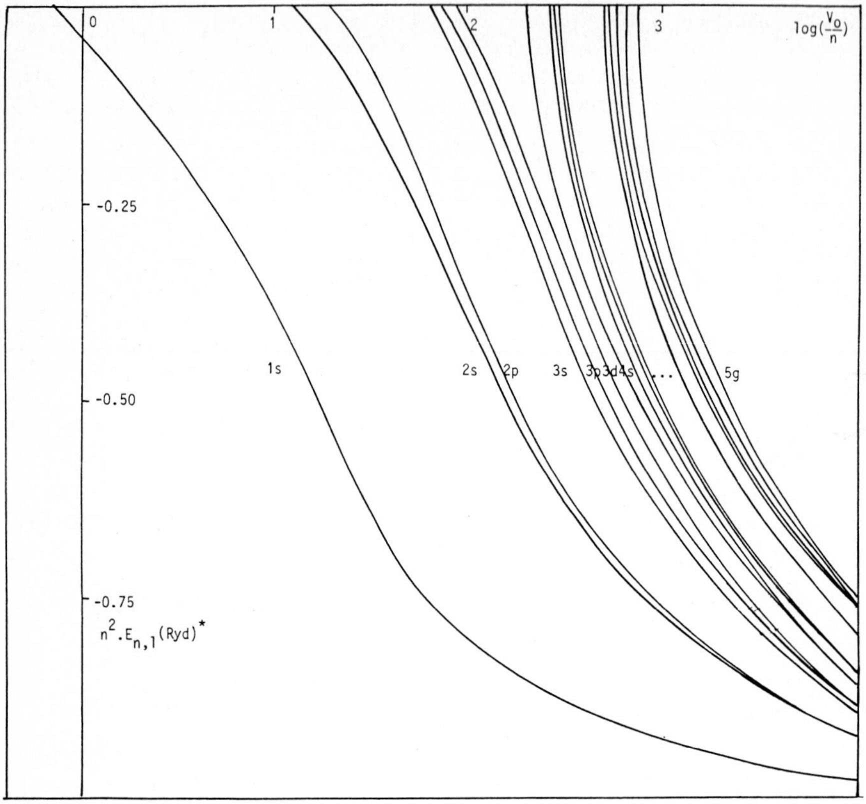

Fig.1 Dependence of $n^2E_{n,1}$ on V_o/N.

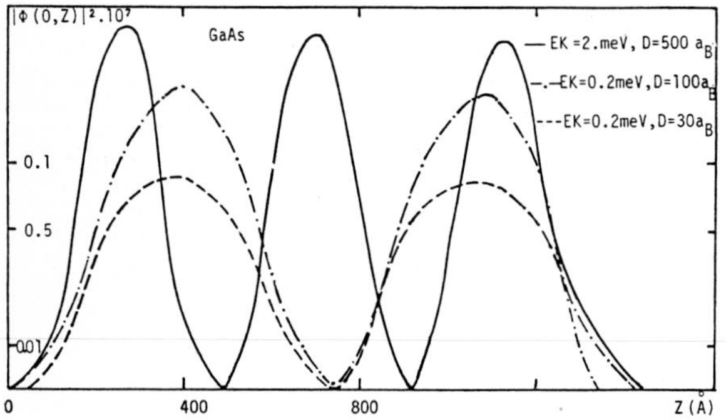

Fig.2 $|\Phi(0,Z)|^2 \times \pi^2 a_B^3$ for GaAs.

$$\Phi(0,Z)= \frac{1}{\sqrt{2}\pi}\Phi_1(0)\left[\exp(-i\, k_z\, Z)+A \exp(ik_z Z) - (1+A) \exp(-PZ)\right]. \qquad (5)$$

The parameter P can be obtained in three ways:

- as an average value of P_n /6/;
- as a fitting parameter with the variational solution (/7/ and this work for reflectivity calculation);
- obtained from

$$\left|\sum_{n,l} C_{n,l} \; \Phi_{n,l}(0)\, e^{-P_{n,l}Z}\right| = e^{-\langle P(Z)\rangle Z} \sum_{n,l} C_{n,l} \; \Phi_{n,l}(0) \quad . \tag{6}$$

We represent in Figs.3 and 4 the curves $\langle P(Z)\rangle Z$ as functions of the center of mass position Z. We see that $\langle P(Z)\rangle Z$ saturates after some depth (corresponding to uniform exciton gas in the bulk). This saturation is mainly due to the limited numerical accuracy of the calculation. We can estimate the extension of the inhomogeneous layer l by considering the function $\langle P(Z)\rangle Z$ that it is quite linear in the region near the surface. The inverse of the slope $\langle P(Z)\rangle Z$ at the origin can be defined as the inhomogeneous layer width. The following results have been obtained for GaAs (center of mass energy E_k=0.2 meV) and for screening length 500 a_B, 100 a_B, 30 a_B equal respectively to 68 Å, 108 Å, 180 Å. For CdS (E_k=1 meV) and for D=500 a_B, l=26 Å and for D=30 a_B, l=69 Å. In the case of CdS the variation of E_k from 1 to 10 meV induces a corresponding decrease of l from 26 Å to 22 Å.

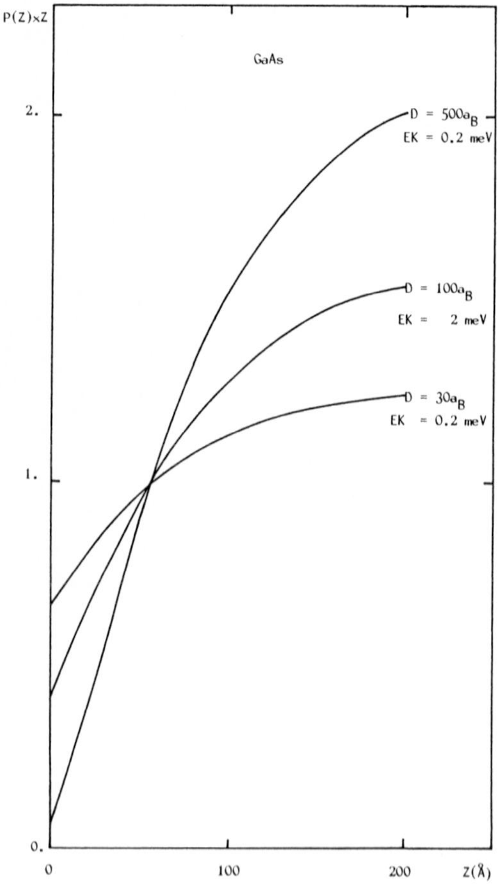

Fig.3 $\langle P(Z)\rangle Z$ for GaAs.

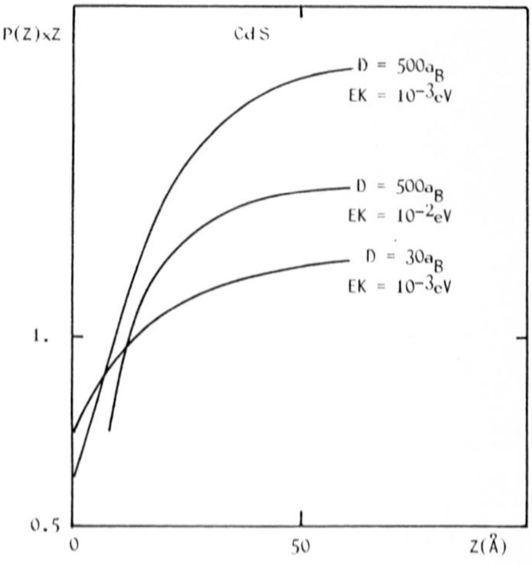

Fig.4 $\langle P(Z)\rangle Z$ for CdS.

5. NORMAL INCIDENCE REFLECTIVITY

In order to calculate the reflectivity coefficient we have used the fitted value of the parameter P. We have calculated the reflectivity for the region near the n=1 exciton resonance frequency with the screened value of the n=1 eigenvalue. For the case of GaAs we see in Fig.5 that the screening effect translates the curve to higher energies by approximately 0.1 meV for the increase of the free e-h concentration by factor 25. In both cases of GaAs and CdS ΔE is lowered with the increasing of the screening effect, ΔR decreases for the GaAs and increases for the case of CdS. See Fig.6.

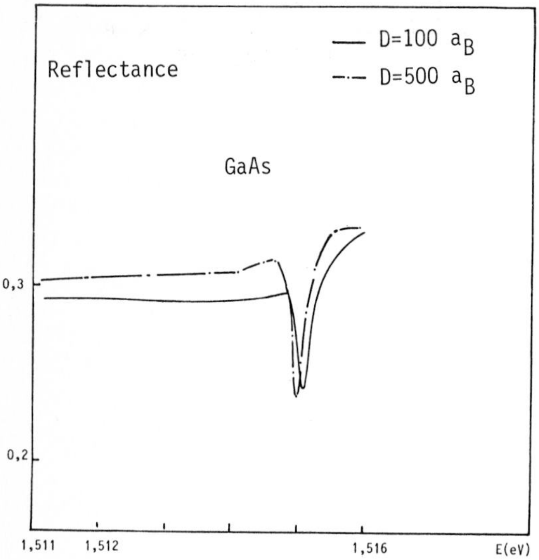

Fig.5 Reflectivity curve for GaAs.

E_g = 1.519 eV
R = 4.1 meV
Γ = 2.10^{-5} eV
$4\pi\alpha_0$ = $1.66\ 10^{-4}$
ε_0 = 12.5
—$P=0.011A^{-1}$, —·—$P=0.0092A^{-1}$

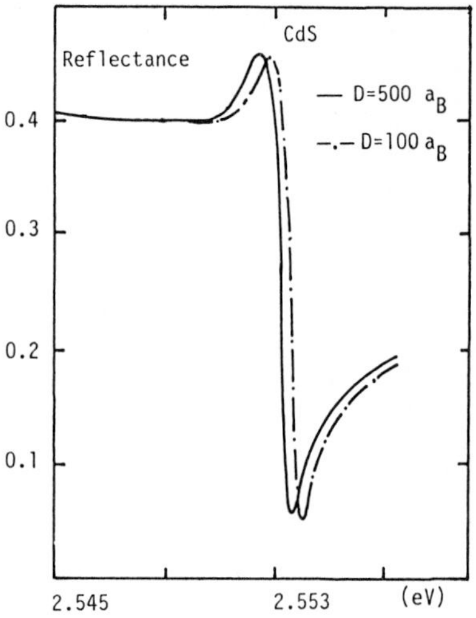

Fig.6 Reflectivity curve for CdS.

E_g = 2.58 eV
R = 28 meV
Γ = $3.\ 10^{-4}$ eV
$4\pi\alpha_0$ = $0.85\ 10^{-2}$
ε_0 = 8.1
—$P=0.038A^{-1}$, —·—$P=0.034A^{-1}$

6. CONCLUSION

We have presented a qualitative study of the effect of weak plasma screening on the exciton gas near the surface. A more accurate analysis needs the study of the effect of the screening on the continuum states and on the values of E_g and ε_0. The study of the excitonic screening will need a many-body treatment of the exciton gas near the surface.

7. ACKNOWLEDGMENTS

We gratefully acknowledge helpful discussions with R.Del Sole and A.D'Andrea.

8. REFERENCES

1. G.Mahler and J.Birman, Phys.Rev B 16, 1552 (1977)
2. R.F.Leheny, H.Shah and G.C.Chiang, Solid State Comm. 25, 621 (1978)
3. J.U.Fishbach, W.Ruhle, D.Dimberg and E.Bauser, Solid State Comm. 18, 1255 (1976)
   J.Lagois, K.Losch, Phys.Rev. B 18, 4325 (1978)
4. D.B.Tran Thoai and H.Haug, J.Luminescence 18/19, 309 (1979)
5. W.D.Kraeft, D.Kremp, W.Ebeling and G.Ropke, Quantum statistics of charged particle systems, Plenum Press, New York (1986)
6. A.D'Andrea and R.Del Sole, Phys.Rev. B 25, 3714 (1982)
7. A.D'Andrea and R.Del Sole, Phys.Rev. B 29, 4782 (1984)
8. F.J.Rogers, A.C.Graboske Jr. and D.Harwood, Phys.Rev. A 1, 1577 (1970)
9. N.Bessis, G.Bessis, B.Dakhel, G.Hadinger, J.Phys. A 11, (1978)

Optical Dephasing of Wannier Excitons in GaAs

L. Schultheis[1,*], *J. Kuhl*[1], *A. Honold*[1], *and C.W. Tu*[2]

[1]Max-Planck-Institut für Festkörperforschung,
Heisenbergstr. 1, D-7000 Stuttgart 80, Fed.Rep.of Germany
[2]AT&T Bell Laboratories, Murray Hill, NJ 07974, USA
*Present address: Brown Boveri Research Center,
CH-5405 Baden/Dättwil, Switzerland

1. Abstract

The optical dephasing of excitons in optically thin GaAs layers is measured by means of time-resolved degenerate four-wave mixing to study the various ultrafast relaxation mechanisms. The temperature dependence of the phase coherence time directly reveals efficient scattering on a picosecond timescale due to acoustic phonons and residual impurities. In addition, the observed ultrafast decay of the optical alignment of the excitons within 7ps indicates fast orientational relaxation of the excitons, corresponding to a short lifetime of the excitonic states.

2. Introduction

The phase coherence of optical transitions is a sensitive probe of the ultrafast scattering processes /1-5/. Time-resolved optical coherent experiments, probing optical dephasing, i.e. the loss of phase coherence of an exciton ensemble, in the model semiconductor GaAs have revealed extremely fast relaxation of a few hundred femtoseconds for free carrier excitation /1,2/, optical dephasing in the picosecond range for delocalized 2D and 3D excitons /3,4,5/ and much longer phase coherence times of the order of a few ten picoseconds for highly localized 2D excitons.

This progress has been enabled by recent advances of the generation of ultrashort optical pulses as well as of materials technology; in particular, the fabrication of high-quality (high-purity as well as strain-free) optically thin GaAs slabs by means of molecular beam epitaxy is crucial to time-resolved nonlinear optical experiments /3/. GaAs layers in the thickness range of 100nm to 200nm have to be exploited to overcome the problem of strong absorption (and therefore reabsorption and high, disturbing, excitation levels) as well as the additional and undesirable polariton transport problems /6/. In addition, cladding of these GaAs platelets with the larger band gap semiconductor GaAlAs significantly improves the optical properties because of carrier confinement and negligible surface recombination /8/.

We describe in this paper the ultrafast relaxation processes of excitons acting initially after optical excitation of the excitonic states. It is this picosecond time regime, in which randomization of the excitonic states starts and optical dephasing occurs, which determines the efficiency of the coherent exciton-photon coupling. The optical dephasing time at low excitation levels probed by the excitonic phase coherence time (or in frequency domain notation.the homogeneous linewidth) is one of the fundamental material parameters which sensitively measures the residual interaction of the exciton with the crystal. A rapid loss of the excitonic phase coherence within 7ps is found, attributed to scattering with impurities and acoustic phonons. These scattering processes also cause fast orientational relaxation of the excitonic states, as confirmed by additional experiments on orientational gratings.

3. Experimental

We used for our experiments high-purity GaAs layers (thicknesses between 100nm and 190nm) cladded by $Al_{0.3}Ga_{0.7}As$ and grown on n^+-GaAs substrates by molecular beam epitaxy. The substrates were polished down to wedges of about 30µm average thickness. The n^+-substrates are transparent below 1.549eV due to the Burstein-Moss shift and are sufficiently thick to avoid strain. The geometrical sample parameters were determined by means of electron transmission microscopy. For most of the experiments, the samples were immersed in superfluid He. A standard synchronously pumped tunable dye laser with Styryl 9 was employed as the excitation source. The autocorrelation width was 3.7 ps, the width of the power spectrum 0.9 meV. Optical characterization of our samples was done by conventional reflectance, transmission and photoluminescence spectroscopy.

4. Characterization in the Frequency Domain

Figure 1 gives typical transmission and photoluminescence spectra in the near band gap region of a sample with a 194nm thick GaAs central layer. The transmission spectrum exhibits a strong absorption peak at 1.51508eV followed on the high energy side by much weaker absorption peaks at 1.5158eV and 1.5165eV.

We attribute these absorption lines to 1s excitonic transitions originating from the quantized electronic subbands of index N. If we take the measured thickness of L_z=194nm, a heavy-hole mass of $0.45m_0$ and an electron mass of $0.067m_0$, and assuming a slightly increased (by 0.17meV) binding energy as compared to the bulk binding energy, the 1s-heavy-hole exciton transitions are determined within an infinite quantum well model as

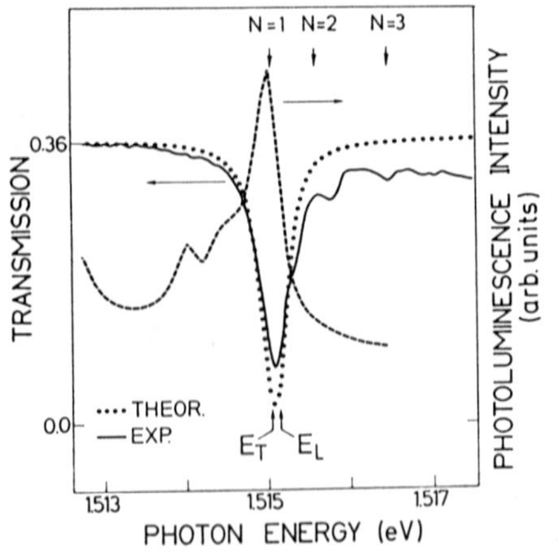

Fig.1: Experimental (solid line) and theoretical (dotted line) transmission and photoluminescence (dashed line) spectra at 2K. The indices N denote the exciton transition belonging to the Nth electronic subband. E_L is the longitudinal and E_T the transverse eigenenergy.

$$E_N = 1.5149eV + 0.17meV*N^2. \quad (1)$$

The very good agreement with the measured absorption peaks demonstrates that the excitonic states can be appropriately described by the quantized (single particle) electron and hole states and not by a quantization of the excitonic translational motion.

In order to obtain the exciton parameters more precisely, in particular the excitonic eigenenergy and the homogeneous linewidth of the N=1 transition, a lineshape analysis has to be performed. The problem of calculating transmission spectra consists of (i) Maxwell's equations and (ii) the calculation of the dielectric response of the exciton in thin slabs, which in turn needs precise knowledge of the exciton wave function to get the local, spatially inhomogeneous exciton response. Because these wave functions are basically unknown for such slabs in the intermediate thickness regime between 2D and 3D bulk systems, we simplified this task by approximating the smooth change of the local excitonic oscillator strength by a simple one layer dielectric oscillator model. The effect of the boundaries leading to the so-called exciton-free surface layer is taken into account assuming an effective layer thickness L_{eff}.

Figure 1 also depicts a theoretical transmission spectrum, calculated by using an effective thickness L_{eff}= 140nm and a longitudinal transverse splitting E_L-E_T= 0.08meV /7/. The agreement between the calculated and the experimental transmission spectrum is satisfactory (as compared to the crude model) and yields the excitonic eigenenergy of E_T=1.51502eV and a homogeneous linewidth of Γ=0.2meV. In addition, the substrate refractive index is incorporated by scaling the theoretical transmission at 1.513eV to the experimentally found value of 36%.

In Fig. 1 we also show a photoluminescence spectrum obtained at a low excitation intensity of 30mW/cm². A narrow emission line at the energy position of the N=1 exciton emission is observed, which exactly coincides with the excitonic eigenenergy. This demonstrates that the exciton in the thin layer can indeed be described as a two-level system. Optical excitation and emission at low exciton densities have to obey the conservation of the wave vector component parallel to the semiconductor layer $k_{\parallel}$. Thus, the continuous excitonic energy distribution $E(k_{\parallel})$ is virtually one single, but highly degenerate, level of typical total width of: $E(k_{\parallel}=2\pi/\lambda_{photon})-E(k_{\parallel}=0)=4\mu eV$.

5. Optical Dephasing of Excitons

Optical dephasing of Wannier excitons is studied by means of a two-pulse degenerate four-wave mixing (DFWM) experiment. A schematical excitation arrangement is shown in Fig. 2. Two optical pulses with center frequencies of $\omega_1=\omega_2=\omega$ and with wave vector components $k_{1\parallel}=-k_{2\parallel}=1.5*10^3cm^{-1}$ are tuned into the excitonic resonance and are focused onto the sample. The first pulse sets up a macroscopic polarization of excitonic dipoles all oscillating in phase, whereas the second and delayed pulse probes the coherent (in phase) part of the macroscopic polarization left from the first pulse. A signal at a center frequency $\omega_3=2\omega_2-\omega_1=\omega$

with $k_{3\parallel}=2k_{2\parallel}-k_{1\parallel}$ (self-diffraction) is emitted from the phased oscillator array. The decrease of the intensity with increasing delay can be used for measuring the excitonic phase coherence time.

The simple qualitative considerations of optical dephasing known since the early work of ABELLA et al. /8/ are restricted to pulse lengths much shorter than any relaxation time. Difficulties in the interpretation of the experimental data arise if the relaxation (dephasing) occurring during excitation can no longer be neglected. In this case, one has to solve the optical Bloch equations of a two-level system taking into account the finite pulse duration /9/. Fortunately we are interested only in the small density regime where bleaching effects (Bloch saturation) are not important. In this limit the Bloch equations can be iteratively solved for an arbitrary pulse shape /9/. The complicated coherent nonlinear interaction is equivalent to a $\chi^{(3)}$-process. All one has to know is the electric field of the pulse, a possible inhomogeneous distribution $g(\omega-\Omega)$, and the phase (T_2) as well as the population (T_1) relaxation time. A simple integration then yields the diffracted intensity as a function of the delay between the two pulses, which can be compared to the experiment.

Our DFWM experiment is equivalent to the commonly employed photon echo experiment /8/. There are only two differences: (i) The DFWM corresponds to the typical photon echo experiment only in the small density regime (no Bloch saturation). (ii) The DFWM signal of a homogeneously broadened line ($g(\omega)=\delta(\omega-\Omega)$) leaves the sample together with the second pulse ($\tau_{signal}=\tau_{12}$), in contrast to the photon echo method on an inhomogeneously broadened line where the signal (photon echo) delay with respect to pulse 1 is two times the delay between pulses 1 and 2 ($\tau_{signal}=2\tau_{12}$). For a detailed discussion of the dynamics of the nonlinear signal, in particular for finite pulse lengths, see reference 10.

Figure 3 depicts a typical diffraction curve versus the delay between the two pulses obtained at a low exciton density $\leq 2*10^{14}cm^{-3}$. The highly asymmetric correlation trace indicates a phase coherence time much larger than the pulse length. In fact, by fitting the correlation trace with a theoretical model we obtain T_2=7ps. The corresponding homogeneous linewidth is Γ=0.18meV and agrees

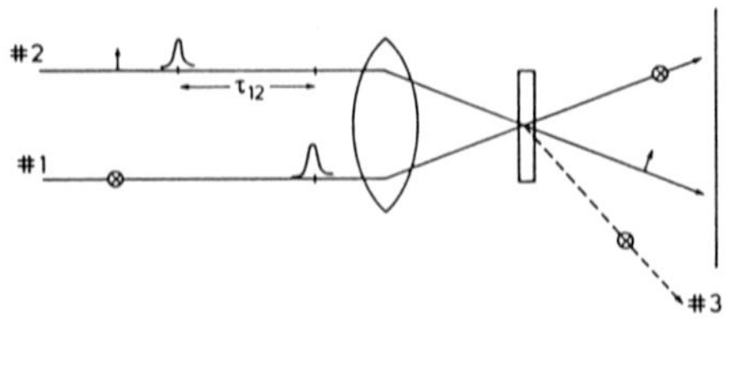

Fig. 2: Schematical excitation arrangement of the two-pulse DFWM experiment

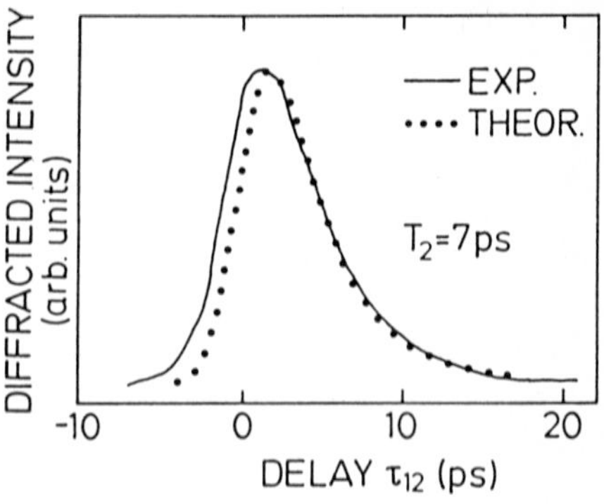

Fig. 3: Experimental (solid line) and theoretical (dotted line) diffraction curves at 2K

well with the homogeneous linewidth from the transmission experiment, substantiating the homogeneously broadened nature of the exciton line.

The DFWM signal is not only highly directional with an extremely small divergence (the same as the laser pulses), but is also highly polarized (polarization ratio up to 10^3) in the direction of the first pulse. This can be expected for (i) a two-level system, which is degenerate for left-handed and right-handed circularly polarized light, in analogy to the photon-echo experiment in ruby /9/ and (ii) also from the tensor properties of $\chi^{(3)}$ known from similar experiments for band-to-band excitation in Germanium and GaAs /11,12/. These experiments, performed in the T_2 =0 limit, reveal also the same symmetry properties of the relevant $\chi^{(3)2}$-elements: χ_{xxxx},χ_{xyyx}.

The optical dephasing of the exciton in GaAs described by the homogeneous linewidth of 0.18meV is by a factor of two larger than the longitudinal-transverse (LT) split of 0.08meV which is a measure of the oscillator strength. The L-T splitting corresponds to a switching time between the exciton and the photon state owing to the coherent exciton-photon interaction. Our results demonstrate that the coherent exciton-photon coupling in GaAs is inefficient due to the much more efficient optical dephasing caused by the residual interaction of the exciton with the crystal. This result strongly questions the polariton concept /13/ employed for GaAs /14/.

The phase coherence time of the Wannier excitons in GaAs is much longer than for free-carrier excitation, which is well below 0.5ps /1,2/. This is basically due to the fact that excitons are neutral particles which are not as efficiently scattered as free carriers with the corresponding long range Coulomb interaction.

6. Temperature Dependence of the Optical Dephasing

Next, we try to answer the question of the phase-relaxing mechanisms. We tackle this question by looking first at the dependence of the optical dephasing on the temperature and second to the lifetime of the excitonic states.

The left part of Fig. 4 shows transmission spectra in the near band gap region obtained for three different temperatures. As compared to Fig. 1 the excitonic absorption line is considerably broadened even at 5K. The broadening increases rapidly with increasing temperature. For temperatures higher than 20K, the excitonic absorption line is only seen as a broad bump on the onset of the bandedge absorption. A lineshape analysis of the transmission data yields the homogeneous linewidth.

The right part of Fig. 4 depicts the two-pulse DFWM diffraction curves for three different temperatures. As can be seen very clearly the excitonic phase coherence strongly decreases with increasing temperature. From the fitting of the correlation traces we obtain the excitonic phase coherence times.

Figure 5 summarizes the results from DFWM and transmission experiments. At low temperatures the homogeneous linewidth increases linearly,with the temperature according to

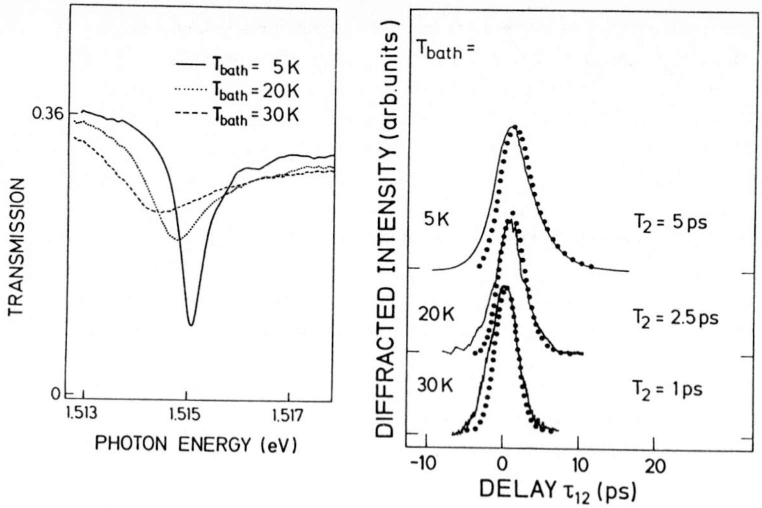

Fig. 4: Transmission spectra (left part) and diffraction curves (right part) for three diffferent bath temperatures

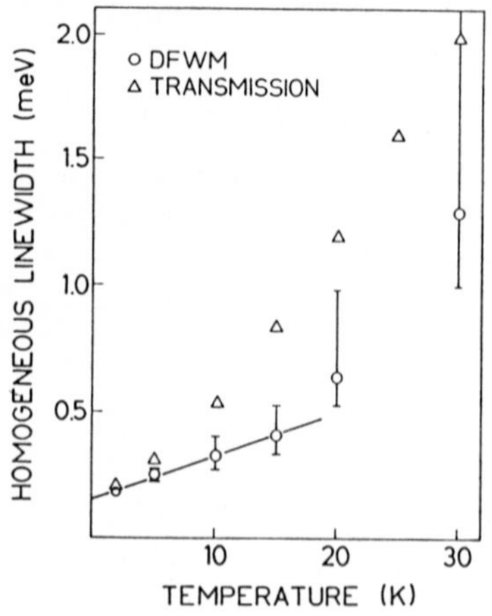

Fig. 5: The homogeneous linewidth from transmission as well as from DFWM experiments versus temperature

$$\Gamma = \Gamma_0 + \gamma * T = 0.146\text{meV} + 17\mu\text{eV/K} * T. \qquad (2)$$

This linear temperature dependence is attributed to anti-Stokes scattering with acoustic phonons, whereas the temperature independent part Γ_0 of the homogeneous linewidth is attributed to scattering with residual impurities.

Deviations from the linear temperature dependence above 15K may be attributed to additional, highly efficient scattering of the excitons with free carriers /15/ originating from thermal dissociation of the excitons.

This assignment implicitly neglects a possible contribution of acoustic phonon scattering involving the modified eigenstate of the slab: the interface polariton. Up to now nothing is known about the slab eigenstate of the coupled exciton-photon, neither the dispersion relation nor the scattering probabilities.

From Fig. 5 it is also seen that the results from DFWM and transmission strongly differ for higher temperatures. A possible explanation for this fact is that a DFWM experiment probes the optical dephasing of well-defined coherent excitonic states, whereas for the transmission also phonon-assisted absorption processes contribute. The surprisingly high efficiency of this mechanism might be explained because the translational symmetry in the z-direction is broken.

7. Orientational Relaxation of Excitons

The orientational relaxation of Wannier excitons is studied by means of a three-pulse DFWM experiment schematically shown in the upper part of Fig. 6. Two pulses with wave vector components $k_{1\parallel}=-k_{2\parallel}$, in temporal overlap are focused onto the sample and are forming a polarization grating (of oriented excitonic dipoles). A third and delayed pulse with a wave vector $k_{3\parallel}$ is diffracted off the grating giving rise to a signal $k_{4\parallel}=k_{3\parallel}+k_{2\parallel}-k_{1\parallel}$, which measures the decay of the orientational grating.

In general, two diffraction mechanisms have to be considered. The most important is the coherent diffraction of the probe pulse by a coherent macroscopic polarization with an interaction governed by the optical Bloch equations. The corresponding population relaxation time T_1 reflects the randomization of the excitonic states originally excited with a well-defined wave vector $k_\parallel$ and polarization. However, after randomization of the excitonic states an incoherent excitonic density modulation grating may still exist /11/. The diffraction then is given by a $\chi^{(1)}$-process, the diffraction off a dielectric grating caused by the density dependence of the excitonic parameters: line shift, linewidth and oscillator strength:

$$\chi^{(1)}(E, N_x) \sim \frac{E_L(N_x)-E_T(N_x)}{E_T(N_x)-E-i\Gamma(N_x)/2}, \qquad (3)$$

where N_x is the periodically varying exciton density. In this case the corresponding decay time is the lifetime of an exciton density grating reflecting spatial diffusion as well as radiative recombination or capture into impurity states.

However, these two mechanisms can be distinguished very elegantly by choosing the proper polarizations for the grating forming pulses. This is also schematically shown in the upper parts of Fig. 6 and 7. For parallel polarization a coherent (polarization) grating as well as an incoherent exciton density grating is formed, whereas for orthogonal polarization only a coherent, orientational grating with periodically changing polarization direction is formed.

This simple picture of the diffraction processes is confirmed experimentally. Figure 6 shows the diffracted intensity versus delay for orthogonal polarization of the grating forming pulses. The orientational grating (periodically varying polarization direction) rapidly decays. A comparison with the theoretical model yields the orientational relaxation time of T_1=7ps.

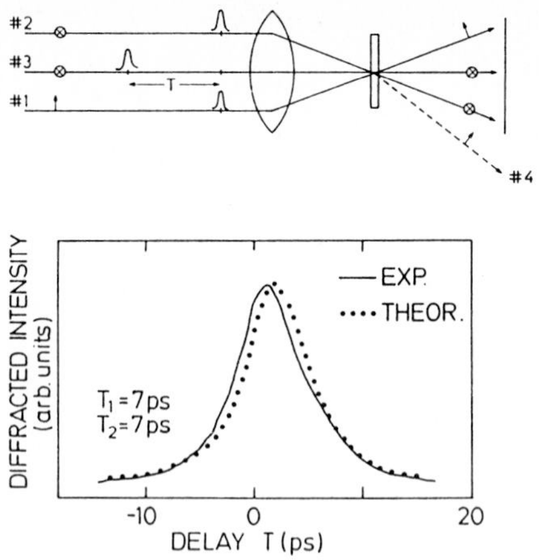

Fig. 6: Schematical excitation arrangement for the orientational grating experiment (upper part) and experimental (strong line) and theoretical (dotted line) diffraction curves (lower part)

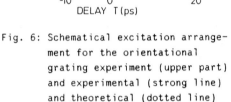

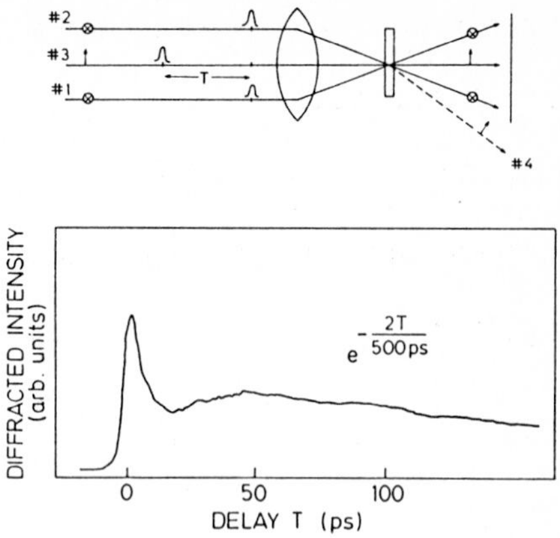

Fig. 7: Schematical excitation arrangement for the population grating experiment (upper part) and experimental diffraction curve (lower part)

Figure 7 depicts the diffracted intensity versus delay for parallel polarization. For small delays ≦10ps a coherent polarization (population) grating is formed which rapidly decays into an incoherent exciton density grating with a time constant below 10ps (not very clearly seen on this time scale). The exciton density grating vanishes with a lifetime of 500ps. For higher exciton densities the density grating lifetime increases, indicating that the grating decay is mainly limited by the lifetime of the excitons due to capture by impurities and not by diffusion processes.

Similiar studies have been reported for free carrier excitation where orientational relaxation times of about 150fs have been reported /1/. The mechanism for the much faster relaxation of free carriers is the same, destroying the phase coherence: efficient carrier-carrier scattering because of the long range Coulomb interaction.

8. Conclusions

Optical dephasing and orientational relaxation of Wannier excitons in optically thin GaAs layers were studied by means of time-resolved DFWM on a picosecond timescale. We found ultrafast optical dephasing within 7ps limited by acoustic phonon and residual impurity scattering. Transient grating experiments revealed also ultrafast orientational relaxation within 7ps.

Further studies have to be made to clarify the dependence of the orientational relaxation time as a function of the temperature and a possible dependence of the optical dephasing on the thickness of the GaAs layers. In particular, we hope that future studies of polariton dephasing can be performed on thick (≧3μm) high-quality GaAs samples, which would greatly enhance our knowledge of the scattering processes which should depend on the polariton wavevector.

9. Acknowledgments

We gratefully acknowledge fruitful discussions with H.J. Polland and we thank H.J. Queisser for critically reading the manuscript and H. Oppolzer, Siemens AG, Munich for determining the geometrical sample parameters.

10. References

1. J.L. Oudar, A. Migus, D. Hulin, G. Grillon, J. Etchepare, and A. Antonetti, Phys.Rev.Lett.53, 384 (1984)
2. J.L. Oudar. D. Hulin, A. Migus, and F. Alexandre, Phys.Rev.Lett.55, 2074 (1985)
3. L. Schultheis, J. Kuhl, A. Honold, and C.W. Tu, Phys.Rev.Lett.57, 1797 (1986)
4. L. Schultheis, M.D. Sturge, and J. Hegarty, Appl.Phys.Lett.47, 995 (1985)
5. L. Schultheis, A. Honold, J. Kuhl, and C.W. Tu, Phys.Rev.B34, 9027 (1986)
6. Y. Masumoto, S. Shionoya, and T. Takagahara, Phys.Rev.Lett.51, 923 (1983)
7. R. Sooryakumar and P.E. Simmonds, Solid State Commun.42, 287 (1982)
8. I. D. Abella, N.A. Kurnit, and S.R. Hartmann, Phys.Rev.141, 391 (1966)
9. T. Yajima and Y. Taira, J.Phys.Soc.Jpn. 47, 1620 (1979)
10. L. Schultheis and J. Hegarty, J. Physique, Colloque C7, 46, 167 (1985)
11. A. Smirl, T.F. Boggess, B.S. Wherrett, G.P. Perryman, and A. Miller, IEEE J. Quantum Electron.19, 690 (1983)
12. J.L. Oudar, I. Abram, A. Migus, D. Hulin, and J. Etchepare, J.Lumin.30, 340 (1985)
13. J.J. Hopfield and D.G. Thomas, Phys.Rev.132, 563 (1963)
14. D.D. Sell, S.E. Stokowski, R. Dingle, and J.V. DiLorenzo, Phys.Rev.B7, 4568 (1973)
15. L. Schultheis, J. Kuhl, A. Honold, and C.W. Tu, Phys.Rev.Lett.57, 1635 (1986)

True Radiative Lifetime of Free Excitons in GaAs

G.W. 't Hooft[1], *W.A.J.A. van der Poel*[1], *L.W. Molenkamp*[1], *and C.T. Foxon*[2]

[1]Philips Research Laboratories, NL-5600 JA Eindhoven, The Netherlands
[2]Philips Research Laboratories, Redhill, Surrey, RH1 5HA, United Kingdom

In this paper we will show that the radiative lifetime of Wannier excitons and the total absorption cross section are not simply related via the oscillator strength per unit cell. For the first time the true radiative lifetime of free excitons in GaAs has been measured. Its value indicates that the coherence volume of the oscillators is much larger than just one unit cell (see also [1]).

Experimentally the transition probability of an exciton line can be deduced from the total absorption cross section. In order to obtain the oscillator strength, however, the density of independent oscillators has to be known. Generally, the density of unit cells is taken. The ensuing oscillator strength is then quite low for most semiconductors. In the case of GaAs we find 7×10^{-5} in this way. Such small values for the oscillator strength have been attributed to the small probability of finding the electron at the position of the hole [2,3]. The transition probability is not only related to the strength of absorption lines but also to the radiative lifetime, τ, of the excited state. Assuming that the oscillator strength per unit cell also governs the radiative recombination process the value for τ is expected to be quite large. For GaAs it amounts to 40 μs. The radiative lifetime of free excitons is difficult to determine from experiment, since free excitons are trapped by defects. According to RASHBA and GURGENISHVILI [4] the oscillator strength of bound excitons is giant, i.e. in the order of unity. Consequently, the radiative lifetime of bound excitons is anticipated to be in the ns time domain. HENRY and NASSAU [5] have verified this for CdS and have found a lifetime of 0.5 ns for donor bound excitons and 1.0 ns for acceptor bound excitons. Therefore, the measured lifetime of free excitons will usually be much shorter due to trapping of the free excitons to impurities and subsequent fast recombination. The measured lifetime is an effective and not a radiative one and will depend on the amount of impurities present in the sample. HWANG [6] found indeed that the effective free - exciton lifetime in GaAs decreased progressively from 2.9 ns to 1.1 ns when the mobility at 77 K decreased from 1.28×10^5 cm^2/Vs to 1.0×10^5 cm^2/Vs.

We have studied the free exciton lifetime in ultra pure GaAs in which the defect-related recombination can be saturated and thus the true radiative lifetime is obtained. The material was grown by molecular beam epitaxy with a Varian GEN II machine. For details on crystal growth see [7]. The structure was grown on a Si-doped (001) oriented substrate and consisted of the following layers. After a 1 μm buffer layer and a 250 period superlattice of 10 monolayers GaAs and 33 monolayers $Al_{0.09}Ga_{0.91}As$ the 1.5 μm thick GaAs active layer was deposited followed by an 80 period superlattice of identical composition as the previous one. The active layer thickness was chosen to be equal to the penetration depth of the exciting light, effectuating homogeneous excitation. The two cladding superlattices prevent the free excitons from diffusing to places with increased defect concentrations like substrates and/or free surfaces. Superlattices also help to minimize carrier spill-over into the active layer from background impurities in the AlGaAs. To a large extent these carriers will now spill over into the GaAs wells of the superlattice. Furthermore, the superlattices ensure that the interfaces around the active layer are of optimum quality [8].

The experiments were performed using a ps dye laser synchronously pumped by a mode - locked krypton laser. The samples were placed in a variable temperature, optical cryostat. The luminescence was dispersed through a 3/4 m monochromator and detected by a cooled photomultiplier with GaAs cathode. Photoluminescence decays were measured via appropriate single - photon - counting equipment with a time resolution of 0.3 ns.

Some of the recorded photoluminescence spectra at 1.7 K are displayed in Fig. 1 for a few excitation densities. The features seen at low excitation density are: the recombination of free excitons in the 2s excited state at 1.518 eV, the free exciton line at 1.5151 eV, the line related to excitons bound to neutral donors at 1.5141 eV, the one related to excitons bound to neutral acceptors at 1.5124 eV, and the so - called g line of KUNZEL and PLOOG [9]. Even at excitation densities as low as 5 mW/cm^2 the free exciton line is dominant and has a full width at half maximum of only 0.36 meV, indicating the very high quality of our material. At excitation densities above 0.2 W/cm^2 an additional luminescence feature can be observed centered at 1.5146 eV. Above 1 W/cm^2 this spectral line broadens and becomes the dominant recombination mechanism. We attribute this line to the recombination of excitonic molecules (biexcitons) on the basis of its spectral position. A number of authors [10,11] have calculated the extra binding energy of the excitonic molecule in units of the excitonic Rydberg as a function of the ratio of the effective masses of electrons and holes. For GaAs the ratio is 0.14, implying that the extra binding energy is 0.08 excitonic Rydberg, *i.e.* approximately 0.35 meV. The actually observed energy difference between the free exciton and the biexciton lines is 0.5 meV. The agreement is reasonably good. Note that the biexciton line cannot be assigned to the formation of an electron - hole liquid or an electron - hole plasma, since the former is not stable and the latter occurs at much lower photon energies [12]. Assuming that biexcitons are only formed when the density of free excitons is so large that their average distance is smaller than their diffusion length, the diffusion length is calculated to be approximately 0.2 μm.

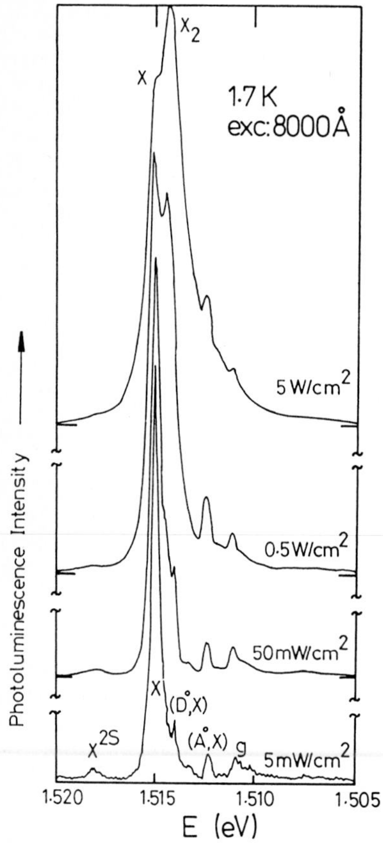

FIG. 1. Photoluminescence spectra of a 1.5 μm *thick GaAs layer at 1.7 K for various excitation power densities. The excitation wavelength was 800 nm. The spectra are scaled to have the same height for the main peak and are shifted for the sake of clarity.*

To verify whether the defect concentration in our sample is low enough, we have measured the dependence of the luminescence on excitation density. Below 0.5 W/cm^2 the free exciton recombination is strictly linear with power density. This is the first time that such a linear behavior for GaAs has been recorded. It demonstrates that the material is ultra pure and that the lifetime of free excitons is governed by radiative processes. For excitation densities above ~ 1 W/cm^2 the biexciton recombination takes over the linear power dependence and the free exciton line more or less saturates.

We measured the lifetime of the free excitons as a function of power density, excitation photon energy and lattice temperature. An example is given in Fig. 2 for resonant excitation at the free exciton line at 1.4 K. We found that the lifetime remains constant for power densities up to ~ 1 W/cm^2. For higher power densities the lifetime progressively decreases, as expected from the appearance of the parallel radiative process of biexcitonic recombination. For excitation photon energies above 1.53 eV the lifetime of the free excitons is found to have increased by some 20 percent. This is probably due to the higher energy of the excited carriers and the slow cooling rate of the carriers when their excess energy is smaller than the optical phonon energy.

The influence of the lattice temperature on the radiative lifetime of the free excitons has been investigated. The results for band edge excitation with 0.1 W/cm^2 are given in Fig. 3. The lifetime exhibits an increase from 3.3 ns at very low temperatures to almost 11 ns at 10 K. The center-of-mass system of the free excitons will have an increased average kinetic energy with increasing lattice temperature. Owing to momentum conservation, only those free excitons close enough to the Brillouin zone center can recombine radiatively. For a parabolic exciton band and an assumed Maxwell - Boltzmann distribution, the fraction, r, of free excitons with a kinetic energy smaller than ΔE is given by

$$r(T) = \frac{2}{\sqrt{\pi}} \int_0^{\Delta E/k_B T} \sqrt{\varepsilon}\, e^{-\varepsilon}\, d\varepsilon. \qquad (1)$$

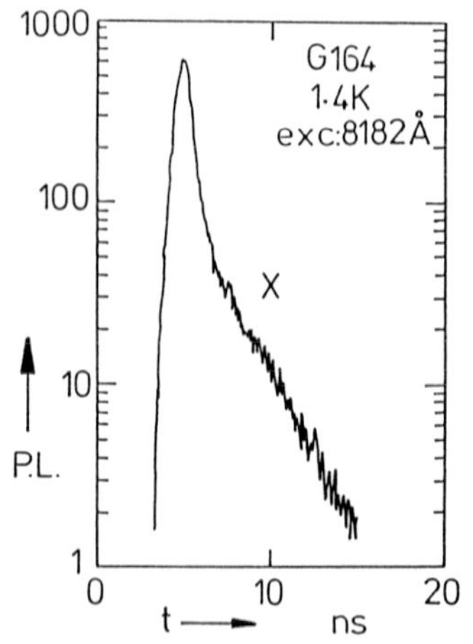

Fig. 2. Photoluminescence decay of the free exciton line at 1.4 K after resonant excitation. The excitation density is 0.1W/cm². The initial decay is due to diffuse scattering of the laser light from the sample surface. The tail of the decay stems from the excitonic transitions.

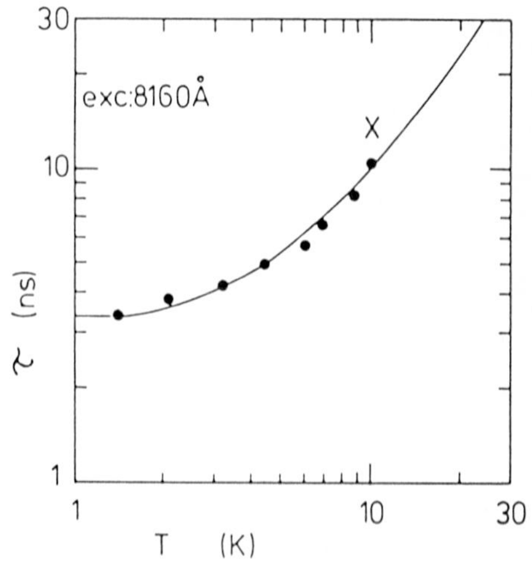

FIG. 3. Double logarithmic plot of the radiative lifetime of the free exciton in GaAs versus temperature. Excitation is resonant at the band edge of GaAs with a power density of ~ 0.1 W/cm².

The temperature dependent radiative lifetime averaged over the ensemble is then

$$\tau(T) = \tau_0/r(T). \tag{2}$$

The theory of (1-2) is fitted to the experimental points of Fig. 3 with τ_0 and ΔE as parameters. The solid curve of Fig. 3 shows a perfect agreement for $\tau_0 = 3.3$ ns and $\Delta E = 0.7$ meV, *i.e.* twice the line width of the free exciton.

The oscillator strength associated with a lifetime of 3.3 ns is of the order of unity. To reconcile this with the total absorption cross section one might take for the density of absorbing entities the concentration of excitons in a closed packed lattice instead of the density of unit cells. This increases the oscillator strength by exactly a factor10^4.The oscillator strength is then per excitonic volume instead of per unit cell. The existence of bound exciton lines in absorption spectra merely reflects the probability of finding an impurity within the volume of the exciton. This probability is proportional to the number of unit cells per excitonic volume. In the limit that the impurity concentration is so large that within every exciton volume an impurity is located, only bound exciton absorption will be observed. At such large concentrations the relation between the oscillator strength of bound excitons and free excitons from RASHBA and GURGENISHVILI [4] breaks down completely, since the ratio of the absorption lines is infinite. To summarize, we suggest it is more fruitful to discuss excitonic transition probabilities in a semiconductor in terms of oscillator strength per excitonic volume instead of per unit cell.

1. G.W. 't Hooft, W.A.J.A. van der Poel, L.W. Molenkamp and C.T. Foxon, Phys. Rev. **B**, (1987), in press.
2. R.J. Elliot, Phys. Rev. **108**, 1384 (1957).
3. R.S. Knox, *Solid State Physics*, suppl. **5**, (ed. F. Seitz and D. Turnbull, 1963), *Theory of Excitons*.
4. E.I. Rashba and G.E. Gurgenishvili, Sov. Phys. Solid State **4**, 759 (1962) [Fiz. Tver. Tela **4**, 1029 (1962)].
5. C.H. Henry and K. Nassau, Phys. Rev. **B1**, 1628 (1970).
6. C.J. Hwang, Phys. Rev. **B8**, 646 (1973).
7. C.T. Foxon and J.J. Harris, Philips J. Res. **41**, 313 (1986).
8. P. Dawson and K. Woodbridge, Appl. Phys. Lett. **45**, 1227 (1984).
9. H. Kunzel and K. Ploog, *Proc. Eighth Int. Symp. on GaAs and Related Compounds (Vienna, 1980)*, Inst. Phys. Conf. Ser. **56**, 519 (1981).
10. O. Akimoto and E. Hanamura, J. Pys. Soc. Japan **33**, 1537 (1972).
11. W.F. Brinkman, T.M. Rice and Brian Bell, Phys. Rev. **B8**, 1570 (1973).
12. O. Hildebrand, E.O. Goebel, K.M. Romanek, H. Weber, and G. Mahler, Phys. Rev.**B17**, 4775 (1978).

Lifetime of Free Excitons in GaAs

W.J. Rappel, L.F. Feiner, and M.F.H. Schuurmans

Philips Research Laboratories, NL-5600 JA Eindhoven, The Netherlands

Recent experiments reported by 'T HOOFT *et al.* [1] yielded a free exciton lifetime in GaAs of 3.3 ns. This was rather surprising because earlier theoretical work of DEXTER [2] and HENRY and NASSAU [3] based on exciton oscillator strength considerations suggested a much longer lifetime of the order of 1 μs. However, HOPFIELD [4] has pointed out that as a result of the coherent (and thereby strong) coupling in a perfect crystal between the free excitons and the photons, neither an exciton nor a photon is a good eigenstate of the system as a whole: one has to consider the mixed-mode excitation, the exciton-polariton. The main purpose of this contribution is to investigate whether a physical interpretation of luminescence in terms of exciton-polaritons and their interaction with phonons can explain the observed lifetime. We have limited ourselves to the lower branch at T=0 K only.

In section I the theory of luminescence in the exciton-polariton picture is given. In section II we present our calculations for both the semi-classical and the rate equations approach. The results and options for future work are discussed in section III.

I Theory

The exciton-polariton (in short polariton) dispersion curve $E(k)$ consists of two branches and is given by

$$\left[\frac{\hbar ck}{E}\right]^2 = \varepsilon_b + \frac{4\pi\beta E_T^2(k=0)}{E_T^2(k) - E^2} \quad , \tag{1}$$

where ε_b is the background dielectric constant which contains contributions from all interactions except the exciton in question, the coupling constant β is the polarizability of the exciton and $E_T(k)$ is the energy of a free exciton with momentum $\hbar\mathbf{k}$, which is assumed to be parabolic, i.e. $E_T = E_0 + \hbar^2k^2/2m^*$, with m^* the effective mass of the exciton. The polarizability β can be determined from the energy splitting ΔE_{LT} of the longitudinal and transverse exciton at $k = 0$. The values of the parameters are given in Table I. Figure 1 shows the dispersion curve for GaAs in the cross-over region, i.e. near $k = k_0$.

Upon absorption photons are converted into exciton-polaritons which thermalize rapidly into the lower branch with $k >> k_0$. There the polaritons are predominantly exciton-like and the probability of being reemitted from the crystal as a photon is small. Due to scattering by phonons the polaritons lose momentum: when k approaches k_0 they rapidly become more photon-like and consequently the radiative escape probability increases. Bottlenecking occurs because energy dissipation takes place rather slowly due to small density of states and to weak polariton-phonon coupling and because the phase-volume for efficient photon escape is small: the effective lifetime is to some extent a confinement time.

In the present investigation we have considered only scattering by longitudinal acoustic phonons, and the electron-phonon interaction is described by the deformation potential. The transition rate for scattering of a polariton with momentum $\mathbf{k}$ and energy E into a polariton with momentum $\mathbf{k}' = \mathbf{k} - \mathbf{q}$ and energy E′ is given by

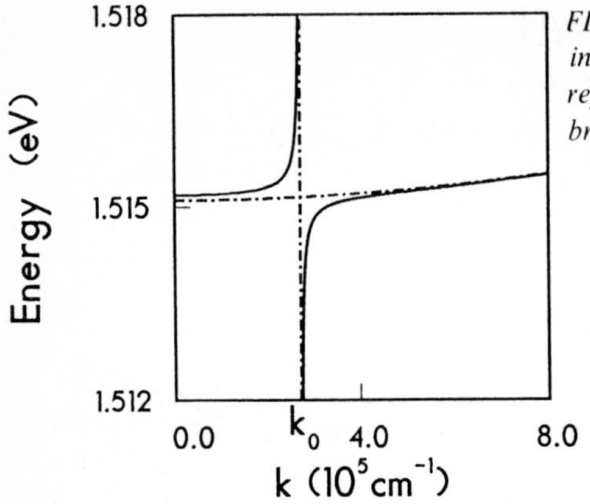

FIG. 1. Dispersion curve of the exciton-polariton in GaAs (solid curves); the dashed curves represent the unperturbed exciton and photon branches

Table I. Parameters used in the calculation

Transverse exciton energy ($E_T(k=0)$)	1.5151 eV
Longitudinal-transverse splitting (ΔE_{LT})	0.08 meV
Polarizability β	$1.06\text{x}10^{-4}$
Background dielectric constant (ε_b)	12.56
Exciton effective mass (m^*)	0.6 m_e
Sound velocity (u)	$4.8\text{x}10^5$ ms^{-1}
Density (ρ)	5.307 gcm^{-3}
Deformation potential strength (D)	7.8 eV
Crystal thickness (L)	1.5 μ m

$$W(\mathbf{k},\mathbf{k}') = \frac{2\pi}{\hbar} | V_{ac}(\mathbf{q}) |^2 \left[\{n(\hbar\omega_q)+1\}\delta(E-E'-\hbar\omega_q) - n(\hbar\omega_{-q})\delta(E-E'+\hbar\omega_{-q}) \right] \quad ,(2)$$

where $V_{ac}(\mathbf{q})$ is the matrix element for scattering by the acoustic phonon with momentum $\hbar\mathbf{q}$ and energy $\hbar\omega_q$:

$$| V_{ac}(\mathbf{q}) |^2 = \frac{\hbar\, |\mathbf{q}|\, D^2}{2V_c\rho u} \quad . \tag{3}$$

Here V_c is the crystal volume, ρ is the density of the crystal, u is the LA sound velocity and D the deformation potential strength (values are given in Table I). In (3) we have ignored the internal structure of the exciton [5].

The radiative decay rate for polaritons with momentum $\hbar\mathbf{k}$ is obtained by multiplying the rate at which these polaritons arrive at the crystal surface by the probability for such a polariton of being transmitted and escaping as a photon:

$$P(\mathbf{k}) = \frac{\upsilon(\mathbf{k})}{L}\int_0^{\theta_r} \sin\theta\cos\theta\; T(\mathbf{k},\theta)d\theta \quad . \tag{4}$$

Here L measures the crystal thickness, $\upsilon(\mathbf{k})$ is the polariton group velocity and $T(\mathbf{k},\theta)$ is the optical transmission coefficient given by

$$T(\mathbf{k},\theta) = 1 - \frac{1}{2}\left[\frac{\sin^2(\theta-\theta')}{\sin^2(\theta+\theta')} + \frac{\tan^2(\theta-\theta')}{\tan^2(\theta+\theta')}\right] . \tag{5}$$

The integration in (4) extends to the critical angle θ_r. The incident angle θ and the outgoing angle θ' are related by Snell's law. Since we are dealing with the case of extremely pure samples [1] we do not include a trapping rate.

II Calculations

a) Semi-classical approach

As a first and rough estimate we have used TOYOZAWA's [5] semi-classical model to calculate the integrated escape probability Y(t), which is a measure of the lifetime of the polariton. In this model one starts with a polariton at a certain k-value, calculates its decay in time while integrating its probability of decay. In other words, we consider only a single polariton and follow it on its way to the bottleneck. Thus the effects of influx from other k-states are neglected and therefore this approach is not capable of describing the process at temperatures different from 0 K. With the help of (4) one can calculate the number of polaritons N(t) from

$$\frac{dN}{dt} = -\,P(k(t))N(t), \tag{6}$$

which results in

$$N(t) = N_0 e^{-\int_0^t P(k(t'))dt'} = N_0 e^{-Y(t)} . \tag{7}$$

In Fig. 2 we have plotted the integrated probability Y(t) *vs.* time. Taking for the lifetime the timespan during which Y(t) is increased to 1 we obtain a lifetime of 17 ns, which is of the right order of magnitude.

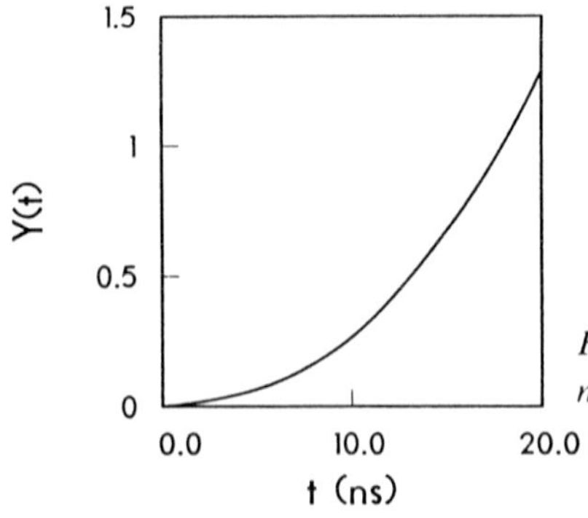

FIG. 2. The integrated escape probability in Toyozawa's model

b) Rate equations

A more sophisticated approach is to calculate the distribution of polaritons in **k** space as a function of time (ASKARY and YU, [6]). Starting from a given initial distribution $\rho(\mathbf{k}, t=0)$, assumed to be isotropic in **k** space, we can determine the distribution $\rho = \rho(\mathbf{k}, t)$ from

$$\frac{d\rho}{dt} = \left[\frac{d\rho}{dt}\right]_{\text{in}} - \left[\frac{d\rho}{dt}\right]_{\text{out}} , \tag{8}$$

where the rates of generation and loss of polaritons are given respectively by

$$\left[\frac{d\rho(\mathbf{k},t)}{dt}\right]_{\text{in}} = \frac{V_c}{(2\pi)^3}\int d\mathbf{k}'\rho(\mathbf{k}',t)W(\mathbf{k}',\mathbf{k}) , \tag{9}$$

and

$$\left[\frac{d\rho(\mathbf{k},t)}{dt}\right]_{\text{out}} = \frac{V_c}{(2\pi)^3}\int d\mathbf{k}' W(\mathbf{k},\mathbf{k}')\,\rho(\mathbf{k},t) + \rho(\mathbf{k},t)P(\mathbf{k}) . \tag{10}$$

Here $W(\mathbf{k}, \mathbf{k}')$ follows from (2). We have solved the rate equations (8) numerically starting from a Gaussian initial distribution of polaritons with energies (on the lower branch) centered at the photon energy, 1.5151 eV, and with a width of 0.5 meV. The parameters used are given in Table I. The results are shown in Figs. 3 and 4. We can see that the polariton distribution rapidly piles up towards the bottleneck. From Fig. 4, which shows the rate of photon emission

$$n(t) = \int P(k)\rho(k,t)4\pi k^2 dk , \tag{11}$$

we estimate the free exciton lifetime to be 8 ns .

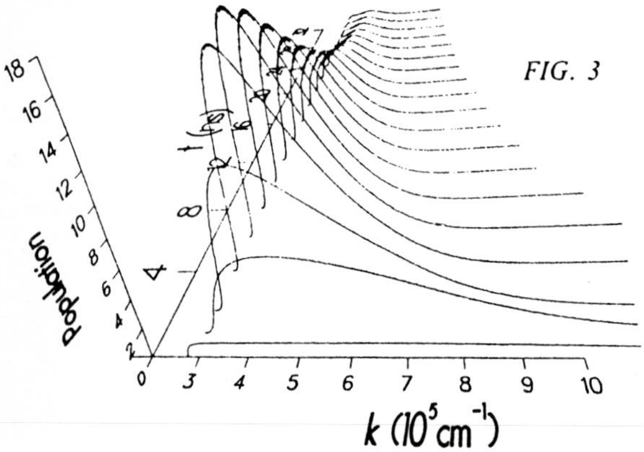

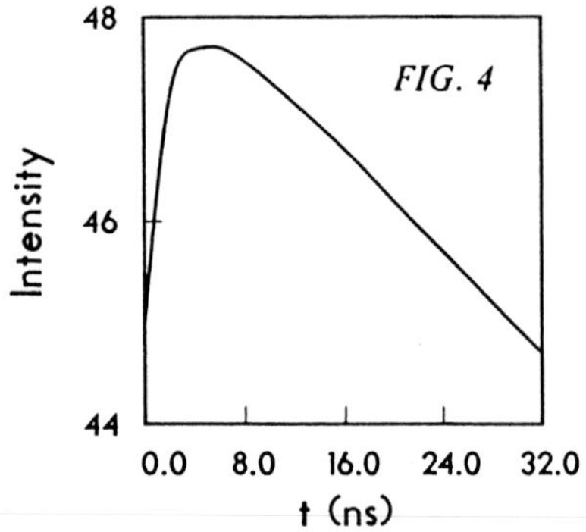

FIG. 3. The polariton population of the lower branch as a function of time and crystal momentum (T=0 K)

FIG. 4. The rate of photon emission corresponding to the populations in Fig. 3.

III Conclusions

We have seen that a description of free exciton luminescence in terms of conversion of exciton-like polaritons into photon-like polaritons by electron-phonon scattering yields a free exciton lifetime 2-3 orders of magnitude smaller than earlier theoretical estimates. Essential in the present work is that we deal with mixed-mode excitations. This is called for by the photon-free exciton coupling being strong in a perfect crystal because of its coherent nature. This aspect was not fully appreciated in the earlier theoretical work.

Our estimated free exciton lifetime of 8 ns is still larger than the experimental value of 3.3 ns in GaAs. However, inclusion of the piezoelectric exciton-phonon interaction will further shorten the theoretical lifetime. The dependence of the lifetime on the temperature is currently under investigation. For this purpose we must include scattering of both the lower and the upper branch polaritons.

Acknowledgements

We thank Dr. G.W.'t Hooft for many valuable discussions.

References

(1) G.W. 't Hooft, W.A.J.A. van der Poel, L.W. Molenkamp and C.T. Foxon, Phys. Rev. B **35**, 8281, 1987
(2) D.L. Dexter, in *Solid State Physics*, Vol. **6** (ed. by F. Seitz and D. Turnbull, 1958), p.353
(3) C.H. Henry and K. Nassau, Phys. Rev. **B1**, 1628 (1970)
(4) J.J. Hopfield, Phys. Rev. **112**, 1555 (1958)
(5) Y. Toyozawa, Prog. Theor. Phys. Suppl. **12**, 111 (1959)
(6) F. Askary and P.Y. Yu, Phys. Rev. B **31**, 6643 (1985)

Exciton Masses in Solids

S.E. Schnatterly[1], *C. Tarrio*[1], *and A.A. Cafolla*[2]

[1]Physics Department, University of Virginia, Charlottesville, VA 22901, USA

[2]Physics Department, Manchester University, Manchester, United Kingdom

An exciton consists of a coherent superposition of a conduction electron and all the electrons in the valence band. To a very good approximation however it can also be thought of as a two-body problem consisting of the conduction electron, a valence hole, and a surrounding passive polarization field. If we accept this reduction to a two-body problem, a question then remains: Are there any surprises? By this we mean do excitons behave in any essential way differently than an ideal two-body problem - i.e. two elementary particles interacting in free space.

According to classical mechanics a composite of two particles in free space can be reduced to two one-particle problems by the use of center of mass and relative coordinates. By Newton's third law the internal forces between the particles cancel and the center of mass point moves under the influence of external forces exactly like a single particle with mass equal to the sum of the two particles' masses. The relative motion reduces to that of a single particle with mass equal to the reduced mass on which a potential given by the interaction between the two particles is acting. These results lie at the very base of our understanding of classical dynamics. Moreover they translate directly to a quantum mechanical treatment of the same problem.

Excitons exist within crystals so they move on a lattice. The dynamics of a single particle on a lattice can be understood in terms of Bloch functions labeled by the crystal momentum in the first Brillouin zone. For motion near a band minimum we can define an effective mass by

$$\frac{1}{m} = \frac{1}{\hbar^2}\frac{\partial^2 E(k)}{\partial k^2},$$

which we assume for simplicity is a scalar. If we define a mass in this way for both the electron and hole then the translational mass of the exciton is expected to be $M = m_e + m_h$ as it would be for two particles in free space.

This result for the exciton mass is only valid in the limit of negligible binding energy, or more precisely, negligible localization of the exciton wave function in space. To see why, consider the opposite limit. A very tightly bound electron and hole have a small wave function. This necessarily means that in momentum space the wave function is very broad, filling the entire Brillouin zone. In this case, moving the center of the momentum space wave function around, i.e. changing its linear momentum, can have no effect on the energy. The exciton bandwidth has collapsed to zero and its translational mass is infinite.

So we have the result that the inertia of a pair of particles on a lattice depends on the interaction between them. This is a novel effect which is not present in the continuum limit. Of course a similar effect occurs at relativistic energies where the inertia is reduced by particle binding. Here we are dealing with much lower energies and the effect is due entirely to quantum mechanics.

Recently Mattis and Gallinar [1] carried out a calculation of this mass enhancement for excitons using a simple cubic lattice with the result

$$M = (m_e + m_h)/\left(1 - \frac{K_n}{\overline{W}}\right),$$

where K_n is the kinetic energy of the exciton relative motion in the nth state and $\overline{W}$ is the average of the electron and hole bandwidths. We will now describe our measurements of this exciton mass enhancement and compare them with this theoretical result.

1. Inelastic Electron Scattering Spectroscopy

To directly test these ideas about the translational mass of excitons we need to be able to create an exciton with definite non-zero momentum. The range of momenta needed is small enough that the energy is still parabolic in q but large enough to determine the mass well (~0.5Å^{-1}). The energy range needed corresponds to what is necessary to create excitons in available materials with reasonably large binding energies so the mass enhancement will be large enough to measure (~ 5-10 eV). The only kind of spectroscopy currently available which can reach these ranges of E and q with adequate resolution is Inelastic Electron Scattering (IES) spectroscopy.

In an IES experiment a beam of fast electrons is passed through a thin film of the material being studied and the inelastic scattering cross section measured. If the incident beam has a high enough energy the Born approximation can be used to describe the cross section with the result [2]

$$\frac{d^2\sigma}{d\omega d\Omega} = \frac{4\hbar}{a_o^2 q^4} S(q,\omega) = \frac{\hbar\gamma^2}{(\pi e a_o)^2 n} \frac{1}{q^2} \mathrm{Im}\left(-\frac{1}{\epsilon(q,\omega)}\right),$$

where $S(q,\omega)$ is the dynamic structure factor and $\epsilon(q,\omega)$ the complex dielectric response function of the material. An exciton appears as a peak below a continuum in these excitation functions.

In our experiments the electron beam energy was 300 kV. The energy and momentum resolutions were approximately 0.1 eV and 0.1Å^{-1} respectively. All samples were prepared by evaporation onto 100Å thick carbon substrates. Sample thickness varied from 150 to 350Å. During measurement the samples were supported on a grid in a vacuum of approximately 10^{-8} Torr.[3]

2. Experimental Results

Figure 1 shows a series of measurements of the valence exciton in CuCl for different momenta. The energy of the exciton clearly increases with q as we expect it to if it is to behave as a composite particle. In order to extract the maximum information from these spectra we shall fit them with a model calculation. Some years ago R. Elliott [4] obtained an analytic expression for the modification of interband optical absorption due to the electron-hole interaction. He used the effective mass approximation so the result is appropriate for Wannier excitons. Figure 2 shows a typical resulting spectrum. Below the interband threshold Eg bound excitons appear. Phonon Coupling broadens the spectrum so individual lines may only be resolved at low temperature. Above Eg the square root density of states is modified into a nearly linear shape. The bound states blend together to match smoothly onto the continuum. In fitting our data we model the bound states with gaussians, and the continuum with a step function with a slope which is convoluted with a gaussian. The solid lines in figure 1 show the resulting fits to our data for CuCl.

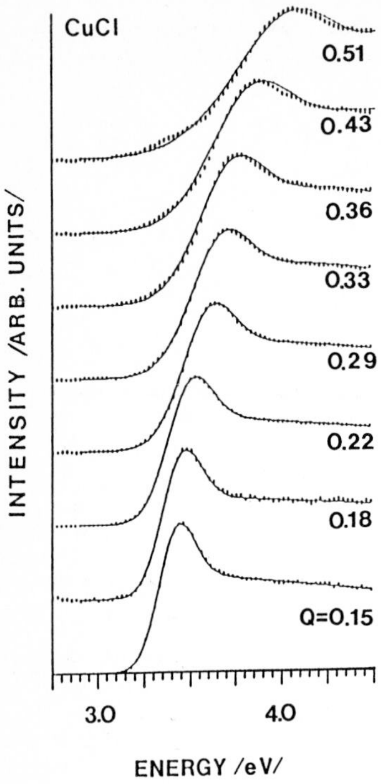

Figure 1 CuCl exciton spectrum measured for various momenta q

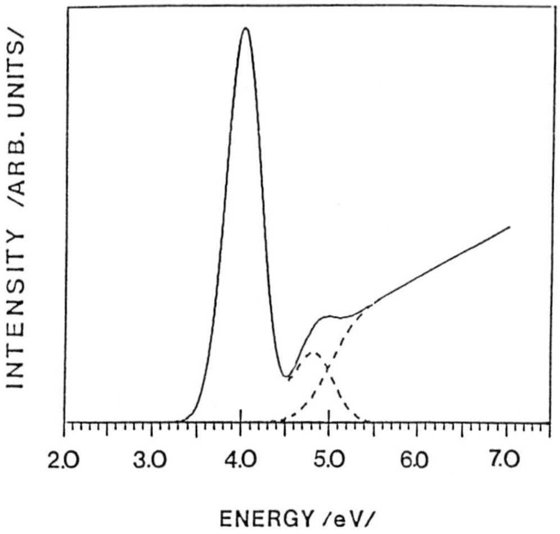

Figure 2 Schematic exciton spectrum according to Elliott [3]

If we now plot Ep, the energy of the n = 1 exciton, versus the square of the momentum, the slope of the resulting line defines the translational mass of the exciton. What we need for comparison is the mass of an unbound electron-hole pair. The energy required to create an unbound electron-hole pair is equal to Eg, the Rydberg limit or continuum threshold in the spectrum. If we plot Eg versus the square of the momentum the slope of the resulting line defines the sum of the electron and hole effective masses. Figure 3 shows these two plots. Both are indeed linear, indicating that the masses are well defined. From the slopes we find the exciton mass, in units of the electron mass, to be 1.23 ± 0.03, and that of the unbound electron-hole pair to be 0.83 ± 0.03. Thus the exciton is heavier than the unbound pair, as expected from the above qualitative arguments and the calculation of Mattis and Gallinar.

Most other mass determinations in CuCl are made near the bottom of the conduction band (or top of the valence band) and so include polaron coupling effects. Our measurements are all carried out well above threshold so we determine bare band masses and therefore a direct comparison with earlier measurements is not possible. This does allow us to directly determine the polaron coupling constant of the exciton however. A measurement near threshold of the mass of the longitudinal exciton has been made with the result 3.14 electron masses. [5] Comparing with our value of 1.23 gives a polaron coupling constant (assuming the simplest form for the mass enhancement $m^* = m(1 + \frac{\alpha}{6})$ of $\alpha = 9.3$. This is a large value, especially since in an exciton the electron and hole partially screen each other from the lattice reducing the polaron coupling.

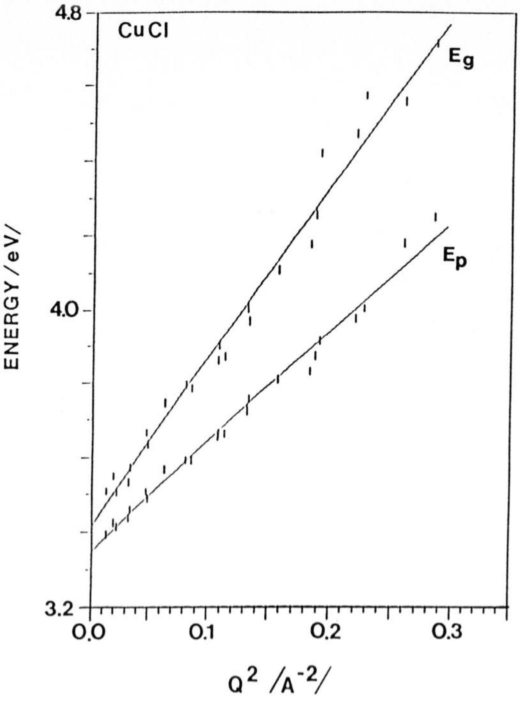

Figure 3 CuCl exciton and continuum energies versus q^2

From a Kramers-Kronig analysis of our data for small q we find a value of 170±10 meV for the binding energy of the transverse exciton. This compares well with optical measurements. Using this value and the optical dielectric constant of ϵ = 3.74 we find an exciton reduced mass of 0.19 electron masses and an effective Bohr radius of 5.4Å. Combining this reduced mass with our measured translational mass we find m_h = 1.0 and m_e = 0.23 electron masses.

We have also carried out measurements like those above for a number of alkali halides. These materials have larger exciton binding energies and correspondingly smaller radii so we first address the question of the validity of the Elliott model to describe these spectra.

Some years ago Hopfield and Worlock [6] carried out two photon absorption measurements of the valence excitons of alkali halides. They found that in spite of the large binding energies the Wannier model allowed them to correlate the one-photon and two-photon spectra, while two Frenkel models did not. In addition, Piacentini [7] found that he could describe the valence exciton spectrum of LiF very well using a Wannier model.

In view of these successes we shall use the Elliott model to describe the alkali halide results, making just one modification. Since the n = 1 exciton is so small, we assume no dielectric screening occurs in this state. All other bound states will be screened in the usual way.

Figure 4 shows several spectra for NaCl. The solid lines are the model fit. Figure 5 shows the exciton and threshold energies plotted versus the square of the momentum. In all the cases we have studied so far, the exciton is heavier than the sum of the electron and hole effective masses.

Table 1 shows a summary of relevant parameters for five materials. E_{ex} is the binding energy of the transverse exciton as determined from these

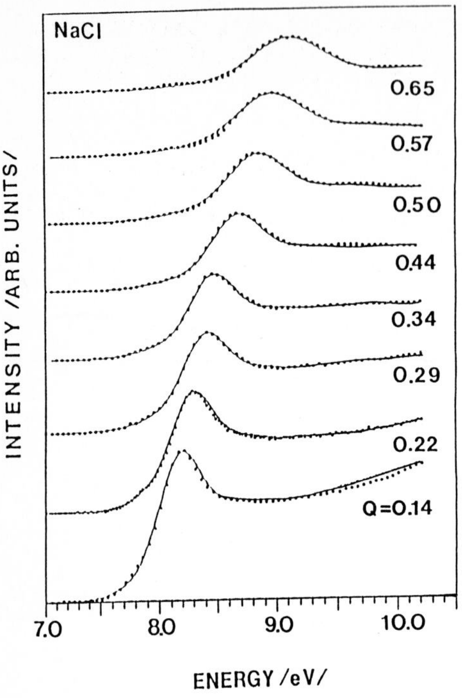

Figure 4 CuCl exciton spectrum measured for various momenta q

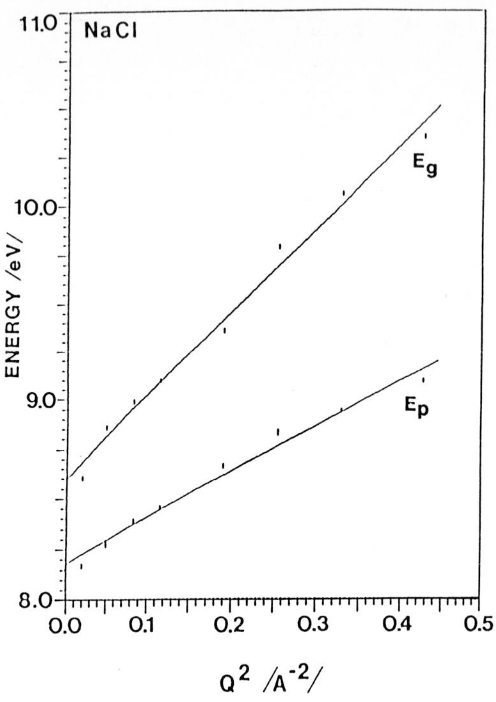

Figure 5 NaCl exciton and continuum energies versus q^2

Table 1 Summary of relevant parameters for the materials studied as described in text.

	M	m_e+m_h	E_{ex}	W	$\frac{E_{ex}}{W}$	R
CuCl	1.23	0.83	0.17	2.0	0.085	0.32
NaCl	1.65	1.13	0.74	2.3	0.32	0.32
KF	7.56	3.26	1.10	2.0	0.55	0.57
NaF	3.40	2.28	1.33	2.4	0.55	0.33
LiF	3.32	2.61	2.04	3.5	0.58	0.21

measurements. These values are close to results obtained optically. W is the width of the valence band of each material obtained by averaging the reported values measured using photoemission. According to the result of Mattis and Gallinar the ratio of the exciton mass to the sum of the electron and hole masses should equal $1/(1 - R)$. The parameter R appearing in the table is obtained from our measured masses using this expression.

The theoretical value of R is the exciton kinetic energy divided by the average of the electron and hole band widths. Using the virial theorem, the exciton kinetic energy is equal to its zero q binding energy E_{ex}. Since the conduction band widths are effectively infinite, we estimate W by the valence band width. Therefore according to the calculation, R should be approximately equal to E/W. The average values of these parameters in Table 1 are remarkably close: $\overline{R} = 0.35$, and $\overline{E}/\overline{W} = 0.42$. Considering the simplicity of the model and the crude estimate used here for W, this is remarkable agreement.

The principal disappointment in the comparison is the disagreement between data and theory on the trend to be expected upon going from the alkali halides to CuCl. The smaller exciton binding energy and similar bandwidth leads us to expect a smaller mass enhancement for CuCl which is not observed.

1. D.C. Mattis and J.P. Gallinar, Phys. Rev. Lett. 53 1392 (1984) See also D.C. Mattis, Revs. Mod. Phys.
2. S.E. Schnatterly in Solid State Physics, H. Ehrenreich, F. Seitz and O. Turnbull, eds. Academic Press, NY 1979 pp 275.
3. A.A. Cafolla, S.E. Schnatterly and C. Tarrio, Phys. Rev. Lett. 55 2818 (1985).
4. R.J. Elliott, Phys. Rev. 108 1384 (1957).
5. T. Mita, K. Sotome and M. Veta, Solid State Comm. 33, 1135 (1980).
6. J.J. Hopfield and J.N. Worlock, Phys. Rev. 137 A1455 (1965).
7. M. Piacentini, Solid State Comm. 17 697 (1975).

Quantum Theory of Polaritons

A. Quattropani, L.C. Andreani, and F. Bassani

Institut de Physique Théorique, EPFL, CH-1015 Lausanne, Switzerland
and Scuola Normale Superiore, I-56100 Pisa, Italy

1 Introduction

In this note we present the quantum mechanical treatment of the coupled exciton-photon system in an infinite dispersive medium. Such a system is described by a quadratic Hamiltonian in the photon and exciton operators and its eigenstates are called polaritons [1], [2], [3].

We first show that the canonical transformation introduced by HOPFIELD [3] can be extended to include spatial dispersion and we obtain the polariton eigenfrequencies which coincide with those of the classical dispersive model.

Second, we discuss the expansion of polariton states in terms of photons and excitons, and we determine analytically the expansion coefficients. We also show that in general a one-polariton state cannot be approximated by a linear superposition of one-photon and one-exciton states [4].

In the case of a *semiinfinite* crystal [5],we expect from the general theory on quadratic Hamiltonians [6] that the structure of the polariton states is similar to the previous case. However, the expansion coefficients of the polaritons in terms of excitons and photons cannot be determined analytically and are solutions of an integral equation which has to be solved numerically.

We conclude this note with some remarks on optical absorption in a crystal in connection with the polariton problem.

2. Quantum Theory of Polaritons

In general, polaritons are mixed modes of the electromagnetic field and elementary excitations of a crystal. Here we consider only the interaction with the polarization field in a narrow frequency region around an exciton resonance ω_0. The other resonances are assumed to give a frequency-independent contribution to the dielectric function.

We take as a starting point the classical equation of motion for the polarization field coupled to the electric field. In order to include spatial dispersion, we assume a wavevector dependence of the resonance frequency $\omega_k = \omega_0 + \gamma k^2$. The constituent equation becomes

$$\frac{1}{\omega_0^2}\ddot{\mathbf{P}} - \frac{2\gamma}{\omega_0}\nabla^2\mathbf{P} + \mathbf{P} = \beta\,\mathbf{E} \qquad (1)$$

We neglect possible higher order terms than ∇^2 ; moreover, we always work to first order in β and γ, since higher order terms would not modify the physical picture. Solving (1) by Fourier transforms, we obtain the dielectric function

$$\varepsilon(\omega, \mathbf{k}) = \varepsilon_\infty + \frac{4\pi\beta\omega_0^2}{\omega_0^2 + 2\gamma\omega_0 k^2 - \omega^2} , \tag{2}$$

which gives the dispersion relations for longitudinal modes

$$\varepsilon(\omega, \mathbf{k}) = 0 \tag{3a}$$

and for transverse modes

$$\varepsilon(\omega, \mathbf{k}) = c^2 k^2 / \omega^2 . \tag{3b}$$

In order to quantize the system, we must have a Hamiltonian formulation. A suitable Lagrangian density for the coupled polarization and radiation system is

$$l = \frac{1}{8\pi}\left(\frac{1}{v}\dot{\mathbf{A}} + \nabla\phi\right)^2 - \frac{1}{8\pi}(\nabla\times\mathbf{A})^2 + \frac{1}{2\beta\omega_0^2}\dot{\mathbf{P}}^2 - \frac{1}{2\beta}\mathbf{P}^2$$

$$+ \frac{\gamma}{\beta\omega_0}\mathbf{P}\cdot\nabla^2\mathbf{P} + \frac{1}{c}\dot{\mathbf{P}}\cdot\mathbf{A} + \phi\,\nabla\cdot\mathbf{P} . \tag{4}$$

We adopt the Coulomb gauge $\nabla\cdot\mathbf{A} = 0$ and disregard the longitudinal part of the polarization, which decouples from the electromagnetic field. Similarly, the scalar potential ϕ can be expressed as a function of $\rho = -\nabla\cdot\mathbf{P}$, and can be eliminated. The canonical momenta are

$$\pi_A = \frac{1}{4\pi v^2}\dot{\mathbf{A}} , \qquad \pi_P = \frac{1}{\beta\omega_0^2}\dot{\mathbf{P}} + \frac{1}{c}\mathbf{A} \tag{5}$$

and the Hamiltonian density is

$$h = 2\pi v^2 \pi_A^2 + \frac{1}{8\pi}(\nabla\times\mathbf{A})^2 + \frac{1}{2}\beta\omega_0^2\left(\pi_P - \frac{1}{c}\mathbf{A}\right)^2 + \frac{1}{2\beta}\mathbf{P}^2 - \frac{\gamma}{\beta\omega_0}\mathbf{P}\cdot\nabla^2\mathbf{P} . \tag{6}$$

The Hamiltonian is quadratic in the fields, and it represents a system of coupled harmonic oscillators, one for each wave vector $\mathbf{k}$ and transverse polarization $\lambda_\perp$ (in the following, we set $k\equiv(\mathbf{k},\lambda_\perp)$ and $-k\equiv(-\mathbf{k},\lambda_\perp)$. We therefore expand the fields in a volume V as follows :

$$\mathbf{A}(\mathbf{x}, t) = \Sigma_k\left(\frac{2\pi h}{k V}\right)^{1/2} e_k\left(a_k e^{i\mathbf{k}\cdot\mathbf{x}} + a_k^\dagger e^{-i\mathbf{k}\cdot\mathbf{x}}\right) , \tag{7a}$$

$$\mathbf{P}(\mathbf{x}, t) = \Sigma_k\left(\frac{h\beta\omega}{2\omega_k V}\right)^{1/2} e_k\left(b_k e^{i\mathbf{k}\cdot\mathbf{x}} + b_k^\dagger e^{-i\mathbf{k}\cdot\mathbf{x}}\right) , \tag{7b}$$

where $\mathbf{e}_k^{(\lambda_\perp)}$ is the unit vector which defines the direction of polarization of the fields.

To quantize the polarization-radiation system, we impose that the expansion coefficients $a_k^\dagger$, a_k, $b_k^\dagger$, b_k satisfy the canonical Bose commutation rules, thus becoming creation and destruction operators for photons and polarization quanta. The volume integral of the Hamiltonian (6) is then expressed in terms of these operators

$$H = \Sigma_k \left[h\, v\, k \left(a_k^\dagger a_k + \frac{1}{2} \right) + h\, \omega_k \left(b_k^\dagger b_k + \frac{1}{2} \right) \right]$$

$$+ \Sigma_k \left[i\, C_k \left(a_k^\dagger + a_{-k} \right) \left(b_k - b_{-k}^\dagger \right) + D_k \left(a_k^\dagger + a_{-k} \right) \left(a_k + a_{-k}^\dagger \right) \right] \quad , \tag{8}$$

where

$$C_k = h\, \omega_0 \left(\frac{\pi\, \beta\, \omega_k}{v\, k\, \varepsilon_\infty} \right)^{1/2} \quad , \qquad D_k = h\, \omega_0 \, \frac{\pi\, \beta\, \omega_0}{v\, k\, \varepsilon_\infty} \quad . \tag{9}$$

The first two terms in (8) describe free photons and free polarization quanta respectively. The third term is of the type $\mathbf{A}\cdot\mathbf{p}$, whereas the fourth term arises from the $\mathbf{A}\cdot\mathbf{A}$ interaction . A complete set in the Hilbert space is given by the tensor products

$$| \{ n_k^{phot} \} \rangle \otimes | \{ m_k^{exc} \} \rangle \quad . \tag{10}$$

The coefficients of the interaction C_k and D_k are $\ll h\, \omega_0$ as $k \gg \beta\, k_0$, but they diverge as $k \to 0$. This is the infrared divergence of quantum electrodynamics, which makes it impossible to treat the interaction perturbatively as $k \leq \beta\, \omega_0 / v$.The quantum theory of the interaction between crystal excitations and electromagnetic waves was first given by FANO [1] and the Hamiltonian (8) was first obtained by HOPFIELD [3]; both authors did not include spatial dispersion ($\omega_k \equiv \omega_0$).

The Hamiltonian (8) is quadratic in creation and annihilation operators, and can be diagonalized by a linear operator transformation first introduced in this context by HOPFIELD [3], which generalizes the well known Bogoljubov transformation to the case of four operators instead of two. We introduce new Bose operators α_{kl}, $\alpha_{kl}^\dagger$ with l=1,2 defined by the linear transformation

$$\begin{bmatrix} \alpha_{k1} \\ \alpha_{k1} \\ \alpha_{-k1}^\dagger \\ \alpha_{-k2}^\dagger \end{bmatrix} = \begin{bmatrix} W_1 & X_1 & Y_1 & Z_1 \\ W_2 & X_2 & Y_2 & Z_2 \\ Y_1^* & Z_1^* & W_1^* & X_1^* \\ Y_2^* & Z_2^* & W_2^* & X_2^* \end{bmatrix} \begin{bmatrix} a_k \\ b_k \\ a_{-k}^\dagger \\ b_{-k}^\dagger \end{bmatrix} \quad . \tag{11}$$

We want to identify $\alpha_{k1}^\dagger$, $\alpha_{k2}^\dagger$ with creation operators for lower and upper polaritons, and therefore require that a, $a^\dagger$ satisfy the Bose commutation rules

$$[\alpha_{kl} , \alpha_{qm}^\dagger] = \delta_{kq}\, \delta_{lm} \tag{12a}$$

and

$$[\alpha_{kl}, \alpha_{qm}] = [\alpha_{kl}^\dagger, \alpha_{qm}^\dagger] = 0 \tag{12b}$$

and moreover, that the Hamiltonian is diagonal in $\alpha^\dagger \alpha$, i. e.,

$$[\alpha_{kl}, H] = h\,\Omega_l(k)\,\alpha_{kl} \tag{13}$$

Conditions (12) - (13) lead to a sort of eigenvalue problem, whose secular equation is

$$\Omega^4 - (v^2 k^2 + \omega_k^2 + 4\pi\beta\,\omega_0^2/\varepsilon_\infty)\,\Omega^2 + v^2 k^2 \omega_k^2 = 0 \tag{14}$$

and coincides with the classical dispersion relation obtained from (2) and (3b). The eigenvectors are the coefficients W, X, Y, Z appearing in (11). Their explicit expressions are given below :

$$W_1 = X_2 = \frac{(\omega_k^2 - \Omega_1^2)(v k + \Omega_1)}{2(v k \Omega_1)^{1/2}[(\omega_k^2 - \Omega_1^2)^2 + 4\pi\beta\,\omega_0^2\,\omega_k^2/\varepsilon_\infty]^{1/2}} \quad , \tag{15a}$$

$$W_2 = X_1 = \frac{-2 i v k C_k / h}{(\omega_k - \Omega_1)(v k + \Omega_1)}\, W_1 \quad , \tag{15b}$$

$$Y_1 = Z_2 = \frac{\Omega_1 - v k}{\Omega_1 + v k}\, W_1 \quad , \tag{15c}$$

$$Y_2 = Z_1 = \frac{\omega_k - \Omega_1}{\omega_k + \Omega_1}\, X_1 \quad . \tag{15d}$$

The fact that the quantum theory gives the same dispersion relations as the classical theory was to be expected, since we have only used the commutator algebra, which is formally equivalent to the algebra of Poisson's brackets. On the other hand, the quantum theory describes particles, whereas this concept is not contained in the classical theory of electromagnetic and polarization fields.

The Hopfield procedure described above gives the polariton eigenvalues and determines polariton creation and annihilation operators, but is not sufficient to obtain the eigenstates. To obtain the eigenfunctions, one has to apply the creation operator $\alpha_{kl}^\dagger$ to the vacuum state of the system; but since the ground state $|\,0'\rangle$ of the Hamiltonian (8) does not coincide with the state with zero photons and excitons $|\,0\,\rangle$, we must first determine the new vacuum. To see this point, note that terms like $a^\dagger a^\dagger$ couple the ground state $|\,0\,\rangle$ of the non interacting system with the state with two photons; hence the interacting vacuum must be a linear combination of states (10).

The polariton vacuum is defined by the relations

$$\alpha_{kl}\,|\,0'\rangle = 0 \qquad , \qquad l = 1, 2 . \tag{16}$$

If we expand the new vacuum in the photon-exciton representation, and substitute for α_{kl} the linear combination (11), (16) gives recursion relations which determine the expansion coefficients uniquely (this point was already recognized by HOPFIELD [3], though it seems to have been overlooked in the literature on the subject). The lowest order expansion, consistent with the fact that the ground state has zero momentum, is

$$| 0' \rangle \cong [1 + \frac{1}{2} \Sigma_k \Sigma_{l,m} G_{lm}(k) A_{kl}^\dagger A_{-km}^\dagger] | 0 \rangle , \quad (17)$$

where we have set $A_{k1} \equiv a_k$ and $A_{k2} \equiv b_k$. The coefficients $G_{lm}(k)$ satisfy

$$W_l G_{11} + X_l G_{21} + Y_l = 0 \quad , \quad l = 1, 2 \quad , \quad (18a)$$

$$W_l G_{12} + X_l G_{22} + Z_l = 0 \quad , \quad l = 1, 2 \quad (18b)$$

and their explicit expressions are

$$G_{11}(k) = G_{22}(k) = - \frac{\Omega_1 + \Omega_2 - \omega_k - v k}{\Omega_1 + \Omega_2 - \omega_k + v k} , \quad (19a)$$

$$G_{12}(k) = G_{21}(k) = -i \left(\frac{\varepsilon_\infty v k \omega_k}{\pi \beta \omega_0^2} \right)^{1/2} G_{11}(k) . \quad (19b)$$

Expansion (17) is equivalent to approximating polariton states with linear combinations of just one photon and one exciton; this approximation is accurate when $| G_{lm} | \ll 1$, which is always satisfied for wave vectors $k \gg \beta k_0$. On the other hand, when $k \to 0$ the interaction blows up and higher order terms cannot be neglected.

It can be proved [4] that (17) is the first order expansion of an exponential; the exact expression for the polariton ground state is

$$| 0' \rangle = \frac{1}{N} \exp [\frac{1}{2} \Sigma_k \Sigma_{lm} G_{lm}(k) A_{kl}^\dagger A_{-km}^\dagger] | 0 \rangle , \quad (20)$$

where N is a normalization constant. The coefficients $G_{lm}(k)$ are the same as given by (19). The proof relies on the important recursion relations

$$A_{kl} | 0' \rangle = \Sigma_m G_{lm}(k) A_{-km}^\dagger | 0' \rangle , \quad l = 1, 2 , \quad (21)$$

which can be directly derived from (20), and which allow one to replace destruction operators acting on the new vacuum by creation operators in (16).

Having obtained the new vacuum state, we can now determine polariton states exactly. If we apply polariton creation operators $\alpha_{kl}^\dagger$ as given by (11) on $| 0' \rangle$ and we use (21), we obtain

$$\alpha_{kl}^{\dagger} | 0' \rangle = [\theta^{l}_{R} a_{k}^{\dagger} + \theta^{l}_{P} b_{k}^{\dagger}] | 0' \rangle \quad , \tag{22}$$

with

$$\theta^{1}_{R} = \theta^{2}_{P} = \frac{W_1}{|X_1|^2 + |W_1|^2} \tag{23a}$$

and

$$\theta^{2}_{R} = \theta^{1}_{P} = \frac{-X_1}{|X_1|^2 + |W_1|^2} \quad . \tag{23b}$$

In (22), the subscript R stands for radiation, while P stands for polarization; moreover, $| 0' \rangle$ represents the linear combination arising from the expansion of (20).

We can now evaluate projections of polariton states on one- and more- particle states (note that only states with an odd number of particles are contained in the expansion of (22), since $| 0' \rangle$ contains only states with an even number of particles). The one-photon components are, for $l = 1, 2$,

$$\langle 0 | a_q \alpha_{kl}^{\dagger} | 0' \rangle = \theta^{l}_{R}(k)\, \delta_{qk} / N \quad , \tag{24a}$$

while the one-polarization components are

$$\langle 0 | b_q \alpha_{kl}^{\dagger} | 0' \rangle = \theta^{l}_{P}(k)\, \delta_{qk} / N \tag{24b}$$

(we remind that l=1 denotes the lower polariton, whereas l=2 denotes the upper polariton).

The coefficient θ_R, θ_P for the lower polariton are plotted in Fig.1 (they are interchanged for the upper polariton). Note that $\theta_P \cong 1$ as $k \gg k_0$, since the lower polariton is almost an exciton for large k; but both coefficients tend to zero for $k \to 0$, showing that non linear terms in (22) become dominant. To check this conclusion, we plot in Fig. 2 the three-photon component $N \langle 0 | a_k a_k a_{-k}^{\dagger} \alpha_{k1}^{\dagger} | 0' \rangle$ of the lower polariton; it is seen that this quantity attains a maximum at $k \approx \beta k_0$.

It can be easily shown from (22) that all projections of polaritons states onto states of the form (10) tend to zero with k; the limit is of course not uniform, since polariton states must remain normalized. This rather unexpected behaviour is another manifestation of the infrared divergence of the photon-polarization interaction at small wave vectors; as a consequence, for $k \leq \beta k_0$ we must consider the exact expression (22) for polariton states, and we are not allowed to keep a finite number of terms.

For a finite crystal, polariton states must be classified according to the photon wavevector $\mathbf{k}_0$ outside the medium [5]. In this case, the interaction between excitons and photons is not as simple as given by (8), but involves an infinite number of normal modes. As a consequence, the canonical transformation analogous to (11) is given by an infinite matrix. The relation between old and new vacuum can again be expressed as an exponential of a quadratic operator like in (20), as follows from the general theory on quadratic Hamiltonians discussed by FRIEDRICHS [6]. However, the linear system (18) has now an infinite number of equations.

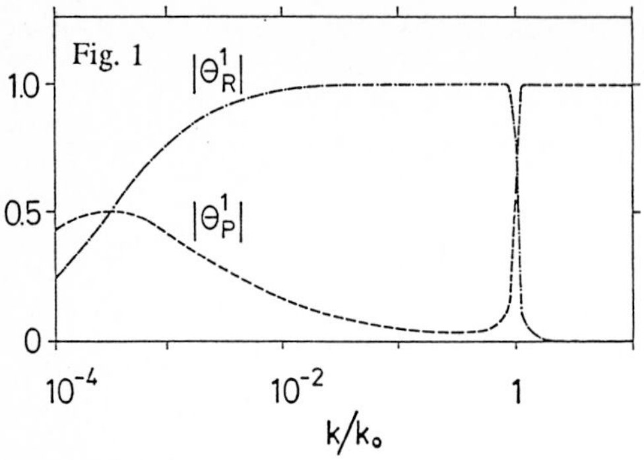

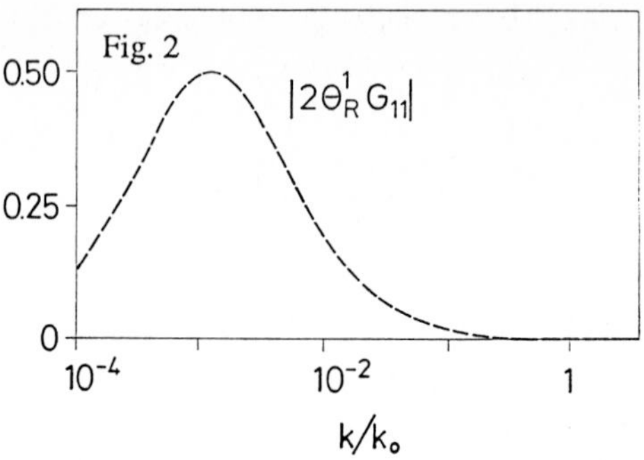

Fig. 1. One-photon and one-polarization components $\theta^1{}_R$ and $\theta^1{}_P$ for the lower polariton. They are interchanged for the upper polariton. Parameters of CuCl have been used: $4\pi\beta=6\cdot10^{-3}$, $\varepsilon_\infty=5$, $k_0=3.6\cdot10^5$ cm^{-1} and $m^*=2.5\ m_0$. From [4].

Fig. 2. Three-photon component $|2\ \theta^1{}_R\ G_{11}|$ for the lower polariton. Parameters are the same as in Fig. 1. From [4].

Finally, it is not proved that in general a quadratic Boson Hamiltonian can be reduced to that of a free Boson system by a canonical transformation like (11). The existence of such a transformation must be examined case by case.

3. Optical Absorption

The concept of polariton was originally introduced in order to formulate a theory of optical absorption below the gap [3]. Exciton-photon interaction in itself does not produce absorption, since a photon interacts with just one exciton because of momentum conservation, giving rise to a stationary polariton state. Polariton decay in a real crystal is due to the interaction with impurities or with phonons. This physical picture is to be contrasted with absorption above the gap, which is due to interband transitions induced by the **A·p** interaction.

From a phenomenological point of view, absorption is usually treated by introducing a damping parameter Γ in the dielectric function as follows :

$$\varepsilon(\omega, \mathbf{k}) = \varepsilon_\infty + \frac{4\pi\beta\omega_0^2}{\omega_0^2 + 2\gamma\omega_0 k^2 - \omega^2 - i\Gamma\omega}, \tag{25}$$

Γ is often taken as a constant, although in principle it depends on wavevector. In fact, a strong energy dependence of Γ was recently measured in a transmission experiment by BROSER [7].

The Ansatz (25) is correct when absorption is mainly due to either the lower or the upper polariton. This condition is verified for most frequencies, except very near the longitudinal frequency. In the above assumption, Γ can be identified with the polariton decay width. A microscopic calculation of Γ proceeds then as follows. First one has to identify the interaction responsible for the decay. Actually, it has been shown by MERLE [8] that in the case of CuCl many different mechanisms are involved, namely piezoelectric and deformation potential interaction with acoustic phonons and Fröhlich and deformation potential interaction with optical phonons. Each of these interactions can be expressed as a trilinear Hamiltonian in exciton and phonon creation and destruction operators:

$$H_{xp} = \Sigma_k \Sigma_q V_q b_k b_{k-q}^{\dagger} c_q^{\dagger} \; + \; h.c. \qquad , \tag{26}$$

where V_q is a q-dependent matrix element. Then one calculates Γ for a given polariton state using Fermi's golden rule, by considering the scattering to all polariton states of lower energy. Since polaritons are zero-order stationary states, the polariton dispersion must be used to compute the density of states. To calculate the matrix elements of H_{xp} between polariton states, the exciton operators b, $b^{\dagger}$ must be expressed as a function of the polariton operators α, $\alpha^{\dagger}$ by the inverse Hopfield transformation. This has first been done by TAIT [9].

Using the Hopfield transformation to express (26) in the polariton representation is an exact procedure, equivalent to considering the infinite expansion (22) of the polariton states. The general theory shows that it is not correct to approximate polariton states as linear combinations of one-photon and one-exciton states in order to calculate matrix elements of (26), particularly near the longitudinal frequency.

References

1. U. Fano: Phys. Rev. 103, 1202 (1956)
2. S. I. Pekar: Crystal Optics and Additional Light Waves (Benjamin-Cummings,Menlo Park,California 1983)
3. J. J. Hopfield: Phys. Rev. 112, 1555 (1958)
4. A. Quattropani, L. C. Andreani and F. Bassani: Il Nuovo Cimento D7, 55 (1986)
5. R. Zeyher, C.-S. Ting and J. L.Birman : Phys. Rev. B10, 1725 (1974).
6. K. O. Friedrichs: Mathematical Aspects of the Quantum Theory of Fields (Interscience Publ., New-York 1953)
7. I. Broser, K. H. Pantke and M. Rosenzweig: Phys. Status Solidi B132, K117 (1985); Solid State Commun. 58, 441 (1986)
8. J. C. Merle: Festkörperprobleme XXV, ed. by P. Grosse (Vieweg, Braunschweig 1985), p.275
9. W. C. Tait and R. L. Weiher : Phys. Rev. 166, 769 (1968).

Influence on Excitons of Band Bending near Surfaces and Interfaces

I. Balslev

Fysisk Institut, Odense Universitet, DK-5230 Odense M, Denmark

The influence on excitonic resonances and translational motion of band bending near semiconductor surfaces and interfaces is discussed. It is shown that both high and low maximum electric fields lead to two-dimensional (2d) exciton states, lower in energy than bulk excitons, as well as pair states consisting of a confined 2d electron and a bulk hole (or vice versa). Relevant experiments on photoluminescence and reflectivity are discussed.

1. Introduction

The electric field near surfaces and heterojunction interfaces in semiconductors can be controlled by external field electrodes, by temperature and by doping. Such fields influence the exciton motion significantly if the penetration depth of the fields is comparable to the Bohr radius a_B of bulk excitons and the total band bending is comparable to the bulk exciton Rydberg E_x. The influence on excitons of homogeneous electric fields has long been known [1-3]: The excitonic resonances are subject to a quadratic (downward) Stark shift of at most 0.1 E_x and to a broadening due to ionization. At very high fields the relevant description is a Franz-Keldysh effect on a band edge without electron-hole interaction. In case of excitons confined in a quantum well a transverse field leads to the so-called quantum confined Stark effect [4].

Near surfaces and interfaces the situation is different. As shown in the examples of Fig. 1, the field is zero in the bulk, but increases gradually towards the interface [5]. The band bending is characterized by a maximum field F_0 and a penetration depth d which is related to the screening length of the material. The electrons and holes also experience in addition to this field the short range forces due to the compositional change. As the slowly varying electric field attracts excitons via the Stark effect, the excitons can only be at rest if this attraction is balanced by repulsive forces from a surface (or from an interface seen from the low-gap side of a heterojunction).

We shall generally neglect polariton effects and concentrate on the energy levels and the kinetics of two-dimensional (2d) excitons, i.e. electron-hole states characterized by a localized wave function perpendicular to the interface

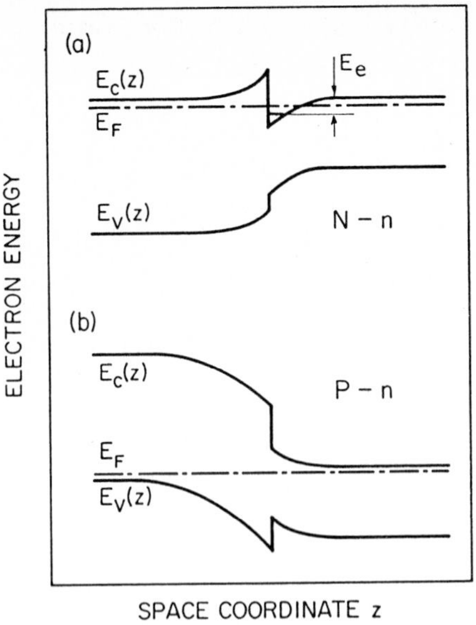

Fig. 1 - Spatial structure of valence and conduction bands in typical heterojunctions. In the N-n junction is shown the level of 2d confined electrons E_e below the bulk conduction band.

and a free motion parallel to the interface. The kinetics of 2d excitons which couple to 3d states can be dominated by either adsorption/desorption:

$$ex^{2d} \leftrightarrows ex^{3d}, \tag{1}$$

or by ionization/association:

$$ex^{2d} \leftrightarrows e^{2d} + h^{3d} \quad \text{or} \quad ex^{2d} \leftrightarrows e^{3d} + h^{2d}. \tag{2}$$

Here ex, e, and h stand for excitons, electrons, and holes, respectively, and 2d, 3d symbolize the degrees of freedom of the translational motion.

Which of the above processes are most important for the kinetics depends primarily on the magnitude of F_0 compared to the ionization field $F_I \equiv E_x/(ea_B)$. The weak field limit is characterized by $F_0 < F_I$. In this case the 2d exciton is only moderately polarized and so the process (1) has the smallest activation energy. The structure and the desorption energy of 2d excitons in this limit is discussed in Sect. 2. In the opposite limit the notch in one of the band edges (see Fig. 1) is able to confine one of the particles and the confinement energy is larger than E_x. Then, as discussed in Sect. 3, the 2d excitons are highly polarized and so the ionization process (2) requires less energy than does desorption.

The system to be considered is an interface in the plane z=0. The short range forces due to a compositional change at z=0 form potential barriers with heights ΔE_c and ΔE_v for electrons and holes, respectively. We assume that both potentials are lower for z > 0 than for z < 0. When dealing with a surface, ΔE_c and ΔE_v can be considered as infinite, while the band offsets are in the range 0-300 mV in typical III-V heterojunctions.

In addition to the short range forces at z=0 we assume forces from an electric field given by the simple expression

$$F(z) = F_0 z/|z|\ e^{-|z|/d}. \qquad (3)$$

The total effective-mass Hamiltonian for an electron-hole pair is then

$$H = E_g + E_{kin} - \frac{e^2}{4\pi\varepsilon_0 \varepsilon r} - e\varphi(z_e) + e\varphi(z_h) + \Delta E_c \theta(-z_e) + \Delta E_v \theta(-z_h), \qquad (4)$$

where the electrostatic potential φ is derived from (3) (with $(\varphi(+\infty)=0)$, θ is the unit step function, E_g is the energy gap for z>0, E_{kin} is the kinetic energy, and r is the magnitude of the relative coordinate $\underline{r} = \underline{r}_e - \underline{r}_h$.

2. The Weak Field Limit

One approximate treatment of (4) is based on the neglect of the center-of-mass kinetic energy when solving for the relative motion. This adiabatic-type approximation is relevant if the influence of the electric field is small compared with the electron-hole interaction and, at the same time, the electron-hole mass ratio is not too close to unity. Then the relative-motion Hamiltonian is (4) with $E_{kin} = -\hbar^2\nabla_r^2/2\mu$ (μ is the reduced mass), and so the center-of-mass depth Z can be considered as a parameter. For example the coordinates z_e, z_h are given by

$$z_e = Z + (m_h/m)z \quad \text{and} \quad z_h = Z - (m_e/m)z\ , \qquad (5)$$

where m is the total mass ($m=m_e+m_h$) and z is the z component of the relative coordinate. The potential formed by the last 5 terms in (4) is shown in Fig. 2 for the case $\underline{r}=(0,0,z)$.

The resulting single particle eigenvalue problem can be studied by standard methods. The ground state is conveniently explored by variational techniques. A suitable trial function in the case ΔE_c and $\Delta E_v >> E_x$ is [6]

$$\psi_{trial} = (1-\exp(-z_e/a_B))(1-\exp(-z_h/a_B))(\psi_{1s} + \alpha\,\psi_{2s} + \beta\,\psi_{2p})$$
$$\text{(for } z_e > 0,\ z_h > 0 \quad \text{and zero otherwise)}\ , \qquad (6)$$

where α and β are variational parameters, and ψ_{1s}, ψ_{2s}, ψ_{2p} are wave functions associated with bulk n=1 and n=2 excitons.

The energy associated with the relative motion $E_{rel}(Z)$ is shown in Fig. 3 for d = 40 a_B , m_e/m_h = 0.2 and $F_0 = \pm F_I$ (from ref. 6). It is seen that $E_{rel}(Z)$ for $0<Z<5a_B$ is dominated by the cut-off forces causing an upshift of the ground state energy. At larger depths the dominant contribution to the energy shift is the quadratic Stark shift

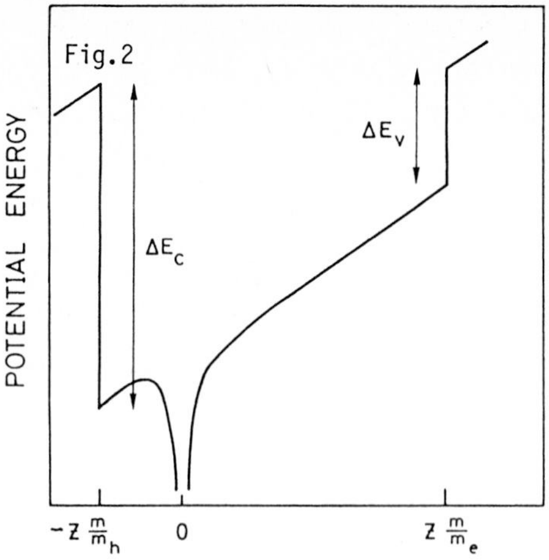

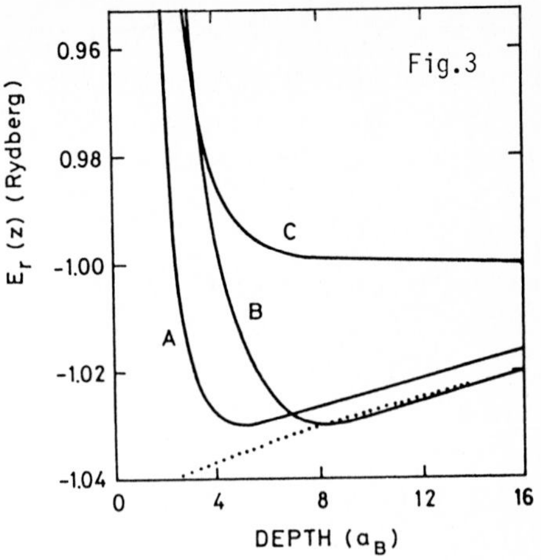

Fig. 2 -Schematic representation of the potential relevant for the motion in relative space for a fixed center-of-mass depth Z.

Fig. 3 - Depth dependence of the energy $E_{rel}(Z)$ of the relative motion. Full curves are calculated for d = 40 a_B and surface fields F_n along the z axis equal to $-0.2F_I$, $0.2F_I$, and 0 for the curves A, B, and C, respectively. The dotted curve is the quadratic Stark shift derived from the local electric field

$$E_{rel} = -E_x(1 + \frac{9}{8}(F(z)/F_I)^2) \tag{7}$$

corresponding to the energy shift obtained in a homogeneous field equal to the local field.

As seen from Fig. 3, the translational motion of 2d excitons in the weak field limit is characterized by confinement in a shallow potential well with a minimum about 5 a_B below the interface. The desorption energy is of the order

$$E_{desorption} \approx E_x(F(z\approx 5a_B)/F_I)^2 . \tag{8}$$

The validity of this estimate is limited to desorption energies considerably less than E_x.

3. High Field Limit

In junctions involving heavily doped materials the fields are often higher than F_I. In this case other approximations than applied in Sect. 2 must be employed. Let us assume that the interface attracts electrons. When this field is strong enough, the two-particle problem given by (4) can be solved in 2 steps. First, the electron-hole interaction is neglected and the part of the Hamiltonian involving the electron coordinate only can be solved. This leads to confined electrons of which the lowest state has energy $-E_e$ (relative to the bulk conduction band edge, see Fig. 1a) and wave function $\Phi_e(z_e)$. The next step is to solve for the hole motion. As E_e is assumed to be larger than E_x, a good excitonic wave function is

$$\psi_{ex}(\underline{r}_e, \underline{r}_h) = \Phi_e(z_e)\, \Phi_h(z_h)\, \exp(-r_{xy}/a_{xy})\, \exp(i\underline{K}.\underline{R}_{xy}), \tag{9}$$

where $\underline{r}_{xy}$,$\underline{R}_{xy}$ are relative and center of mass coordinate in the xy plane, $\underline{K}$ is the translational wave vector and a_{xy} and $\psi_h(z)$ are to be varied to give minimum total energy. Instead of a variational method one can also derive an effective potential for the z-motion of the hole [7] and from this determine $\Phi_h(z)$. A situation typical for GaAs-(AlGa)As heterojunctions is shown in Fig. 4. Note that the average electron-hole separation is $\approx$ 600 Å corresponding to $\approx 5a_B$. The ionization energy of the 2d exciton is determined by the balance between the repulsion of the hole by the "macroscopic" electric field and the electron-hole interaction, and it approaches zero when the penetration length of the field i.e. the screening length, becomes large compared to the bulk Bohr radius. In the example in Fig. 4, this ionization energy is $\approx$1 meV or $\approx 0.25\, E_x$.

The above description is in several respects oversimplified. This is particularly true if a large fraction of the confined electron states are occupied as is the case of the N-n junction (see Fig. 1a). Such a situation requires a complicated calculation of the selfconsistent field $\varphi(z)$ [7-8]. In the presence of a high density 2d electron gas in the well, the screening of the electron-hole interaction and the band renormalization need to be considered in detail [9]. Furthermore, at high excitation levels the interface may attract so many excitons that the gas of highly polarized excitons most likely develops into a 2d hole gas and a 2d electron gas separated about 3-5 a_B from each other [10, 11].

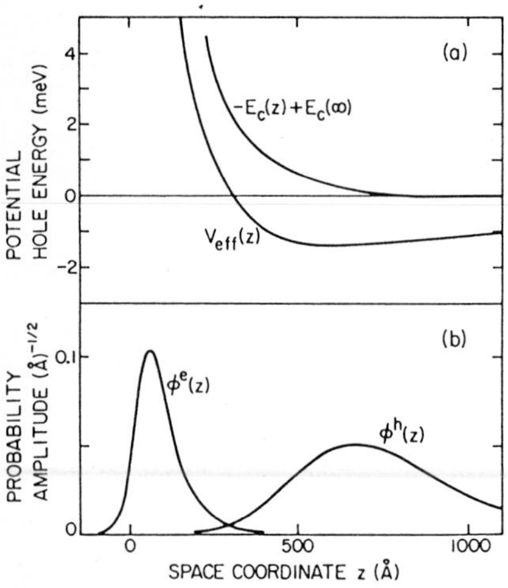

Fig. 4 - (a) Electrostatic potential energy $E_c(z)$ and total effective potential $V_{eff}(z)$ experienced by the hole near a N-n GaAs-(AlGa)As heterojunction. (b) Transverse wave functions of electron and hole for a 2d exciton in the high field limit.

4. Experiments on Excitons in the Band Bending Region

The first experimental evidence for the redshift of the exciton resonance a few Bohr radii below a surface is reported by LAGOIS [12] who studied in detail the attenuated total reflection and the normal incidence reflection in various II-VI compounds. He fitted to the experiments a 4-layer model with freely adjustable exciton energies. In ZnO it was found that the resonance had a clear blue shift for $0 < z < 6a_B$ and a red shift for $6a_B < 14a_B$ in agreement with the calculated behaviour in Fig. 3. Related to these results is the observation of anomalous reflectivity spectra of epitaxial GaAs [13]. As discussed in ref. 14, the oscillatory spectra observed for certain surface treatments can be explained by band bending with a surface field of $\approx 0.3\ F_I$.

In photoluminescence experiments it is expected that bulk excitons diffuse or drift to interfaces and surfaces, and that the 2d exciton states, having lower energy bulk excitons, are probably important channels for radiative recombination. A candidate for this interpretation is the H-line observed in GaAs-(AlGa)As heterojunctions by YUAN et al. [15]. The line cannot be assigned to bulk impurities and is observed so much below the bulk exciton line that the high field limit is applicable [7]. KOTELES and CHI [16] also assign an observed line to the GaAs-(AlGa)As interface. Its spectral position is characteristic for the low field limit. In both the above cases the lines disappear at temperatures above 10-15 K. This indicates an activation energy of about $0.25\ E_x$, a value which is not contradicted by the theoretical arguments in Sects. 2-3.

5. Concluding Remarks

It is shown that for a wide range of band bending parameters (F_0 and d) there exist 2d exciton states which are stable against both desorption into bulk excitons and ionization into a bulk hole and a 2d electron (or vice versa). However, the corresponding activation energies cannot be more than a small fraction of the exciton Rydberg. In this respect the 2d excitons confined by band bending and a compositional halfspace are very different from the excitons confined in quantum wells [4, 17].

The experimental verifications of the considerations in Sects. 2-3 are rather sparse. This situation calls for further experiments. Among the theoretical challenges of the interface excitons are the inclusion of polariton effects, a treatment of many-body effects in a dense 2d electron gas, and studies of the influence of strong illumination.

Acknowledgements

The author is indebted to J.L. Merz for inspiring discussions and hospitality during his visit at Electrical and Computer Engineering, University of California, Santa Barbara, in the spring of 1986.

References

1. W. Franz, Z. Naturforschung 13a, 484 (1958)
2. L. V. Keldysh, Zh. Eksperim. i. Teor. Fiz. 34, 1138 (1958) [Engl. transl.: Sov. Phys. - JETP 7, 788]
3. J.D. Dow and D. Redfield, Phys. Rev. B1, 3358 (1973)
4. D.A.B. Miller, D.S. Chemla, T.C. Damen, A.C. Gossard, W. Wiegmann, T.H. Wood, and C.A. Burrus, Phys. Rev. B32, 1043 (1984)
5. H. Kroemer, in Proc. NATO Advanced Study Institute on Molecular Beam Epitaxy and Heterostructures, Erice, Sicily, L.L Chang and K. Ploog, eds., (M. Nijhoff, The Netherlands, 1984).
6. L. Schultheis and I. Balslev, Phys. Rev. B28, 2292 (1983)
7. I. Balslev, to be published in Semiconductor Science and Technology (1987)
8. G. Barraf and J. Appelbaum, Phys. Rev. B5, 475 (1974)
9. S. Schmitt-Rink, D.S. Chemla, and D.A.B. Miller, Phys. Rev. B32, 6601 (1985).
10. P.D. Altukhov, A.V. Ivanov, Yu.N. Lomasov and A.A. Rogachev, Pis'ma Zh. Eksp. Teor. Fiz. 39, 432 (1984) [Engl. transl. JETP Letters 39, 523]
11. A.A. Rogachev. P.D. Altukhov, V.M. Asnin, A.A. Bakun, A.M. Monakhov, and V.I. Stepanov, in Proc. 18. Int. Conf. Phys. Semicond., Stockholm 1986, O. Engstrom, ed. (World Scientific, Singapore, 1987) p. 1441
12. J. Lagois, Phys. Rev. B23, 5511 (1981)
13. J. Lagois, E. Wagner, W. Bludau, and K. Losch, Phys. Rev. B18, 4324 (1978).
14. A. Stahl and I. Balslev, in Electrodynamics of the Semiconductor Band Edge, Springer Tracts of Modern Physics, Vol. 110 (Springer, Berlin, 1987)
15. Y.R. Yuan, M.A.A. Prudensi, G.A. Wawter abd J.L. Merz, J. Appl. Phys. 58, 397 (1985)
16. E.S. Koteles and J.Y. Chi, J. Superlatt. and Microstruct. 2, 421 (1986)
17. R. Dingle, in Festkörperprobleme - Advances in Solid State Physics, H.J. Queisser, ed. (Pergamon/Vieweg, Braunschweig, 1975) Vol. XV, p. 21.

Part II

Excitons in Thin Films

Exciton-Polaritons in a Slab; ABC and ABC-Free Theory

K. Cho and H. Ishihara

Faculty of Engineering Science, Osaka University,
Toyonaka 560, Japan

§1 Introduction

Among various excited states of solids, exciton-polaritons in non-metallic crystals have provided a peculiar field of very precise spectroscopic studies. One of the peculiar features is the so-called spatial dispersion effect, which arises from the dependence of exciton energy on its translational wave vector. Because of this effect, the Maxwell equations for this type of bulk medium allow two (or more) different plane wave solutions for a given frequency, and consequently the existence of several modes gives rise to a problem of additional boundary condition (ABC) at the boundary of the medium, which essentially determines the relative amplitudes of the different plane waves in the spatially dispersive medium. Thus, the two most important aspects in the study of exciton-polaritons have been [A] the multi-mode dispersion relation $\{k = k_j(\omega)\}$, and [B] the form of ABC /1-3/.

Among others, a thin film geometry provides one of the most valuable systems to study the two important aspects [A] and [B], since the interference patterns in reflectance and transmittance spectra are, in spite of their simple nature, quite sensitive means to investigate [A] and [B]. The measured spectra of CdS, CdSe /4-6/ and CuCl /7/ clearly show the existence of two branches in polariton dispersion, as manifested by several different periods in the interference pattern, and the theoretical analyses /4-8/ of the measurements allow detailed arguments about the form of polariton dispersion curves and ABC to be used.

As to the theoretical calculation of fields inside spatially dispersive medium, a first principle theory requires the following three steps: [I] to prepare the eigenvalues and eigenfunctions of a system in consideration of its ground and excited states, whereby the boundary condition for the wave functions must be properly considered, [II] to calculate the system's susceptibility $\chi(R,R';\omega)$ in site representation, and [III] to solve the Maxwell (Helmholz) equation

$$\mathbf{rot\ rot}\ E(R) - (\omega/c)^2\ E(R) - Q^2 \int dR'\chi(R,R')\ E(R') = 0\ , \quad (1)$$

where the variable ω in E and χ is omitted and

$$Q = \sqrt{4\pi}\ \omega/c\ . \quad (2)$$

In the last step [III], there is an alternative formulation based on integral equation /9/, but it is essentially equivalent to solving eq.(1). Until very recently, the solution has always been written as a linear combination of bulk polariton waves (plus surface localized components, if necessary) which contain more free parameters than necessary, and it is accompanied by additional relation(s) between the free parameters, which play the role of ABC. Before the full step [I – III] calculations /10, 11, 5/ were demonstrated, it had been noticed in terms of model susceptibilities that the eq.(1) contains all the necessary informations including ABC /9, 12,13/. But the model susceptibilities in such works are just truncated bulk susceptibilities where the existence of boundary is not considered quantum mechanically.

All the ABC's derived from the full step calculations are based on particular models in the first step [I]. There has been no explicit argument, to the authors' knowledge, about the way to solve eq.(1) without depending on the specific form of $\chi(R,R';\omega)$. Very recently we have developed a theory to solve eq.(1), for a general form of χ, without referring to ABC /14/: The solution $E(R,\omega)$ contains a minimal necessary number of free parameters, and the usual Maxwell boundary conditions are enough to fix them uniquely. In order to show that this kind of ABC-free treatment is generally possible, it is sufficient to use the general form of χ expected from linear response theory. Therefore the method applies not only to semi-infinite systems but also to any confined systems including quantum wells. Since the ABC and ABC-free theories are just different ways to solve eq.(1), they should give the same physical result. This equivalence is also seen for the particular exciton model in this paper for a slab of arbitrary thickness. The exciton wave function employed for this theory is the type of D'ANDREA – DEL SOLE (DA-DS) /11/ with explicit consideration of size quantization of translational wave vector. Since the full presentation of the theory would need much more space than allowed, we will give the outline of the whole story in this paper.

§2 Model

We consider normal incidence of light on a slab. The z-axis is taken perpendicular to the surface, and the slab thickness is d. The bulk medium is assumed to be isotropic having a single exciton branch as resonant levels. The (analytic) model of DA-DS describes the distortion of exciton wave function in terms of a single evanescent wave ($\exp[-PZ]$) for translational motion multiplied by a linear combination of relative motion wave functions for higher ($n \geq 2$) excited states. Together with the no-escape boundary condition for both electron and hole, the model leads to the following matrix element of transition dipole moment density (at site Z) for the exciton state with translational wave number K /8/:

$$M_K(Z) = (1s,K|\ \boldsymbol{P}(Z)\ |0)$$

$$= \mu\, N_K \{\ e^{-iKZ} + A_K\, e^{iKZ} - (1 + A_K)\, e^{-PZ} - (e^{-iKd} + A_K\, e^{iKd})\, e^{-P\bar{Z}}\ \} \quad , \qquad (3)$$

where μ is a constant with the dimensions of dipole moment, and

$$N_K = [2d - 4(P^2-K^2)/P(P^2+K^2)]^{-1/2} \quad , \qquad (4)$$

$$A_K = (-P+iK)/(P+iK) \quad , \qquad (5)$$

$$\bar{Z} = d - Z \quad . \qquad (6)$$

The allowed values of K are the roots of

$$Kd = n\pi + 2\tan^{-1}(K/P) \quad , \quad (n = 1,2,3,\cdots) \tag{7}$$

and the susceptibility in site representation is

$$\chi(Z,Z';\omega) = \chi_b\delta(Z-Z') + \frac{1}{V_0}\sum_K \frac{M_K(Z)^* M_K(Z')}{E_{1s}(K) - \hbar\omega - i\gamma} , \tag{8}$$

where V_0 is the volume of a unit cell, and all the non-resonant terms are put into the background (χ_b), γ is the phenomenological damping constant, and

$$E_{1s}(K) = \hbar\omega_0 + \hbar^2K^2/2m_x \tag{9}$$

is the exciton energy. The summation in (8) must be carried out for the discrete values of K according to the quantization condition (7).

§3 ABC Theory

This model has previously been treated by CHO and KAWATA (CK) /8/ with the following assumptions: [a] Pd >> 1 , [b] P >> q_0 , where q_0 is the wave number of the exciton with energy hω, i.e.,

$$\hbar q_0 = [2m_x(\hbar\omega - \hbar\omega_0 + i\gamma)]^{1/2} \quad , \quad (\mathrm{Im}[q_0] > 0) \quad . \tag{10}$$

Based on these assumptions, CK linearized (7) (near $K = q_0$) as

$$Kd = n\pi + 2K/P \tag{11}$$

and carried out the summation in (8) for this set of (equidistant values of) K. The error in this procedure was estimated, by rewriting the omitted part of the summation into an integral, to be O(1/Pd). The resultant susceptibility has the form (29), and the solution of the Maxwell eq.(1) {$(R,R') \to (Z,Z')$, **rot rot** $E \to -d^2E(Z)/dZ^2$} is given in the form

$$E(Z) = \sum_j (W_j e^{ik_jZ} + \bar{W}_j e^{ik_j\bar{Z}}) + g e^{-PZ} + \bar{g} e^{-P\bar{Z}} , \tag{12}$$

where j runs over the two bulk polariton modes (k_1, k_2), and (g, $\bar{g}$) are known linear combinations of the four constants {W_j, $\bar{W}_j$}, which satisfy two simultaneous linear equations. These equations are essentially the ABC of DA-DS on each surface, except that the polariton amplitudes contain those reflected from the other surface.

Though the result of CK applies rather well to the analysis of the interference pattern of CuCl [Z_3]-exciton for a d=1500 Å sample /7,8/, the assumptions [a] and [b] may put limitation to the application of the theory to thinner samples or to other materials. To get rid of the constraint, it is necessary to carry out the summation in (8) with the exact quantization condition (7). This is possible, if one converts the summation into an integral in the complex K-plane according to

$$\sum_K \cdots = \frac{1}{2\pi i} \int_C \cot[Kd-\Delta(K)]\cdot[d-\Delta'(K)] \cdots \; dK \, , \qquad (13)$$

where

$$\Delta(K) = 2 \tan^{-1}(K/P) \qquad (14)$$

and the contour C picks up all the poles of $\cot[Kd-\Delta(K)]$ on the real axis, which are nothing but the allowed values of quantized K in (7). If one deforms the contour so that it encircles the whole complex plane, the integral can be evaluated at the complex poles of the whole integrand. The expression (8) contains the following two types of summation:

$$\sum_K N_K^2 \cos(KX) \,/\, (K^2 - q_0^2) \quad , \quad (|X| \leqq d) \qquad (15)$$

$$\sum_K N_K^2 (-1)^n \cos(KX) \,/\, (K^2 - q_0^2) \quad , \quad (|X| \leqq d) \, . \qquad (16)$$

They can be evaluated by the integral in the complex K-plane as mentioned above: The complex poles are $\pm\, q_0$, $\pm\, i\bar{P}$, and $\pm\, iP_\pm$, where the latter two sets arise from N_K^2 and $\cot(Kd-\Delta)$, respectively. Throughout the whole evaluation, we have used a single assumption

$$\exp[-Pd] \cong 0 \quad . \qquad (17)$$

The resultant expression of the susceptibility is

$$\chi(Z,Z';\omega) = (\alpha_0\, \omega_0\, m_x \,/\, \hbar)$$

$$\times\, [S_0\, J_1 + 2S_1(J_2 - J_4) - 2S_2(J_2 - J_3)], \qquad (18)$$

where

$$\alpha_0 = 4e^2|P_{vc}|^2|\phi_1(0)|^2/m_0^2\omega^2\hbar\omega_0 \qquad (19)$$

$$S_0 = [1-\delta' P^2/(P^2+q_0^2)]/[\{1-\delta'(P^2-q_0^2)/(P^2+q_0^2)\}\; 2q_0 \sin(q_0 d-\delta_0)] \qquad (20)$$

$$S_1 = \delta'[(1-\delta')/(1+\delta')]^{1/2}/[4P(1+q_0^2/P^2)\{1- \delta'(P^2-q_0^2)/(P^2+q_0^2)\}] \qquad (21)$$

$$S_2 = S_1\, (\sqrt{1+\delta'} + \sqrt{1-\delta'})^2/2\delta' \qquad (22)$$

$$\delta' = 2/Pd \qquad (23)$$

$$\delta_0 = 2 \tan^{-1}(q_0/P) \qquad (24)$$

$$J_1 = -\cos[q_0(d-|Z-Z'|)-\delta_0] + \cos[q_0(\bar{Z}-Z')]$$

$$+ [\cos q_0 d - \cos(q_0 d-\delta_0)](e^{-PZ-PZ'} + e^{-P\bar{Z}-P\bar{Z}'})$$

$$+ (1-\cos\delta_0)(e^{-PZ-P\bar{Z}'} + e^{-P\bar{Z}-PZ'})$$

$$+ [\cos(q_0\bar{Z}-\delta_0) - \cos q_0\bar{Z}]\, e^{-PZ'} + [\cos(q_0 Z-\delta_0) - \cos q_0 Z]\, e^{-P\bar{Z}'}$$

$$+ [\cos(q_0\bar{Z}'-\delta_0) - \cos q_0\bar{Z}']\, e^{-PZ} + [\cos(q_0 Z'-\delta_0) - \cos q_0 Z']\, e^{-P\bar{Z}} \qquad (25)$$

$$J_2 = -e^{-PZ-PZ'} + e^{-PZ-\bar{P}Z'} + e^{-PZ'-\bar{P}Z}$$

$$- e^{-P\bar{Z}-P\bar{Z}'} + e^{-P\bar{Z}-\bar{P}\bar{Z}'} + e^{-P\bar{Z}'-\bar{P}\bar{Z}} \qquad (26)$$

$$J_3 = e^{-\bar{P}Z-\bar{P}Z'} + e^{-\bar{P}\bar{Z}-\bar{P}\bar{Z}'} \qquad (27)$$

$$J_4 = e^{-\bar{P}|Z-Z'|}\,. \qquad (28)$$

This should be compared with CK's result;

$$\chi(Z,Z';\omega)_{CK} = (\alpha_0\, \omega_0\, m_x / \hbar)\, S_0\, J_1 \Big|_{\substack{\delta_0 \to 2q_0/P \\ \delta' \to 0}} \times (1-\delta')\quad , \qquad (29)$$

namely, the exact summation leads to (i) δ_0 instead of $2q_0/P$ in the S_0J_1 term, and (ii) the additional terms, (J_1, J_2, J_3). Since

$$S_1 = O(1/Pd)\, O(q_0/P)\, S_0 \quad , \qquad (30)$$

$$S_2 = O(q_0/P)\, S_0 \quad , \qquad (31)$$

$$J_2 - J_3 = O(1/Pd) \quad , \qquad (32)$$

the omitted part in CK's calculation is of the order $O(1/Pd) \times O(q_0/P)$ of the retained term. Thus, under the assumptions, $Pd>>1$, $P>>q_0$, the result of CK is assured, and, in the limit of $d \to \infty$, χ of (18) is shown to become the susceptibility of DA-DS.

Next step is to solve eq.(1) with χ of (18). For that purpose, we try to reduce the integro-differential equation to higher order differential equation

/11-13/. Because of the J_4 term in χ, we operate

$$(d^2/dZ^2 - \bar{P}^2)\,(d^2/dZ^2 + q_0^{\,2}) \tag{33}$$

on eq.(1) instead of $(d^2/dZ^2 + q_0^{\,2})$ alone as in DA-DS and CK. This reduces eq.(1) to a sixth order differential equation

$$\mathcal{D}\,E(Z) + B\,(P^2-\bar{P}^2)(P^2+q_0^{\,2})(H\,e^{-PZ} + \bar{H}\,e^{-\bar{P}Z}) = 0\,, \tag{34}$$

where

$$\mathcal{D} \;=\; (d^2/dZ^2-\bar{P}^2)(d^2/dZ^2+q_0^{\,2})(d^2/dZ^2+\varepsilon_b\omega^2/c^2)$$

$$+\; B[\{4S_1\bar{P} - 2S_0q_0\sin(q_0d-\delta_0)\}(d^2/dZ^2-\bar{P}^2) + 4S_1\bar{P}(\bar{P}^2+q_0^{\,2})] \tag{35}$$

$$B = 4\pi\,\omega^2\,\alpha_0\,\omega_0\,m_x\,/\,c^2\,h \tag{36}$$

$$\varepsilon_b = 1 + 4\pi\,\chi_b \tag{37}$$

and $(H, \bar{H})$ are the constants defined by integrals of the form

$$\int_0^d E(Z)\,\xi(Z)\,dZ\,, \tag{38}$$

ξ being a known function (plane wave or evanescent wave). The solution of the homogeneous equation $\mathcal{D}\,E(Z) = 0$ are plane waves, $\exp[i\lambda Z]$. From the form of $\mathcal{D}$, (35), we get three pairs of solutions

$$\lambda = \pm\,q_1,\; \pm\,q_2,\; \pm\,q_3\;;\quad (\;\mathrm{Im}[q_j] > 0\;)\qquad. \tag{39}$$

Two of them, q_1 and q_2, correspond essentially to bulk polaritons, and the rest, q_3, to the evanescent wave, $\exp[-PZ]$. For finite d, there is a certain modification, but in the limit of $d \to \infty$, the correspondence is exact. The general solution of eq.(34) is now written as

$$E(Z) = \textstyle\sum_{j=1}^{3}\,(W_j\,e^{iq_jZ} + \bar{W}_j\,e^{iq_j\bar{Z}}) + g\,e^{-PZ} + \bar{g}\,e^{-P\bar{Z}}, \tag{40}$$

where the coefficients $(W_j, \bar{W}_j, g, \bar{g})$ are to be determined. By substituting (40) in eq.(1), there appear the terms with different Z-dependences on the left-hand side, the exponents of which are $\pm\,iq_j$, $\pm\,P$, $\pm\,iq_0$, and $\pm\,\bar{P}$. From the requirement that each of the coefficients should be zero, we get [a] $(g, \bar{g})$ as linear combinations of $(W_j, \bar{W}_j)$, which allows us to rewrite (40) as

$$E(Z) = \textstyle\sum_{j=1}^{3}[W_j(e^{iq_jZ} + h_j\,e^{-PZ}) + \bar{W}_j(e^{iq_j\bar{Z}} + \bar{h}_j\,e^{-P\bar{Z}})] \tag{41}$$

and [b] four linear simultaneous equations for $(W_j, \bar{W}_j)$, which are the ABC's in this case. This solution can be shown to reduce to that of CK for Pd>>1, and DA-DS for $d \to \infty$: For Pd>>1, the terms with coefficients W_3 and $\bar{W}_3$ in (41) tend to vanish, other terms to those of CK, and the two of the four ABC's approach to those of CK' (see (54) below) which do not contain W_3 and $\bar{W}_3$, while the other two of the ABC's retain W_3 and $\bar{W}_3$ in themselves, thus being irrelevant relations in this limit. In this way, the solution (40) together with the four ABC's are the exact result (up to the assumption, $\exp[-Pd] \cong 0$) for a slab of finite thickness within the DA-DS model of exciton wave function. Comparison between this and other results will be discussed later.

§4 ABC-Free Theory

If one eliminates four of the six constants $(W_j, \bar{W}_j)$ in (41) by using the four ABC's, the electric field E(Z) is expressed in terms of two free amplitudes. It is to this form of solution that ABC-free formalism should lead directly. Its general framework /14/ is based on the characteristic form of $\chi(R,R';\omega)$: The linear response theory leads generally to an expression where the R and R' dependence is described as a sum of the products of respective functions of R and R', as in (8).

By substitution of (8), eq.(1), for normal incidence, can be rewritten as

$$d^2E(Z)/dZ^2 + q^2\, E(Z) + Q^2 \sum_K \bar{\chi}_K(\omega)\, \rho_K(Z)^*\, F_K = 0 \quad , \tag{42}$$

$$q = \sqrt{\varepsilon_b}\, \omega/c \quad , \tag{43}$$

$$\rho_K(Z) = M_K(Z) \,/\, \mu \quad , \tag{44}$$

$$\bar{\chi}_K(\omega) = (\mu^2/V_0) \,/\, [E_{1s}(K) - \hbar\omega - i\gamma] \quad , \tag{45}$$

$$F_K = \int_0^d \rho_K(Z)\, E(Z)\, dZ \quad . \tag{46}$$

The general solution of (42) is

$$E(Z) = E_1\, e^{iqZ} + E_2\, e^{iq\bar{Z}} - \sum_K D_K(Z)\, F_K \quad , \tag{47}$$

$$D_K(Z) = Q^2\, \bar{\chi}_K\, [-\frac{e^{iKZ} + A_K^*\, e^{-iKZ}}{K^2 - q^2} + \frac{(1+A_K^*)e^{-PZ} + (e^{iKd}+A_K^*\, e^{-iKd})e^{-P\bar{Z}}}{P^2 + q^2}]\, N_K \tag{48}$$

with two arbitrary constants, E_1 and E_2. To determine the values of $\{F_K\}$, we put (47) into (46). Using the relation

$$\int_0^d (e^{-iKZ} + A_K e^{iKZ})(e^{iK'Z} + A_{K'}^* e^{-iK'Z})\, dZ = \delta_{KK'}(2d - \frac{4P}{P^2 + K^2}) , \tag{49}$$

we can rewrite the equations for $\{F_K\}$ as

$$F_K = C_K \int_0^d \rho_K(Z)\, (E_1 e^{iqZ} + E_2 e^{iq\bar{Z}})\, dZ$$

$$- C_K N_K [(1+ A_K) B_1 + (e^{-iKd} + A_K e^{iKd}) B_2] , \tag{50}$$

$$B_1 = Q^2 \sum_K N_K \bar{X}_K(\omega) (1+ A_K^*) F_K , \tag{51}$$

$$B_2 = Q^2 \sum_K N_K \bar{X}_K(\omega) (e^{iKd} + A_K^* e^{-iKd}) F_K , \tag{52}$$

$$C_K^{-1} = 1 - \frac{N_K^2 Q^2 \bar{X}_K(\omega)}{K^2 - q^2} (2d - \frac{4P}{P^2 + K^2}) . \tag{53}$$

Equations (50-52) provide a closed set of inhomogeneous linear equations for B_1 and B_2, which gives B_1 and B_2 (and consequently F_K, too) as linear combinations of E_1 and E_2. In this way, we see that the electric field in the medium E(Z) is obtained in terms of two arbitrary constants. The Maxwell boundary conditions at Z = 0 and Z = d provide four equations. Since the unknown amplitudes are those of incident (E_i), reflected (E_r), transmitted (E_t) waves, E_1 and E_2, the four equations are sufficient to uniquely determine the ratio of these waves. In the actual expression of E(Z), there appear various types of summations over size quantized K. These summations have been evaluated by the method of contour integral mentioned in §3.

To compare the results of ABC- and ABC-free theories with each other, the coefficients of the terms with different exponents in the expression of E(Z) have been calculated numerically, and shown to be exactly the same as each other, as expected.

§5 Comparison Between "Exact" and Approximate Theories

The CK-scheme /8/ is based on the assumptions $Pd \gg 1$ and $P \gg q_0$, while the present result (which we call CI hereafter) relies just on $\exp[-Pd] \cong 0$ alone. Though CI is certainly superior to CK, the latter is mathematically much simpler than the former. Therefore, it is of practical interest to see where in the parameter space the two schemes begin to show difference in physical quantities. For the comparison, we include also a modified version of CK (CK', hereafter), where we replace ABC as.

$$ABC(CK') = ABC(CK)\Big|_{\cot(q_0/P) \to P/q_0} , \tag{54}$$

while keeping the expression for E(Z) unchanged. This is a scheme already proposed in CK to relax the constraint $P>>q_0$. (§3 of CK's paper is formulated according to CK'.) We compare the reflectance spectra calculated from CK, CK', and CI to see the critical region of parameter space where the conditions $Pd>>1$ and $P>>q_0$ begin to be invalid appreciably.

In the case of CuCl $[Z_3]$-exciton, where $P=0.3$ Å^{-1} according to the original DA-DS theory, the three cases cannot be distinguished in the plotted spectra of reflectance within the studied region of d (≧ 100 Å). This is an additional support to the validity of CK analysis in ref.8. For $P=0.0154$ Å^{-1}, which would be appropriate for GaAs, the calculated spectra are shown in Fig.1 (a-d) for d = 1000, 500, 300 Å. The parameter values used for the calculation are

$$\hbar\omega_0 = 1.515\ [\mathrm{eV}]\ , \quad \varepsilon_b = 12.6\ , \quad m_x = 0.298\ m_0\ ,$$

$$\gamma = 0.035\ [\mathrm{meV}]\ , \quad 4\pi\alpha_0 = 0.0022\ . \qquad (55)$$

The difference between CK and CI is already seen for d=2000 Å, but the deviation of CK' from CI is not seen until d≃400 Å. This means that the difference between CK and CI for d≧500 A is caused mainly by the breakdown of the condition $P>>q_0$. This is also seen from the fact that the difference is larger for higher ω, where q_0 becomes larger. As to the resonant energy

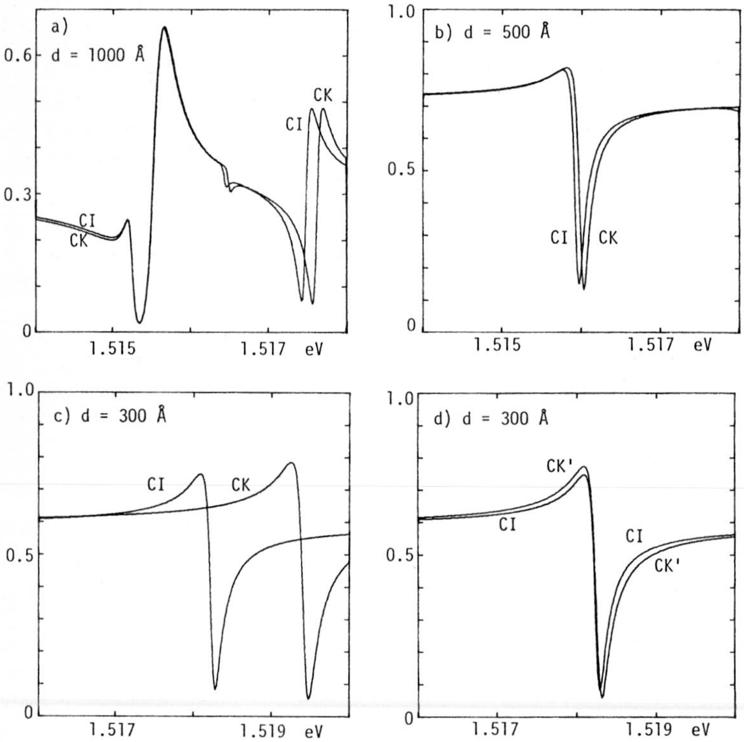

Fig.1 Reflectance spectra according to CI, CK, and CK' for d = 1000, 500, 300 A

positions, there is almost no difference between CK' and CI even for d=300 Å. Their difference occurs in the value of reflectivity, which is due to the breakdown of Pd>>1. (Pd=6.16, 4.62 for d=400, 300 Å, respectively) But the difference is still at most several % even for d=300 Å, which is almost in the quantum-well regime. Therefore, the conclusion is that the scheme of CK (eqs.(2.40), (2.41), §3 of ref.8) is an excellent approximation in most of the parameter space except in and very near the quantum-well regime.

§6 Exciton Quantization and Polariton Interference

According to a naive picture of light absorption, one might expect minima in transmission spectrum at the levels of quantized exciton states, since they "absorb" light. This looks rather plausible in resonant region ($\omega \geqq \omega_0$), but clearly it does not apply to the minima in the lower energy region ($\omega < \omega_0$), since there is no exciton level. On the other hand, the classical picture of Fabry-Perot interference well explains the reflectance and transmittance spectra in the off-resonant region, where the medium can be described by a (weakly) ω-dependent dielectric constant $\varepsilon(\omega)$. In the exciton resonance region, the interference theory must include the effect of spatial dispersion, namely (in correspondence with the k-dependence of $\varepsilon(\omega)$ in the case of a bulk) the existence of a series of poles in $\varepsilon(\omega)$ at the frequencies of size-quantized exciton levels in the case of a slab.

What is shown in CK and CI is that the explicit consideration of the size quantization of exciton's translational motion does lead to the picture of multi-mode polariton interference. In their final forms, the effect of size quantization becomes implicit, and physical quantities are calculated from the amplitudes of multi-mode polaritons with or without explicit reference to ABC. In terms of this theory, it is possible to unify the two simple pictures mentioned above: In the low ω, off-resonant region, upper polariton is a strongly damped mode, and therefore only lower polariton contributes to optical spectra, the result of which resembles very much to a single mode Fabry-Perot interference. On the other hand, in the resonant region, there are a series of short period interference patterns due to the lower polariton. Its spikes and/or dips correspond well to the quantized exciton levels. In addition to these aspects which are reminiscent of the simple pictures, there are other, more complicated features due to the presence of multi-modes: They can produce [a] various different periods of interference, [b] a change in the spectral shape from what is expected in a single mode theory, and sometimes [c] a "shift" of the position of interference fringe. An example is shown in Fig.2. Tick marks represent the energies of exciton (EXC) and upper and lower polaritons (UP, LP) corresponding to the size quantized wave numbers. They correspond to some structure in reflectance and transmittance in higher $\hbar\omega$ region. But, in the neighborhood of $\hbar\omega_0$, there is no such correspondence, which is a consequence of the multi-mode interference. The line shape for each "exciton resonance" is usually a superposition of absorptive and dispersive curves. Thus, a better description may be obtained by $A(\omega) = 1 - R(\omega) - T(\omega)$, where R and T are reflectance and transmittance respectively: The peaks in $A(\omega)$ correspond well to the quantized exciton levels in higher $\hbar\omega$ region. This will be discussed in more detail elsewhere /15/.

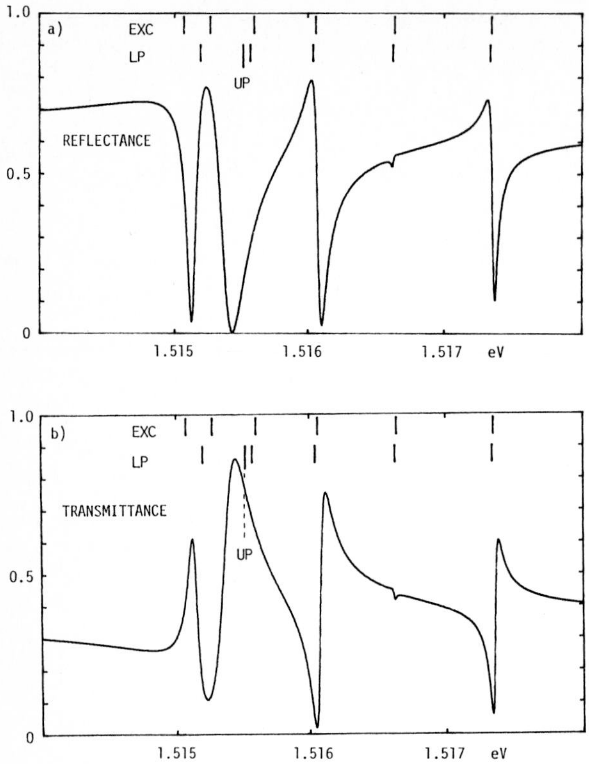

Fig.2 Reflectance (a) and transmittance (b) spectra calculated for d = 1500 Å, P = 0.0154 /Å, and γ = 0.001 eV. Other parameters are those in (55). Tick marks are the positions of quantized exciton levels (EXC), and single mode (lower and upper) polariton interference (LP, UP).

Acknowledgments

The authors are grateful to Professor A. Yoshimori for useful discussions and support.

References

1. S.I. Pekar: Sov. Phys. JETP **6**, 785 (1957)
2. J.L. Birman: *Excitons*, ed. E.I. Rashba and M.D. Sturge (North Holland, 1982) p.27
3. V.M. Agranovich and V.L. Ginzburg: *Crystal Optics with Spatial Dispersion, and Excitons* (Springer Verlag, 1984)
4. V.A. Kiselev, B.S. Razbirin, and I.N. Uraltsev: Phys. Status Solidi (b)**72**, 161 (1975)
5. I.V. Makarenko, I.N. Uraltsev, and V.A. Kiselev: Phys. Status Solidi (b)**98**, 773 (1980)
6. V.A. Kiselev, I.N. Uraltsev, and I.V. Makarenko: Solid State Commun. **53**, 591 (1985)
7. T. Mita and N. Nagasawa: Solid State Commun. **44**, 1003 (1982)
8. K. Cho and M. Kawata: J. Phys. Soc. Jpn. **54**, 4431 (1985)
9. J.L. Birman and J.J. Sein: Phys. Rev. **B6**, 2482 (1972)

10. R. Zeyher, J.L. Birman, and W. Brenig: Phys. Rev. **B6**, 4613 (1972)
11. A. D'Andrea and R. Del Sole: Phys. Rev. **B25**, 3714 (1982)
12. G.S. Agarwal, D.N. Pattanayak, and E. Wolf: Phys. Rev. Letters **27**, 1022 (1971); Phys. Rev. **B8**, 4768 (1973)
13. A.A. Maradudin and D.L. Mills: Phys. Rev. **B7**, 2787 (1973)
14. K. Cho: J. Phys. Soc. Jpn. **55**, 4113 (1986)
15. A. D'Andrea, R. Del Sole, H. Ishihara, and K. Cho: in preparation

Wave Functions and Optical Properties of Excitons in a Slab

A. D'Andrea[1] *and R. Del Sole*[2]

[1]Ist. Metodologie Avanzate Inorganiche, C.N.R., I-00016 Monterotondo Staz. Roma, Italy
[2]Dip. di Fisica, II Univ. di Roma "Tor Vergata", Via O. Raimondo, I-00173 Roma, Italy

1. Introduction

It is well known that, if we want to extract the properties of Wannier excitons from reflectance measurements, the situation is complicated by two different effects: the spatial dispersion and the surface potential. The spatial dispersion arises from the finite value of the total exciton mass M and generates two polariton branches, whose relative amplitude is determined by the so-called additional boundary condition (ABC). The surface repulsive potential, which occurs even if the total mass is infinite, causes the so called dead layer effect, a zone at the surface where the exciton polarization is vanishingly small.

It is now clear that both properties, namely the ABC and dead layer, can be derived from the knowledge of exciton wave functions/1/. Moreover, a link has been established, within a realistic model of the exciton-surface interaction, between dead layer depth and ABC/2/. In spite of its analytically complicated form/3/, the latter bears a physical meaning identical to that of Pekar's ABC/4/, namely the vanishing of the exciton polarization at the surface, which in turn is a direct consequence of the no-escape boundary conditions imposed on exciton wave function. The depth of the dead layer (or, speaking more precisely, of the inhomogeneous transition layer where the exciton wave function is smaller than in bulk) comes out to be smaller than the exciton radius/5/. However, it is supposed to be affected by the image potential and by surface electric fields, both neglected in the calculation. On the other hand, its determination from experiments carried out on thick samples is hampered by the poor experimental sensitivity and by the occurrence of extrinsic dead layers/6/. We think that optical measurements on semiconductor slabs are a sensitive tool for checking the validity of the microscopic model adopted/1,2,5,6/ and for determining the dead layer depth. The quantization of the center-of-mass motion is actually strongly dependent on the slab dimension L and dead-layer depth d, while the ABC's are essentially dependent on the chemical nature of the interfaces between the active slab and the barriers. Therefore a measurement of the exciton quantization as a function of L should give direct information on the dead layer depth. On the other hand, the exciton quantization in thin films is an interesting topic per se, in view of the recent unexpected experimental findings/7/ of excitons following the electron and hole quantization (rather than that of the center-of-mass motion) in GaAs films as thick as $18a_B$, where a_B is the exciton radius.

Reliable exciton wave functions are available only in the quantum well limit, $L \simeq a_B$ /8,9/ and for $L >> a_B$ /10/. The first purpose of this work is to close this gap, by constructing an exciton wave function which is reliable in the range $L > a_B$. We will also show that the quantization of the lowest exciton level in the range $L > 10\ a_B$ is determined by the center-of-mass motion, and therefore by the exciton mass M. The third purpose is to show that it is possible to shed some light on the dead layer problem by performing optical measurements on semiconductor slabs, in the range $10\ a_B < L < \lambda/2$, where λ is the light wave length.

Section 2 is devoted to the study of the exciton wave function in a slab, while Section 3 deals with the optical response. The conclusions are drawn in Section 4.

2. Exciton wave functions in a slab.

In order to find a suitable approximation to the exciton wave function in a "thick" quantum well ($L > a_B$), we start from the wave function appropriate to the semi-infinite crystal/5/. In the case of a slab of thickness L an analogous treatment leads to the wave function/11/

$$\Psi(r,z,Z)=N[\cos(KnZ)-Feven(z)\cosh(PZ)+Fodd(z)\sinh(PZ)]\exp(-r/a), \quad (1)$$

where Z is the center-of-mass distance from the central plane of the slab, $\vec{r}$ is the electron-hole distance, z its component along the slab axis and N the normalization constant; furthermore

$$Feven(z)=[\sinh(PZ1)\cos(KnZ2)-\sinh(PZ2)\cos(KnZ1)]/\sinh[P(Z1-Z2)], \quad (2)$$

$$Fodd(z)=[\cosh(PZ1)\cos(KnZ2)-\cos(PZ2)\cos(KnZ1)]/\sinh[P(Z1-Z2)], \quad (3)$$

where $Z1=L/2-m_hZ/M$, $Z2=-L/2+m_eZ/M$, and m_e and m_h are the electron and hole mass respectively. The exciton momentum Kn is given by the quantization condition

$$Kn\,tg(KnL/2) + P\,tgh(PL/2)=0. \quad (4)$$

P is the inverse transition layer depth. The wave functions (1) are even for reflection with respect to the central plane of the slab. Since we are mainly interested in the lowest exciton state n=1, which is even, we do not quote the expressions of odd wave functions. Cho/10/ has already treated the particular case of very thick slabs, where the interaction of the two surfaces, proportional to exp(-PL), can be neglected. When the interaction between the two transition layers is considered the values of P and a may be different from the values appropriate to semi-infinite crystals. Therefore they are considered here as variational parameters, to be adjusted in order to minimize the n=1 exciton energy. Their values as functions of L are shown in Figs. 1 and 2 for the case of CdS, by using $\varepsilon_0=8.1$, M=0.95 m, R*= 28 meV, μ=0.135 m. While a reaches its bulk value a_B already for $L= 4a_B$, $d \equiv 1/P$ shows a slower convergence. It is about constant, equal to $1.5a_B$, for $L>18a_B$, in contrast with the value found by numerical calculations in the semi-infinite case ($d \simeq 0.7\ a_B$). This discrepancy ,which is not explained so far,might be

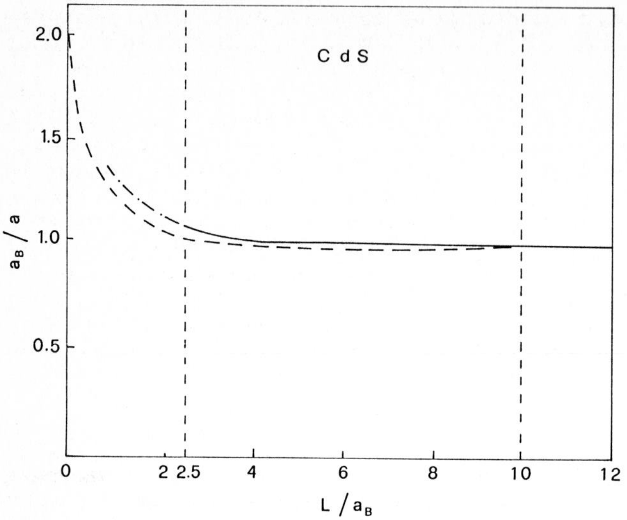

Fig.1 The variational parameter a shown as a function of the slab thickness L. The exciton parameters are: ε_0 = 8.1, M=0.94 m, μ=0.135 m. Dashed curve: Bastard et al.'s results/8/. Solid curve: present calculation.

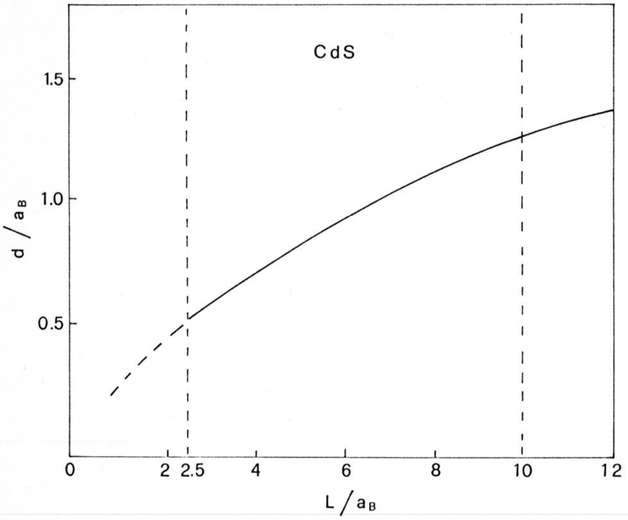

Fig.2 The variational parameter d = 1/P shown as a function of the slab thickness L. Exciton parameters chosen as in Fig.1.

due to the different methods of solving the Schroedinger equation used in the two cases. It will be the object of further research.

We have checked the reliability of (1) for small L by comparison with the wave function of Bastard et al./8/:

$$\psi(\varrho, z_e, z_h) = No \cos(\pi z_e/L)\cos(\pi z_h/L) \exp(-r/a^0), \qquad (5)$$

where a^0 is the variational parameter, No the normalization constant, z_e and z_h are the electron and hole distances from the central plane of the slab, and ϱ is the component of $\vec{r}$ in the surface plane. The comparison between the lowest exciton state energies computed according to (1) and (5) is shown in Fig.3. For $L < 2.5a_B$ our wave function is completely unable to give sensible values of the exciton ground-state energy, yielding a negative binding energy. For $L > 2.5\ a_B$ the computed energies apparently coincide, showing the reliability of both wave functions. As L tends to infinity, the binding energy of the wave function of Bastard et al. is $R^* - \hbar^2\pi^2/(6\mu L^2)$, while the binding energy of our wave function is $R^* - \hbar^2\pi^2/[2M(L-2/P)^2]$. The larger exciton binding energy of our approach is related to its increasing reliability for thick slabs.

The square modulus of the lowest-exciton wave function given by (1) and by (5) for vanishing electron-hole distance, that is relevant for optical properties/3/, is shown in Figs.4 (L=500Å) and 5(L=100Å) for the case of CdS. The wave function of Bastard et al. shows greater localization of the exciton in the middle of the slab than ours, but the same value of oscillator strength, which is proportional to the integral value of the curves. In the sample of 100 Å thickness the Bastard et al.'s wave function intensity is systematically lower than ours. Since for this thickness spatial dispersion is negligible the optical spectra computed by our wave function show greater oscillator strengths (about 20%).

Concluding, in CdS we can distinguish three different ranges of thicknesses, namely: a) from zero to $2.5a_B$ we have the quantum well zone. In this zone the electron and the hole are quantized independently along Z direction. b) from $2.5a_B$ to $10a_B$, the overlap zone, where spatial dispersion is still negligible and the exciton shows only one bound state, and finally c) from $10a_B$

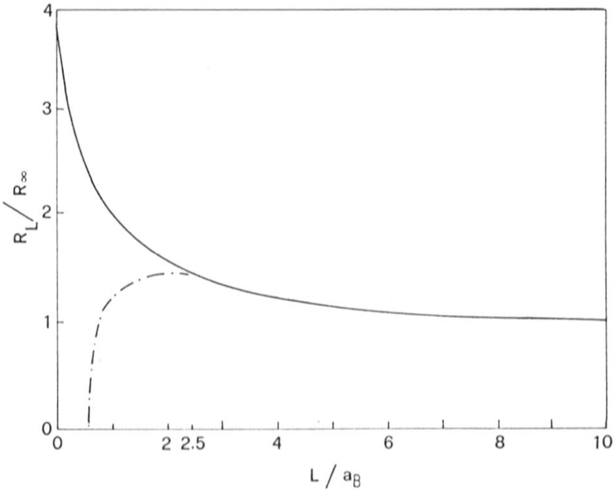

Fig.3 Binding energy R_L of the lowest exciton state for CdS. Solid curve: Bastard et al./8/. Dotted-dashed curve: present calculation.

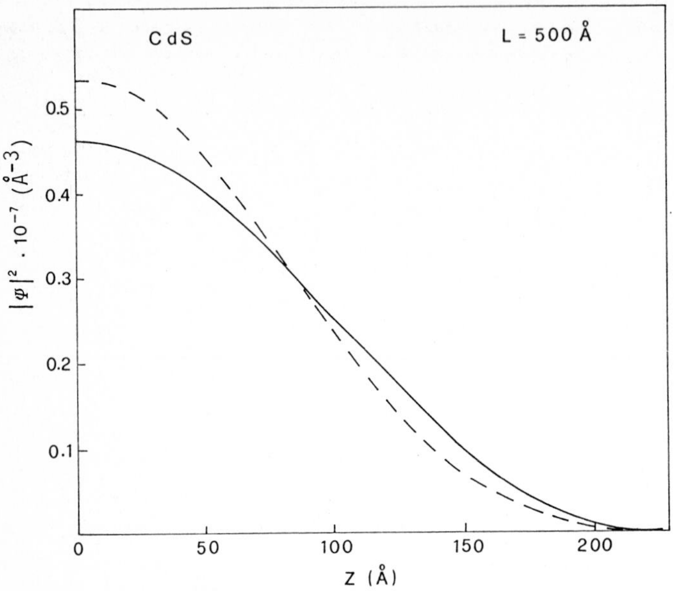

Fig.4 $|\psi(0,Z)|^2$ for a 500 Å thick slab of CdS. Dashed curve: computed according to the formulation of Bastard et al./8/; Solid curve: our calculation, with $P = 0.022$ Å^{-1} and $a = 32$ Å.

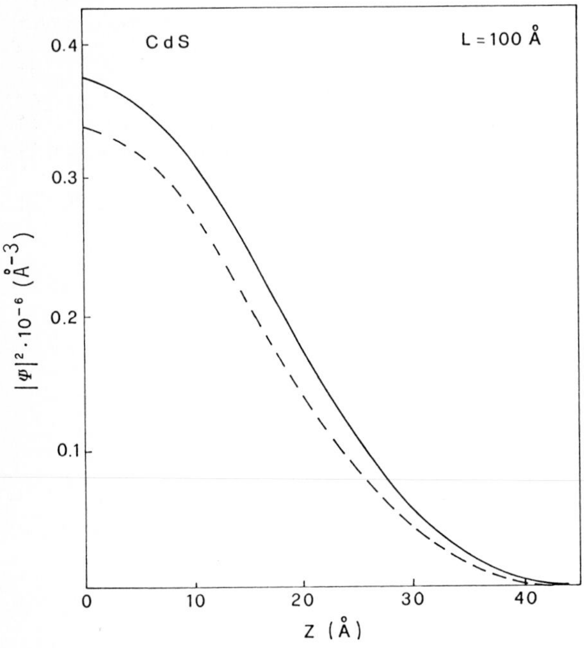

Fig.5 The same as Fig.4 for L = 100 Å. Our calculation uses $P = 0.054$ Å^{-1} and $a = 31$Å.

to semi-infinite samples, the bulk exciton scheme is restored and the center-of-mass motion is quantized. The exciton parameters reach bulk parameters as limit values.

3. The optical response.

In this Section we consider reflection and transmission of light normally incident on a self-sustained semiconductor slab of length L. The method of solution of Maxwell's equations is similar to that of Ref./12/, involving the inversion of a matrix whose size is determined by the number of quantized exciton states considered. The details of the calculation will be given elsewhere /11/. The number of quantized states to be taken into account in order to yield well converged results in an energy range of the order of a few meV around the lowest level depends of course upon L. We find that including up to 50 levels yields converged results for L up to 2000 Å in CdS. (Our results are quite general. We consider CdS rather than GaAs to avoid the complications arising from the valence band degeneracy.) For L< 200 Å, only the lowest state is important, whose energy was actually minimized by adjusting a and P. For $L>18a_B$ (~ 500Å in Cds) the bulk value of P is reached, so that it is reasonable to associate the same P value to all quantized states, as in the case where the surfaces do not interact/10/. In the intermediate case, i.e. for 200 Å < L < 500 Å, we have arbitrarily used the same P-value determined for the lowest state also for the higher states. This makes our results not completely reliable in this range.

The reflectance of CdS slabs as a function of the thickness L is shown in Fig.6. For L = 100 Å a single structure appears, associated with the n = 1 state. For L = 300 Å, also the n = 2 level shows up as a small peak at 2.5538 eV. For L = 500 Å, also the structure related to n = 3 can be seen. The peculiar features of polariton interference do not yet appear, since we are in the range L< λ/2. The sensitivity of reflectance to the dead layer depth is shown in Fig. 7, for L = 300 Å. By changing d from 25 (semi-infinite crystal result/5/) to 40 Å (L→ ∞ limit of the slab calculation) there is a clear shift of the small peak associated with the n=2 level. The experimental observation of this peak should however be difficult because of its small intensity.

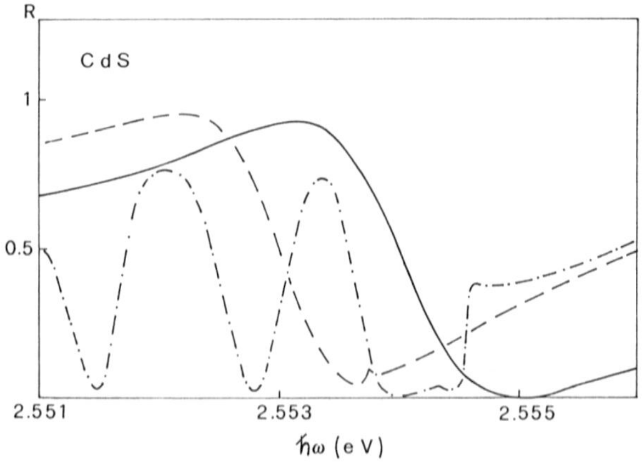

Fig.6 Normal incidence reflectivity of CdS slabs. The exciton parameters are: ε_0=8.1; M=0.94 m, $\hbar\omega_0$ = 2.5515 eV, $4\pi\alpha$ = 0.013 and γ = 0.1 meV. Solid curve: L = 100 Å; dashed curve: L=300Å; dotted - dashed curve : L = 500 Å.

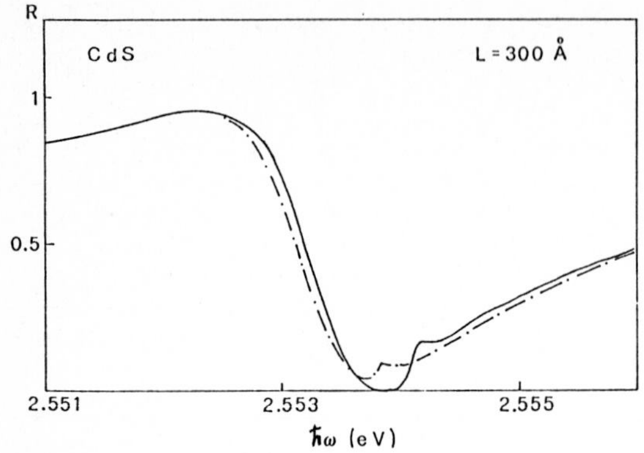

Fig.7 Normal incidence reflectivity for CdS. Solid curve: present calculation with P=0.025 $Å^{-1}$; dotted-dashed curve: present calculation with P=0.04 $Å^{-1}$.

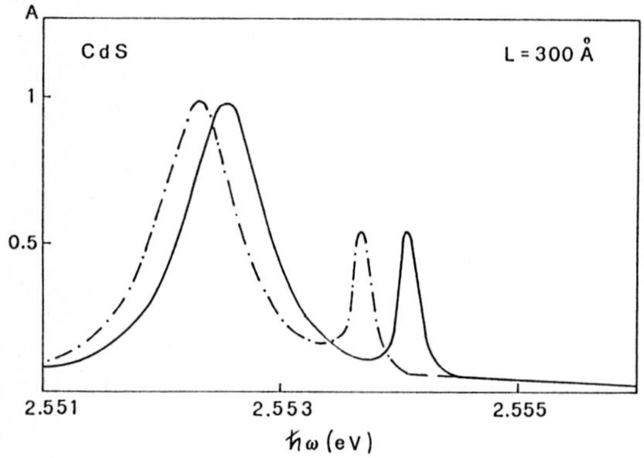

Fig.8 Absorption spectrum of a CdS slab. Solid curve: present calculation with P = 0.022 $Å^{-1}$; dotted-dashed curve: present calculation with P = 0.04 $Å^{-1}$.

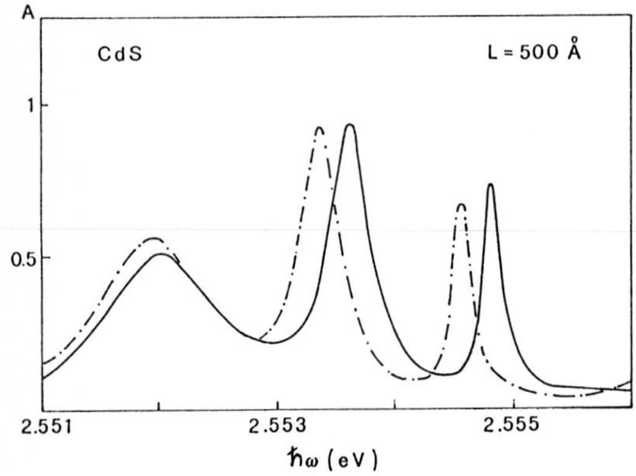

Fig.9 Absorption spectrum of a CdS slab. Solid curve: present calculation with P = 022 $Å^{-1}$; dotted-dashed curve: present calculation with P=0.04 $Å^{-1}$.

Absorption measurements result to be more sensitive to the dead layer depth. The transmitted intensity T, normalized to the light intensity entering the sample 1-R, is shown in Figs. 8 and 9 for L = 300 Å and L = 500 Å respectively. Two values of P are considered, as in Fig. 7. The peaks associated with n = 2 and n= 3 result to be quite sensitive to the value of P, so that absorption measurements in the range $10\ a_B < L < \lambda/2$ can determine it. The lower bound on L is determined by the required occurrence of the n = 2 level, while the upper bound is to avoid the appearance of polariton interference in the slab, which would make more difficult the comparison between theory and experiment.

4. Conclusions

The exciton wave function derived in this work is valid for slab thickness L larger than $2.5\ a_B$. The concept of transition layer firstly introduced in the case of semi-infinite crystals/13/ and later on extended to thin films/10/, is generalized to the case of "thick" quantum wells. Its value 1/P reduces by a factor 3 going from $L = \infty$ to $L = 2.5\ a_B$.

We find that for $L > 10\ a_B$, the lowest-state exciton energy is quantized according to the center-of-mass motion involving the exciton mass M. A meaningful comparison with experiment /7/, which gives information on the energy differences between the lowest exciton states, requires however the calculation of all their energies, not only the lowest-state energy.

As a final result, we point out the possibility of a direct measurement of the dead-layer depth by absorption measurements performed on very thin films, in the thickness range where the n=2 state shows up in the spectrum but thin enough to avoid the occurrence of polariton interference patterns.

References

1) R.Del Sole and A.D'Andrea, these Proceedings.
2) A.D'Andrea and R.Del Sole, Phys.Rev.B 29, 4782(1984)
3) A.D'Andrea and R.Del Sole, Phys.Rev.B 25, 3714(1982)
4) S.I.Pekar, Sov.Phys.Jept 6, 785(1957)
5) A.D'Andrea and R.Del Sole, Phys.Rev.B 32, 2337(1985)
6) A.D'Andrea and R.Del Sole, to be published.
7) L.Schultheis, J.Kuhl and A.Honold, these Proceedings.
8) G.Bastard, E.E.Mendez, L.L.Chang and L.Esaki, Phys.Rev.B 26, 1974(1982)
9) Y.Shinozuka and M.Matsuura, Phys.Rev.B 28, 4878(1982)
10) K.Cho, J.Phys.Soc. of Japan 54, 4431(1985)
11) A.D'Andrea and R.Del Sole, to be published.
12) K.Cho, J.Phys.Soc. of Japan 55, 4113(1986)
13) A.D'Andrea and R.Del Sole, Sol.St.Commun. 30, 145(1979)

Wannier Excitons at GaAs Surfaces and in Thin GaAs Layers

L. Schultheis[1,*], *K. Köhler*[1,+], *and C.W. Tu*[2]

[1]Max-Planck-Institut für Festkörperforschung, Heisenbergstr. 1, D-7000 Stuttgart 80, Fed. Rep. of Germany

[2]AT&T Bell Laboratories, Murray Hill, NJ 07974, USA

1. Abstract

The optical properties of excitons at GaAs surfaces and in thin GaAs layers are studied by means of reflectance, transmission, and photoluminescence spectroscopy. Free GaAs surfaces exposed to air are found to affect not only the reflectance but also the luminescence lineshape of the excitonic line due to a strong electric surface field. Highly parallel GaAs layers reveal quantization of the excitonic polariton for layer thicknesses of a few µm as well as quantization of excitonic energy levels for ultrathin GaAs layers.

2. Introduction

The optical properties of excitons in the model semiconductor GaAs have been the subject of extensive studies in the past /1-7/. Excitonic parameters like eigenenergy /1/, binding energy /2/ and oscillator strength /3/ have been studied in detail. However, by employing epitaxial GaAs samples grown by liquid phase epitaxy or vapour phase epitaxy, the excitation geometry does not correspond to the model of the semi-infinite crystal commonly considered in the theoretical analyses, in particular with respect to the surface condition as well as to the microscopic surface geometry. Surface preparations, on the other hand, such as oxygen exposure, polishing, and etching are known to strongly modify the excitonic properties near the surface /4-7/.

Recent progress in material technology, however, has now led to growth of well-defined and parallel layers of a controlled thickness by molecular beam epitaxy, not only on GaAs substrates but also on or clad with $Al_xGa_{1-x}As$, a semiconductor which is perfectly lattice-matched to GaAs, but has a higher bandgap (typical 2eV for x=0.3). Problems with the surface preparation as well as an unknown surface geometry can be overcome successfully, preventing the formation of extrinsic surface states and thus electric surface fields /7/.

The purpose of this paper is to emphasize the importance of well-defined and properly chosen GaAs layer for optical investigations of the excitonic properties.

---

* present address: Brown Boveri Research Center
CH 5405 Baden/Dättwil, Switzerland

\+ present address: Fraunhofer Institut für Angewandte Festkörperforschung
D-7800 Freiburg, Federal Republic of Germany

In the first part, we study the influence of an electric surface field to the excitonic emission and reflectance properties. The second part is concerned with the impact of the finite slab geometry to excitonic polaritons as well as to excitons. Studies by reflectance, transmission and photoluminescence spectroscopy on GaAs layers with thickness ranging from 20nm to 4.8µm reveal the characteristic features of quantized excitonic polaritons as well as of quantized excitons.

3. Experimental

We used for our experiments GaAs-$Al_{0.3}Ga_{0.7}As$ heterostructures grown by molecular beam epitaxy on GaAs substrates. All of our experiments were performed in superfluid helium. For reflectance and transmission experiments we used a tungsten iodine lamp as the excitation source. Photoluminescence was excited with the 752.5nm line of a Kr^+-laser. The reflectance and transmission signals were detected by a single grating monochromator with a (nonintensified) optical multichannel analyzer. The photodiode array was cooled down to -40°C. The spectral resolution was 0.025nm.

The advantage of this system is the high sensitivity to small changes of the signal. This sensitivity is achieved by the long integration time of 20s for a given channel (0.025nm), which also corresponds to the total measuring time. In contrast, a standard photomultiplier set-up reaches the same signal-to-noise-ratio after about 6 hours measuring time!

Such a high spectral sensitivity (0.1% amplitude change!), however, can be only obtained after eliminating the strongly varying sensitivity of the <u>single</u> diodes by dividing the measured spectra by a reference spectrum (usually from an AlGaAs sample in the case of reflectance and a white light spectrum in the case of transmission). Interferences of the cryostat windows can also be eliminated. The absolute value of the reflectivity and the transmission is obtained by a separate measurement with a narrow-line dye laser at 1.513eV.

4. Electric Field Effects at GaAs Surfaces

Figure 1 depicts the schematic energy versus wave vector dispersion curves of the exciton and the photon as well as of their combined eigenstate: the excitonic polariton. Well below the excitonic resonance, only photon-like polariton states exist. Above the excitonic eigenenergies, a photon-like and an exciton-like state are both simultaneously excited. The ratio of both amplitudes is determined by the so-called additional boundary condition /8/. Near the excitonic resonance, both polaritons are equally important for energy transport, and therefore for the optical spectra, whereas above and below the resonance the photon-like polaritons dominate the optical spectra as well as the energy transport /9/.

Since the early work of SELL et al. /1/, optical spectra have been discussed in the framework of the polariton picture. Anomalies in the optical spectra which seemed to be characteristic for the model semicondutor GaAs, like a reflectance spike and a luminescence dip at the longitudinal eigenenergy, have been the subject of extensive investigations and have been interpreted controversially within the polariton concept /1,5,6,10/.

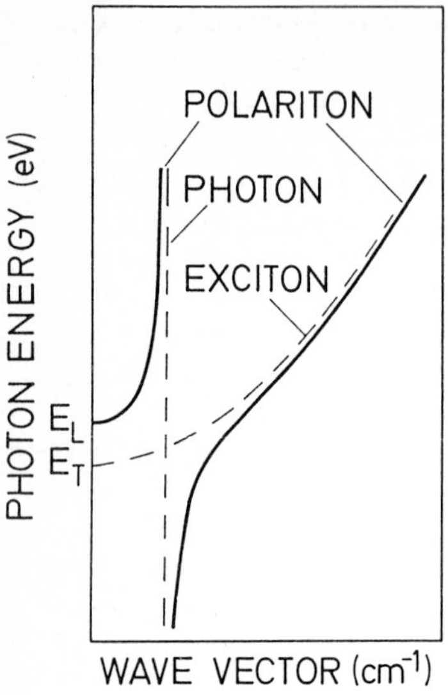

Fig. 1: Schematical plot of the polariton, exciton and photon energy vs wave vector dispersion relation

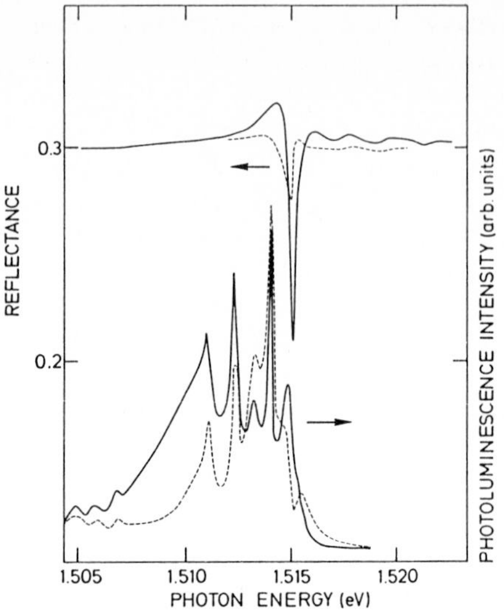

Fig. 2: Reflectance and photoluminescence spectra of a GaAs sample clad (full line) and unclad (dashed line) with a protective AlGaAs layer

Figure 2 shows reflectance and photoluminescence spectra of a GaAs sample, approximately 5µm thick, for two different surface conditions. In the case of cladding the GaAs with a protective 100nm thick AlGaAs layer, we see a strong reflectance structure in the excitonic energy range. The photoluminescence exhibits a single excitonic emission line centered at the excitonic eigenenergy (1.515eV) and additional impurity-related luminescence peaks are observed on the low energy side.

Removing the cladding by etching produces a much weaker reflectance signal with a changed reflectance lineshape. In particular, a small spike at 1.5151eV is seen. This reflectance feature is typical for free GaAs surfaces exposed to air; it indicates the presence of an electric surface field /5/, because of the formation of extrinsic, charged surface states /6/.

The prominent influence of an electric surface field is confirmed by the photoluminescence experiments, depicted in Fig. 2. Whereas the excitonic luminescence of the GaAs sample clad with AlGaAs exhibits a single emission line /7/, a doublet structure with a dip at the longitudinal eigenenergy is observed which is typical for the usual, air-exposed GaAs surfaces /1,6/. This luminescence feature can be explained within the exciton framework by the presence of a narrow (few hundred nm) region near the surface in which excitons can be created but radiatively annihilated because of rapid field ionization /7/. The surface layer simply acts as a narrow-band filter with a transmission characteristic given by the excitonic absorption.

These observations demonstrate that reproducible optical spectra of excitons in GaAs (and probably in other III-V compounds) can only be obtained in a controlled manner on clad, heterostructure samples avoiding exposure to air and subsequent formation of electronic surface states with the corresponding strong electric surface fields.

However, by using clad layers the unambiguous interpretation of the optical (reflectance) spectra is impeded. The reflectance lineshape, for example, strongly depends on the dielectric properties (refractive index and thickness) of the cladding AlGaAs layer which are not known. In order to answer fundamental questions concerning the so-called "additional boundary conditions" /8/ or potentials at the crystal boundary /8/ unambiguously the dielectrics of the AlGaAs layer have to be determined precisely.

5. Quantization of Polaritons and Excitons

The energy spectrum of the excitonic polaritons is a continuous band as seen from Fig. 1. Every wave vector perpendicular to the surface $k_{\perp}$ corresponds to an excitonic polariton state due to the infinity of the bulk semiconductor. A real semiconductor sample (platelet or slab), however, is characterized by two boundaries. This limit allows only eigenstates with wave vectors satisfying the quantum mechanical boundary conditions: $k_{\perp}=N\pi/L_z$, where L_z is the thickness of the slab. In the limit that the electronic properties have not changed very much, the corresponding energy states are simply given by $E= E(k_{\parallel},N\pi/L_z)$, where $E(k)$ denotes the continuous dispersion of the elementary excitation considered. The energy spectrum now is split into discrete states. From another and more practical viewpoint, this quantization is the result of constructive and destructive interference of the plane-wave-type wave functions.

The observability of the discrete energy levels E_i depends on the homogeneous linewidth Γ. Only for the case $\Delta E=E_{i+1}-E_i \geq \Gamma_i$, discrete spectral features can be resolved. This criterion is basically equivalent to the condition that the mean free path exceeds the layer thickness.

The effect of the quantization on the optical properties of the excitons can be classified by two intrinsic lengths: the photon wave length in the medium λ ($\approx$230nm) and the excitonic Bohr-radius a_0 ($\approx$13nm).

(i) $L_z \gg \lambda$

For slab thicknesses in the range of a few µm, the electronic and excitonic properties are slightly changed because the distribution of the discrete levels is quasi-continuous. However, the photon-like part of the polariton is strongly affected. As it can be seen in Fig. 3a for L_z= 3µm, only a few photon-like polariton states are allowed. Thus, a Fabry-Perot-type oscillating interference pattern of the optical spectra well above and below the excitonic resonance should be observed. The energy levels of the quantized excitonic polariton are given by

$$E(k_{\parallel})=\hbar c\sqrt{(k_{\parallel}^2+N^2\pi^2/L_z^2)/\varepsilon(k_{\parallel},N\pi/L_z)}. \qquad (1)$$

(ii) $a_0 \ll L_z \approx \lambda$

For a slab thickness below 1µm (shown in Fig. 3b for L_z=300nm) the excitonic part of the polariton becomes now really quantized. The photon-like states of the polariton are well separated by more than 200meV because the layer thickness is in the same range as the polariton wave length of the photon-like branch. Thus, polariton effects are expected to play a minor role as compared to true exciton effects. (From another viewpoint: they are simply reduced to excitonic effects). The energy levels are characterized by the excitonic dispersion and not by the polariton dispersion:

$$E(k_{\parallel})=\hbar^2(N^2\pi^2/L_z^2+k_{\parallel}^2)/2M, \quad (2)$$

where M is the translational excitonic mass.

In the same intermediate thickness range the quantization already starts to modify the electron and hole wavefunctions. Therefore the optical spectra of excitons (in particular for thinner layers!) are additionally modified due to the change of the electronic states. This may impede an unambiguous assignment to exciton quantization.

(iii) $a_0 \approx L_z \ll \lambda$

A slab thickness of L_z=30nm simplifies the picture: The quantization has completely changed the electronic properties: The valence band is split into a heavy-hole and a light-hole band and the energy separations between the states are typically in the meV range as depicted in Fig. 3c. The excitonic state is now composed from energetically well separated, quantized electron and hole states and the binding energy is increased. The usual hydrogenic exciton energy series is lost because the exciton becomes two-dimensional. The energy states are rather given by

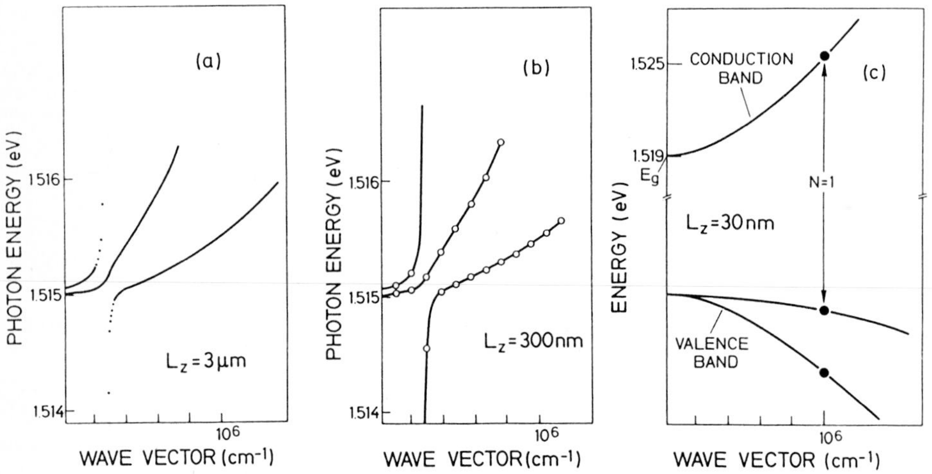

Fig. 3: Schematical energy vs wave vector dispersion relation for polaritons in a slab of L_z=3µm (a), polaritons in a slab of L_z=300nm (b) and electrons and holes in a quantum well of L_z=30nm (c)

$$E(k_{\|}) = \hbar^2 (N^2 \pi^2 / L_z^2 + k_{\|}^2)/2\mu, \tag{3}$$

where μ is the reduced mass.

In Fig. 4 we show reflectance and photoluminescence spectra of an unclad GaAs sample with a thickness of 4.8μm. At energies well below and well above the excitonic eigenenergy, an oscillating reflectance pattern is seen, having its origin in the interference of polarization waves (quantization of polaritons). Near excitonic resonance, the reflectance structure is basically unchanged as expected for $\Delta E < \Gamma \approx 0.2$meV /11/. The small spike at 1.5151eV indicates the presence of an electric surface field typical for such unclad samples which mainly affects the resonance region /1,5/. This assignment is confirmed by the luminescence spectrum. The typical field-induced luminescence doublet structure is observed with a dip centered at the longitudinal eigenenergy /1,7/.

Figure 5 depicts reflectance and photoluminescence spectra of a GaAs-AlGaAs heterostructure sample consisting of two GaAs layers of 450nm thickness separated by a central AlGaAs layer of 7nm thickness and clad with an AlGaAs layer of 100nm thickness. The polariton is only slightly affected by the 7nm thick AlGaAs barrier and the effective thickness for the polariton quantization is 900nm whereas the excitonic wave functions are quantized in each layer separately.

The reflectance spectrum of the thin GaAs layer is strongly altered near the resonance. New, sharp structures at the excitonic resonance arise due to the discrete excitonic states /12/. A sharp photoluminescence peak at 1.515eV, the transverse excitonic eigenenergy, is seen, indicating the characteristic intrinsic excitonic polariton emission.

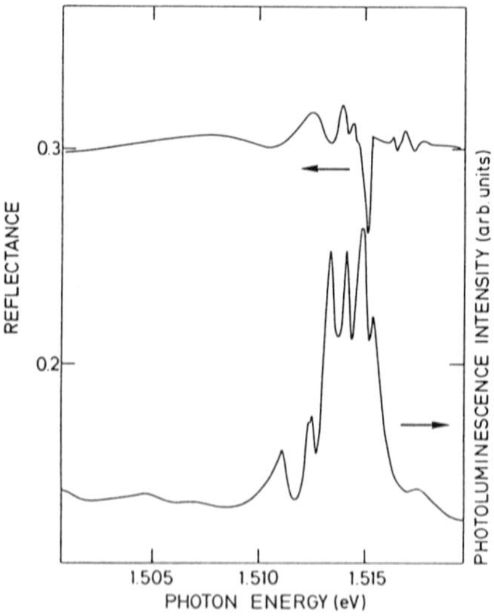

Fig. 4: Reflectance and photoluminescence spectra of a GaAs layer of 4.8μm thickness

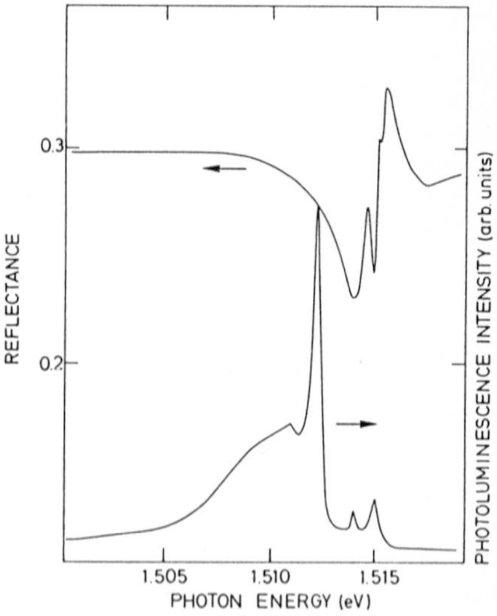

Fig. 5: Reflectance and photoluminescence spectra of a GaAs-AlGaAs heterostructure with two GaAs layers of 450nm thickness

In Fig. 6 and 7 we depict reflectance, transmission and photoluminescence spectra of the samples with 194nm (130nm) and 27.7nm (21.4nm) separated by AlGaAs barrier layers of 14nm (24nm).

The reflectance spectra of the two optically thin GaAs layers of 194nm and 130nm thickness show sharp structures above the 1s exciton resonance located at 1.515eV. Transmission spectra confirm the assignment to the discrete 1s exciton levels belonging to the quantized electronic subbands of index N:

$$E=E_0 + E_c*N^2, \tag{4}$$

with E_0= 1.51483eV (1.51462eV) for L_z=194nm (130nm) and the confinement energy E_c calculated to E_c=0.17meV (0.38meV) by using an electron mass of $0.066m_0$ and a hole mass of $0.45m_0$.

The luminescence spectra of the two optically thin GaAs layers exhibit a single narrow emission line centered at 1.515eV, (the excitonic eigenenergy of the bulk /1/) and in close vicinity to the transmission minima. This confirms that the exciton in such thin (100nm-200nm) GaAs layers can be described as a two-level system with highly degenerate energy levels. In addition, the binding energy has not changed very much (E_0 is close to the bulk value of 1.515eV) and the exciton has retained its 3D properties.

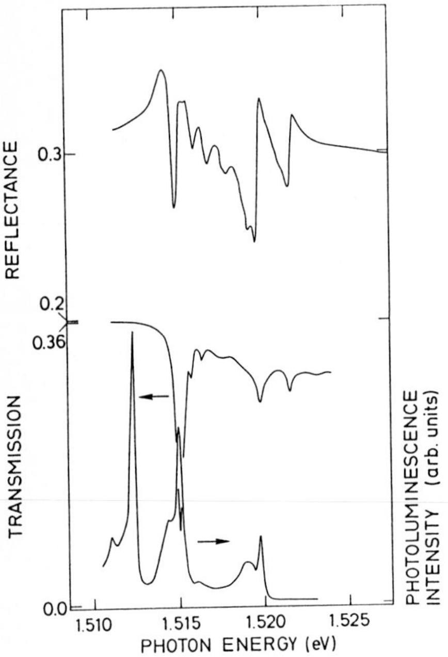

Fig. 6: Reflectance, transmission and photoluminescence spectra of a sample with 194nm and 27.7nm GaAs layers separated by 14nm AlGaAs

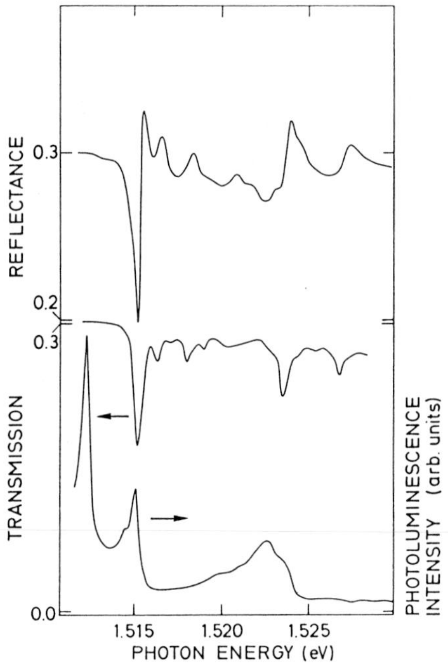

Fig. 7: Reflectance, transmission and photoluminescence spectra of a sample with 130nm and 21.4nm thick GaAs layers separated by 24nm AlGaAs

In contrast to the optically thin GaAs layers, the excitonic reflectance and transmission features of the two GaAs quantum well (QW) layers are considerably shifted toward the blue. In addition, the exciton line is split because the valence band degeneracy is lifted. The confinement energies E_c of the 2D excitons are calculated as

$$E_c^{hh}=8.3\text{meV}, \qquad E_c^{lh}= 12.9\text{meV} \quad \text{for } L_z = 27.7\text{nm},$$

$$E_c^{hh}=14.0\text{meV}, \qquad E_c^{lh}= 21.6\text{meV} \quad \text{for } L_z = 21.4\text{nm},$$

where a heavy-hole (hh) mass of $0.45m_0$ and a light-hole (lh) mass of $0.088m_0$ are used.

The predicted shifts of the excitonic transitions agree roughly with the experiments. A detailed analysis incorporating the increase of the excitonic binding energy with decreasing layer thickness should give a much better agreement.

The luminescence spectra exhibit strong emission lines due to 2D exciton transitions which coincide with the transmission minima. Thus, the 2D exciton can be described simply as a two-level system. For resonant excitation as well as for emission there is just one energy level on a large area ($\approx 3*10^{-4}\text{cm}^2$), and not an inhomogeneous distribution of local exciton energy levels, indicating the superior quality of our samples.

6. Conclusions

We have investigated the optical properties of excitons at GaAs surfaces and in thin GaAs layers. We find a strong electric surface field causing anomalies in reflectance as well as in photoluminescence spectra on free, oxygen-exposed surfaces and truly intrinsic optical features for GaAs surfaces clad with AlGaAs. Reflectance studies on GaAs slabs in the thickness range of a few µm reveal interference of excitonic polaritons. For much thinner GaAs layers quantization of excitons is observed by means of reflectance, transmission, and photoluminescence experiments. Our studies demonstrate that the use of high-quality (high-purity, highly parallel) GaAs-AlGaAs heterostructures with excellent interfaces not only allows to discriminate extrinsic effects but is also essential for optical study of the various excitonic quantization effects.

7. Acknowledgments

We gratefully acknowledge helpful discussions with J. Kuhl. We thank H.J. Queisser for critically reading the manuscript and H. Oppolzer, Siemens AG, Munich for determining the geometrical sample parameters.

8. References

1. D.D. Sell, S.E. Stokowski, R. Dingle, and J.V. DiLorenzo, Phys.Rev.B7, 4568 (1971)
2. D.E. Hill, Phys.Rev.B1, 1863 (1970)

3. R.G. Ulbrich and C. Weisbuch, Phys.Rev.Lett.38, 865 (1977)
4. F. Evangelisti, A. Frova, and F. Patella, Phys.Rev.B10, 4253 (1974)
5. L. Schultheis and I. Balslev, Phys.Rev.B28, 2292 (1983)
6. B. Fischer and H.J. Stolz, Appl.Phys.Lett.40, 56 (1982)
7. L. Schultheis and C.W. Tu, Phys.Rev.B32, 6978 (1985)
8. J.J. Hopfield and D.G. Thomas, Phys.Rev.132, 563 (1963)
9. R.G. Ulbrich and G.W. Fehrenbach, Phys.Rev.Lett.43, 963 (1979)
10. C. Weisbuch and R.G. Ulbrich, Phys.Rev.Lett.39, 654 (1977)
11. L. Schultheis, J. Kuhl, A. Honold, and C.W. Tu, Phys.Rev.Lett.57, 1797 (1986)
12. L. Schultheis and K. Ploog, Phys.Rev.B29, 7058 (1984)

Part III

Excitons in Superlattices and in Quantum Wells

Key Aspects of Molecular Beam Epitaxy and Properties of $Ga_{0.47}In_{0.53}As/Al_{0.48}In_{0.52}As$ Superlattices

K. Ploog

Max-Planck-Institut für Festkörperforschung,
Heisenbergstr. 1, D-7000 Stuttgart 80, Fed. Rep. of Germany

1. INTRODUCTION

The recent advances in the crystal growth technique of molecular beam epitaxy (MBE) allow the engineering of semiconductors on an atomic scale / 1 / . The development of sophisticated artificially layered structures made of III-V compound semiconductors has opened up new horizons in semiconductor research / 2 / . The interfaces between ultrathin epitaxial layers of different composition are used to confine electrons (and holes) to two-dimensional (2D) motion (electrical and optical confinement). Excitons play a more significant role in the quasi-2D systems than in the corresponding bulk material / 3 / . In this review paper we first briefly outline the key aspects of MBE of III-V semiconductors. Then we discuss the structural and optical properties of $Ga_{0.47}In_{0.53}As/Al_{0.48}In_{0.52}As$ superlattices (SL) and multi-quantum-well heterostructures (MQWH) lattice-matched to InP, which emit in the wavelength range $1.3 < \lambda < 1.6$ µm. The ternary materials $Ga_xIn_{1-x}As$ and $Al_xIn_{1-x}As$ are thus important for application in optoelectronic devices used in fiber communication systems. The growth and the optical properties of $Ga_xIn_{1-x}As/Al_xIn_{1-x}As$ SL and MQWH are strongly affected by the ternary nature of this materials system. We show that the recent improvements in the quality of MBE grown material now allows the investigation of intrinsic electronic properties of $Ga_{0.47}In_{0.53}As/Al_{0.48}In_{0.52}As$ superlattices.

2. MBE GROWTH OF III-V SEMICONDUCTORS

The unique capability of MBE to create a wide variety of complex compositional and doping profiles in semiconductors arises from the conceptual simplicity of the growth process, which is schematically shown in Fig. 1. The particular merits of MBE are that thin films can be grown with precise control over thickness, alloy composition, and doping level / 4 - 6 / . The basic process consists of a co-evaporation of the constituent elements of the epitaxial layer (Al, Ga, In, P, As, Sb) and dopants (mainly Si for n-type and Be for p-type doping) onto a heated crystalline substrate where they react chemically under ultra-high vacuum (UHV) conditions. The composition of the layer and its doping level depend on the relative arrival rate of the constituent elements which in turn depends on the evaporation rate of the appropriate sources. Accurately controlled temperatures (to within $\pm$ 0.1^{o} at

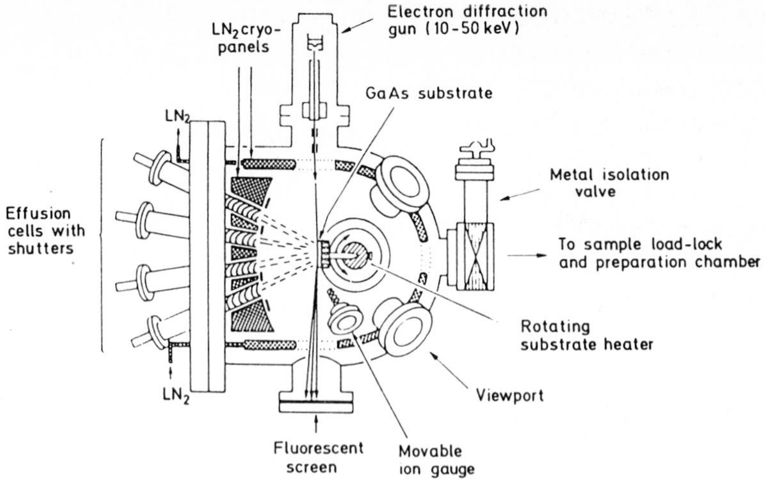

Fig. 1 Schematic cross-section (top view) of MBE growth chamber illustrating the basic evaporation process under UHV conditions (i.e. base pressure of 2×10^{-11} Torr).

100 ^{o}C) have thus a direct, controllable effect upon the growth process. The group-III-elements are always supplied as monomers by evaporation from the respective liquid element and have a unity sticking coefficient over most of the substrate temperature range used for film growth (e.g. 500 - 650 ^{o}C for GaAs). The group-V-elements, on the other hand, can be supplied as tetramers (P_4, As_4, Sb_4) by sublimation from the respective solid element or as dimers (P_2As_2, Sb_2) by dissociating the tetrameric molecules in a two-zone furnace. The growth rate of typically 0.5 - 1.5 µm/hr is chosen low enough that migration of the impinging species on the growing surface to the appropriate lattice sites is ensured without incorporating crystalline defects. Simple mechanical shutters in front of the evaporation sources are used to interrupt the beam fluxes to start and stop deposition and doping. Due to the slow growth rate of 1 monolayer/s, changes in composition and doping can thus be abrupt on an atomic scale. The transmission electron (TEM) micrograph of a GaAs/AlAs superlattice displayed in Fig. 2 demonstrates that this independent and accurate control of the individual beam sources allows the precise fabrication of artificially layered semiconductor structures on an atomic scale.

The stoichiometry of most III-V semiconductors during MBE growth is selfregulating as long as excess group-V-element molecules are impinging on the growing surface. The excess group-V-species do not stick on the heated substrate surface, and the growth rate is essentially determined by the arrival rates of the group-III-elements. A good compositional control of III-III-V alloy films can be achieved by supplying excess group-V-species and adjusting the flux densities of the impinging group-III-beams, as long as the substrate temperature is kept below the congruent evaporation

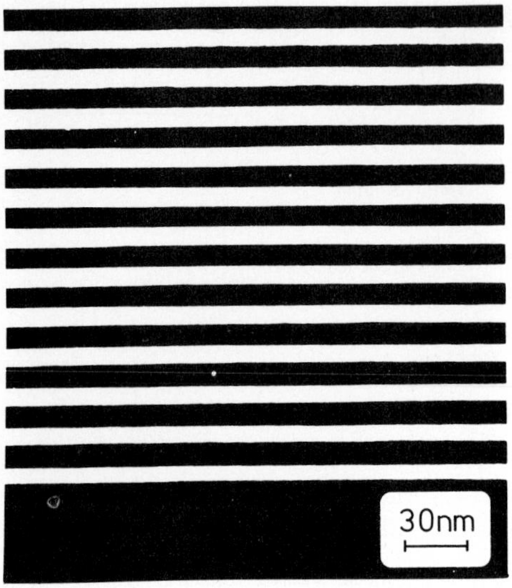

Fig. 2 (110) cross-sectional TEM of a GaAs/AlAs superlattice.

limit of the less stable of the constituent binary III-V compounds (e.g. GaAs in the case of $Al_xGa_{1-x}As$). At higher growth temperatures, however, preferential desorption of the more volatile group-III-element (i.e. Ga from $Al_xGa_{1-x}As$) occurs so that the final film composition is not only determined by the added flux ratios but also by the differences in the desorption rates. To a first approximation we can estimate the loss rate of the group-III-elements from their vapour pressure data. This assumption is reasonable because the vapour pressure of the element over the compounds, i.e. Ga over GaAs, is similar to the vapour pressure of the element over itself. The results are summarized in Table 1. The surface of alloys grown at high temperatures will thus be enriched in the less volatile group-III-element. As a consequence, we expect a loss of In in $Ga_xIn_{1-x}As$ films grown above 550 ^oC and a loss of Ga in $Al_xGa_{1-x}As$ films grown above 650 ^oC. An independent detailed calibration based on measured film composition is thus recommended for accurate adjustments of the effusion cell temperatures. The growth of ternary III-V-V alloy films

TABLE 1 Approximate loss rate of group-III-elements in monolayer per second estimated from vapour pressure data

Temperature (^oC)	Al	Ga	In
550	-	-	0.03
600	-	-	0.3
650	-	0.06	1.4
700	-	0.4	8
750	0.05	2	30

(e.g. GaP_yAs_{1-y}) by MBE is more complicated, because even at moderate substrate temperatures the relative amounts of the group-V-element incorporated into the growing film are not simply proportional to their relative arrival rates.

The simplicity of the MBE process allows composition control from x = 0 to x = 1 in $Al_xGa_{1-x}As$, $Ga_xIn_{1-x}As$ etc. with a precision of $\pm$ 0.001 and doping control, both n- and p-type, from the 10^{14} cm^{-3} to the 10^{19} cm^{-3} range with a precision of a few percent. The accuracy is largely determined by the care with which the growth rate and doping level were previously calibrated in test layers. The most common dopants used during MBE growth are Be for p-type doping and Si for n-type doping. Be behaves as an almost ideal shallow acceptor in MBE grown III-V compounds for doping concentration up to 1×10^{19} cm^{-3}. Each incident Be atom produces one ionized impurity species providing an acceptor level 29 meV above the valence band edge in GaAs. The group-IV-element Si is primarily incorporated on Ga sites during MBE growth under arsenic-stabilized conditions, yielding n-type material of low compensation. Up to $n = 1 \times 10^{19}$ cm^{-3} the observed doping concentration is simply proportional to the dopant arrival rate provided care is taken to minimize the H_2O and CO level during growth.

Advanced MBE systems consist of three basic UHV building blocks (the growth chamber, the sample preparation chamber, and the load-lock chamber) which are separately pumped and interconnected via large diameter channels and isolation valves. High-quality layered semiconductor structures require background vacuums in the low 10^{-11} Torr range to avoid incorporation of impurities into the growing layers. Therefore, extensive LN_2 cryoshrouds are used around the substrate to achieve locally much lower background pressures of condensible species. For MBE growth of III-V semiconductors the starting materials are evaporated in resistively heated effusion cells made of pyrolytic BN which operate at temperatures up to 1400 ^oC.

In general MBE growth of III-V semiconductors is performed on (001) oriented substrate slices about 300 - 500 µm thick. The preparation of the growth face of the substrate from the polishing stage to the in-situ cleaning stage in the MBE system is of crucial importance for epitaxial growth of ultrathin layers and heterostructures with high purity and crystal perfection and with accurately controlled interfaces on an atomic scale. The substrate surface should be free of crystallographic defects and clean on an atomic scale with less than 0.01 monolayer of impurities. Various cleaning methods have been described for deposition of III-V semiconductors / 5 - 8 / . The first step always involves chemical etching, which leaves the surface covered with some kind of a protective oxide. After insertion in the MBE system this oxide is removed by heating. This heating must be carried out in a beam of arsenic or phosphorus.

The most important method to monitor in-situ surface crystallography and kinetics during MBE growth is reflection high energy electron diffraction (RHEED) operated at 10 - 50 KeV in the small glancing angle reflection mode. The diffraction pattern on the fluorescent screen contains information from the topmost nanometer of the deposited material that can be related to the topography and structure of the growing surface. The specific surface reconstruction can be identified and correlated to the surface stoichiometry which is an important growth parameter. In addition, the temporal intensity oscillations observed in the features of the RHEED pattern are used to study MBE growth dynamics and the formation of heterointerfaces in multilayered structures / 9, 10 / . The periodic intensity oscillations in the specularly reflected beam of the RHEED pattern shown in Fig. 3 provide direct evidence that MBE growth occurs predominantly in a two-dimensional (2D) layer-by-layer growth mode. The period of the intensity oscillations corresponds exactly to the time required to grow a monolayer of GaAs (i.e. a complete layer of Ga plus a complete layer of As), AlAs, or $Al_xGa_{1-x}As$. To a first approximation we can assume that the oscillation amplitude reaches its maximum when the monolayer is completed (maximum reflection). Although the fundamental principles underlying the damping of the amplitude of the oscillations are not completely understood, the method is now widely used to monitor and to calibrate absolute growth rates in real time with monolayer resolution.

The oscillatory nature of the RHEED intensities provides direct real-time evidence of compositional effects and growth modes during interface formation. As for the widely used $GaAs/Al_xGa_{1-x}As$ heterointerface, the sequence of layer growth is critical for compositional gradients and crystal perfection, which in turn is important for optimizing 2D transport properties. When the Al flux is switched at the maximum of the intensity oscillations, the first period for the growth sequence from ternary alloy to binary compound corresponds neither to the $Al_xGa_{1-x}As$ growth rate nor to the steady-state GaAs rate, but shows some intermediate value. For the growth sequence from binary compound to ternary alloy or between the two binaries

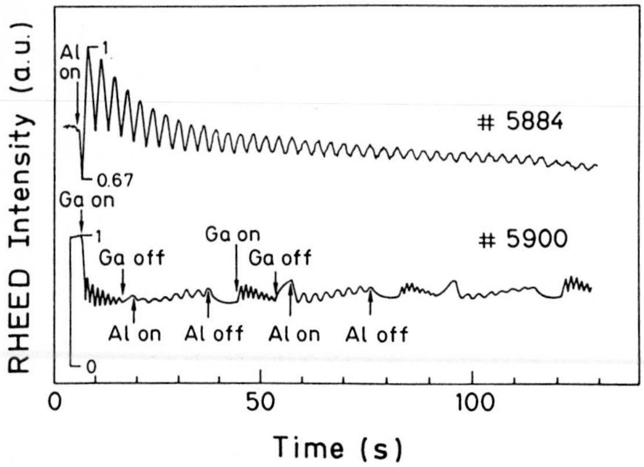

Fig. 3 RHEED intensity oscillations of specularly reflected beam from (001) surface in [100] azimuth during growth of GaAs/AlAs heterojunctions

an intermediate period does not exist. A possible explanation for this phenomenon can be found in the relative surface diffusion lengths of the group-III-elements Al and Ga, which were estimated to be $\lambda_{Al} \cong 3.5$ nm and $\lambda_{Ga} \cong 20$ nm on (001) surfaces under typical MBE growth conditions / 10 / . These differences in cation diffusion rates have striking consequences on the nature of the interface. While a GaAs layer should be covered by smooth terraces of 20 nm average length between monolayer steps, those on an $Al_xGa_{1-x}As$ layer would be only 3.5 nm apart. The important result of this qualitative estimate is that the GaAs/$Al_xGa_{1-x}As$ interface is much smoother on an atomic scale than the inverted structure. Direct experimental evidence for this distinct difference in binary-to-ternary layer growth sequence is obtained from the high-resolution TEM investigations of Suzuki and Okamoto / 11 / . Their lattice image of a $Al_{0.2}Ga_{0.8}As$/AlAs superlattice shows clearly that the heterointerface is abrupt to within one atomic layer only when the ternary alloy is grown on the binary compound but not for the inverse growth sequence. Since the nature of the heterointerface is critical for optimising excitonic as well as transport properties in quantum wells, various attempts have been made to minimize the interface roughness (or disorder) by modified MBE growth conditions. The most successful modification is probably the method of growth interruption at each interface. Growth interruption allows the small terraces to relax into larger terraces via diffusion of the surface atoms. This reduces the step density and thus simultaneously enhances the RHEED specular beam intensity which can be used for real-time monitoring. The time of closing both the Al and the Ga shutter (while the As shutter is left open) depends on the actual growth condition. Values ranging from a few seconds to several minutes have been reported / 12, 13 / .

3. STRUCTURAL AND OPTICAL PROPERTIES OF $Ga_xIn_{1-x}As/Al_xIn_{1-x}As$ SUPERLATTICES AND MULTI-QUANTUM-WELL HETEROSTRUCTURES

3.1 Structural Properties

The examination of the structural properties of $Ga_xIn_{1-x}As/Al_xIn_{1-x}As$ SL and MQWH is extremely important for the investigation of their optical properties because they are lattice matched to the InP substrate only for a given composition and because of the large differences of the lattice constant and bandgap for the constituent binary compounds AlAs, GaAs, and InAs (Fig. 4). High-angle X-ray diffraction is a very useful non-destructive technique to determine the strain profile, the composition (periodicity), and the interface quality of $Ga_xIn_{1-x}As$/ $Al_xIn_{1-x}As$ MQW and SL, if a detailed analysis of the experimental diffraction curve is performed / 14 - 16 / . In Fig. 5a we show the X-ray diffraction pattern of a $Ga_{0.47}In_{0.53}As/Ga_{0.48}In_{0.52}As$ SL taken in the vicinity of the (004) reflection of InP. The average lattice mismatch of the SL to the (001) substrate surface is obtained from the angular spacing between the substrate peak and the main epilayer

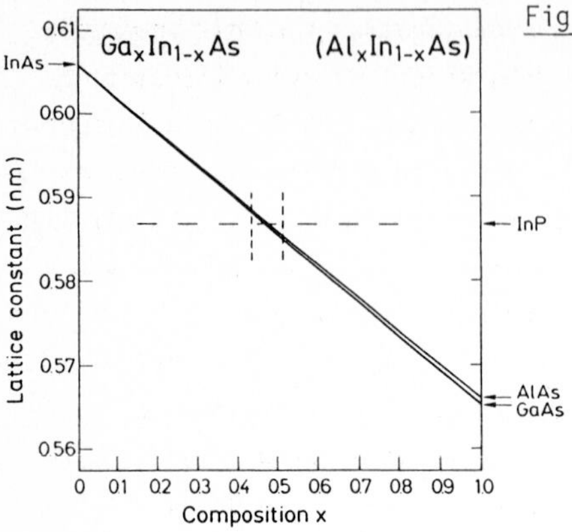

Fig. 4 Variation of lattice constants of $Ga_xIn_{1-x}As$ and $Al_xIn_{1-x}As$ as a function of composition.

peak "0" by applying Bragg's equation. For the samples of the present study this lattice mismatch normal to the (001) growth surface for the tetragonally distorted epilayers on InP is less than / 2.8×10^{-3} / , i.e. smaller than in the GaAs/AlAs system. The angular separation $\Delta\Theta^{\pm n}$ between the main epitaxial peak and the satellite peaks ($\pm n$) is given by the period length T of the SL, according to $T = n\lambda/(2\Delta\Theta^{\pm n}$ $\cos\Theta_B$, where λ is the X-ray wavelength and Θ_B the kinematic Bragg angle. The Pendellösung fringes observed between the main diffraction peak and the satellite peaks "-1" and "+1" of Fig. 5a demonstrate the excellent thickness and composition homogeneity normal and parallel to the crystal surface.

For a quantitative description of the structure of superlattices composed of ternary compounds a comparison between the theoretical diffraction pattern and experimental data is required. We have used a semikinematical approach of the dynamical theory of X-ray diffraction for distorted crystals to calculate the theoretical diffraction pattern / 15 / . In Fig. 5b we show a theoretical diffraction pattern for a perfectly lattice-matched $Ga_{0.468}In_{0.532}As/Al_{0.477}In_{0.523}As$ SL with $L_Z = L_B = 10.6$ nm. It should be noted that the satellite peaks "-1" and "+1" have almost disappeared. This result is in contrast to that observed in strained-layer SL where

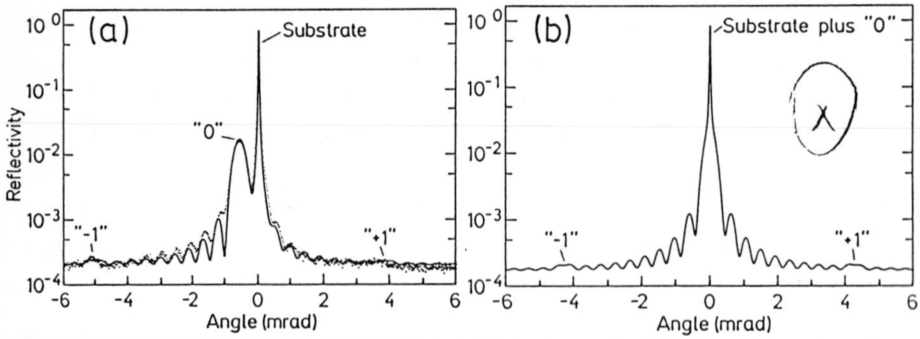

Fig. 5 (a) $CuK\alpha_1$ (004) diffraction pattern of $Ga_xIn_{1-x}As/Al_xIn_{1-x}As$ SL on (001) InP with $L_Z = L_B = 10.2$ nm (... exp., —— theor.); (b) Theoretical diffraction pattern of a perfectly lattice-matched superlattice on (001) InP.

a strain periodicity produces strong satellite peaks. The low-intensity satellite peaks and the Pendellösung fringes in Fig. 5b are caused only by the periodicity of the structure factors in the SL. However, if the lattice strains of the constituent $Ga_xIn_{1-x}As$ and $Al_xIn_{1-x}As$ layers are of the same magnitude, also a weak intensity of the satellite peaks is observed.

3.2 Optical Properties

Until recently, intrinsic excitonic emission, which dominates the low-temperature photoluminescence (PL) spectra of $GaAs/Al_xGa_{1-x}As$ SL, has been difficult to identify in the PL spectra of $Ga_{0.47}In_{0.53}As/Al_{0.48}In_{0.52}As$ SL / 17, 18 / . In Fig. 6 we show the low-temperature PL spectra obtained from bulk-like $Ga_{0.47}In_{0.53}As$ and $Al_{0.48}In_{0.52}As$ layers and from four SL with different well widths L_Z. With decreasing well width we observe a pronounced shift of the quantum well luminescence to higher energies. As the quality of the $Al_xIn_{1-x}As$ barrier material is critical, we have investigated the effect of the barrier configuration on the edge luminescence of $Ga_{0.47}In_{0.53}As$ quantum wells (QW) / 19 / . The $Ga_{0.47}In_{0.53}As$ QW are confined either by the ternary $Al_{0.48}In_{0.52}As$ alloy or by a short-period superlattice (SPS) composed of alternating 2.5 nm thick $Ga_{0.47}In_{0.53}As$ and $Al_{0.48}In_{0.52}As$ layers. The PL and PLE (photoluminescence excitation) spectra shown in Fig. 7 clearly demonstrate the superior quality of the SPS confined QW. The E_{1h} excitonic peak in the PLE spectrum is narrower (3.1 meV HWHM) and the Stokes shift of the PL peak relative to the E_{1h} PLE resonance is reduced from 11.3 meV to 6.5 meV in the SPS confined $Ga_{0.47}In_{0.53}As$ QW.

The emission peak of the SPS confined $Ga_{0.47}In_{0.53}As$ MQWH does not shift in energy when the excitation density of the incident laser is increased by three orders of magnitude (Fig. 8b) The low-energy shoulder of the luminescence line, attributed to a band-to-acceptor transition, saturates with enhanced excitation

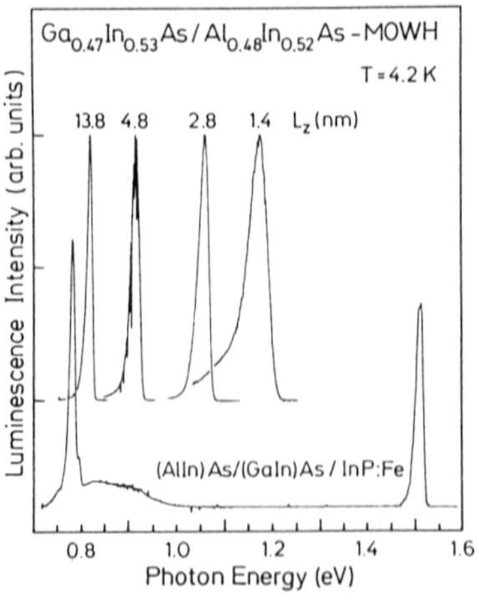

Fig. 6 PL spectra of thick bulk-like $Ga_{0.47}In_{0.53}As$ and $Al_{0.48}In_{0.52}As$ layers (lower trace) and of $Ga_{0.47}In_{0.53}As/Al_{0.48}In_{0.52}As$ SL of different well widths L_Z (upper trace).

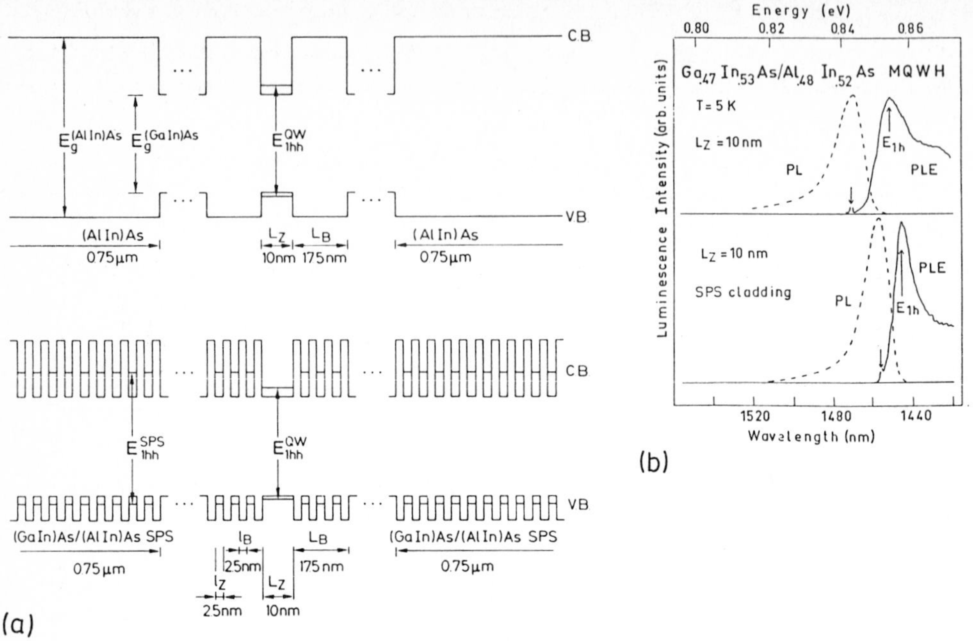

Fig. 7 $Ga_{0.47}In_{0.53}As$ quantum wells confined by ternary $Al_{0.48}In_{0.52}As$ alloy (top) or by $Ga_{0.47}In_{0.53}As/Al_{0.48}In_{0.52}As$ SPS (bottom). (a) Schematic real-space energy band diagram and (b) PL and PLE spectra obtained at 5 K.

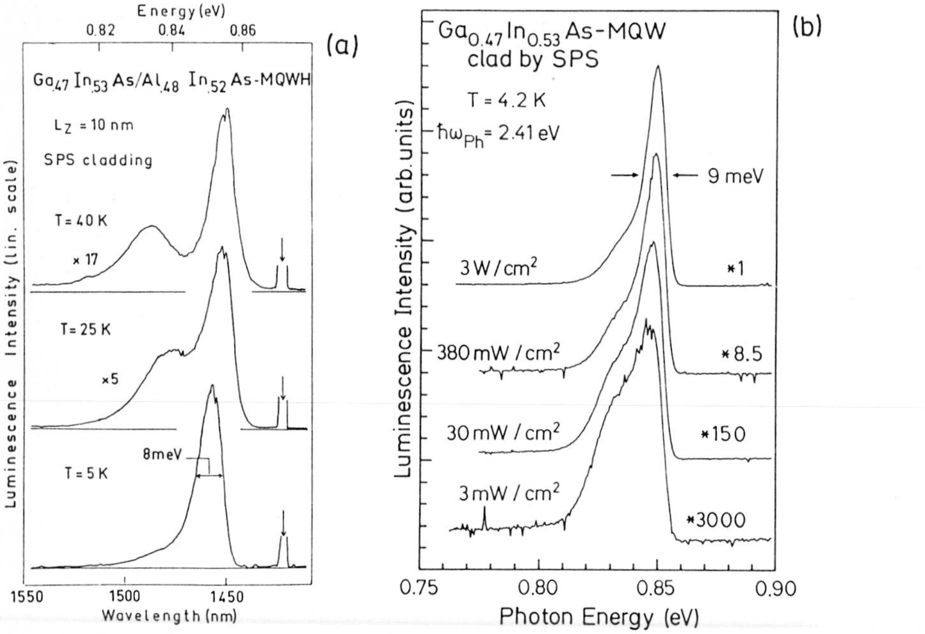

Fig. 8 $Ga_{0.47}In_{0.53}As$ quantum wells confined by $Al_{0.48}In_{0.52}As/Ga_{0.47}In_{0.58}As$ SPS obtained (a) at different sample temperatures and (b) at different excitation densities.

density. In the whole temperature range $5 < T < 40$ K the narrow (8meV FWHM) emission peak located 6.5 meV below the E_{1h} PLE peak (at 5 K) is the dominant feature in the PL spectrum (Fig. 8a). We attribute this emission line to intrinsic excitonic recombination. The intensity of the band-to-acceptor recombination appearing in the temperature range $15 < T < 45$ K is considerably lower. When the sample temperature is increased from 5 to 50 K the excitonic emission shifts slightly to higher energy and the Stokes shift between PL and PLE spectra decreases to 1.5 meV at 50 K. The observed Stokes shift / 20 / of 6.5 meV at 5 K is caused partly by statistical alloy fluctuations of the well material / 21 / , which account for about 3 - 4 meV, and partly by monolayer fluctuations of the well width, which also accounts for about 3 - 4 meV using $\Delta L_Z = 0.3$ nm for $L_Z = 10$ nm. The improvement of the optical properties of SPS confined $Ga_{0.47}In_{0.53}As$ QW is due (i) to a removal of substrate defects by the SPS layers / 22 / , (ii) to an amelioration of the interface between well and barrier, and (iii) to a modification of the dynamics of injected carriers in the SPS barrier / 23, 24 / . The first two items are based on strain modulation in the SPS which has not yet been evaluated quantitatively.

The features of free-exciton absorption in $Ga_{0.47}In_{0.53}As/Al_{0.48}In_{0.52}As$ SL and MQW can be observed also at room temperature / 25 / . In Fig. 9a we show the low-temperature absorption spectra for three samples with different well widths / 26 / . The step-like behaviour of the absorption coefficient and thus of the density of states in these quasi-2D structures is clearly observed. The strong resonances at the subband edges are due to discrete excitonic transitions. In Fig. 9b the energy shift of the subband transitions as a function of well width is compared with theoretical calculations based on two different models / 26 / . The dashed line results from the single band envelope function approximation (EFA) and the full line from the coupled 6-band EFA. For the lowest subband transitions the results of both models agree to within 5 - 10 meV and are thus represented by common dashed lines. For the higher transitions a reduction in energy occurs when the nonparabolicity of the conduction and light-hole valence band is taken into account. This gives a better agreement between theory and experiment.

The application of a magnetic field B normal to the well layers, i.e. in z-direction, leads to a complete quantization of the electron and hole subbands into Landau levels. The allowed optical transitions (ΔN) between the electron and heavy-hole Landau levels have a transition energy of $\Delta E_n = E_{1h} + (N + 1/2)\,\hbar eB/\mu_{1h}$, where μ_{1h} is the reduced effective mass of electron and heavy hole. In the perpendicular magnetic field the exciton continuum splits into discrete excited excitonic states. / 27 / . In Fig. 10 we show that the optical transitions between these states experience the expected linear energy shift with magnetic field. The discrete ground-state exciton transition, on the other hand, shows a weaker energy dependence on the magnetic field as long as $\hbar\omega_c/2B_{1h} \ll 1$, where ω_c is the cyclotron frequency

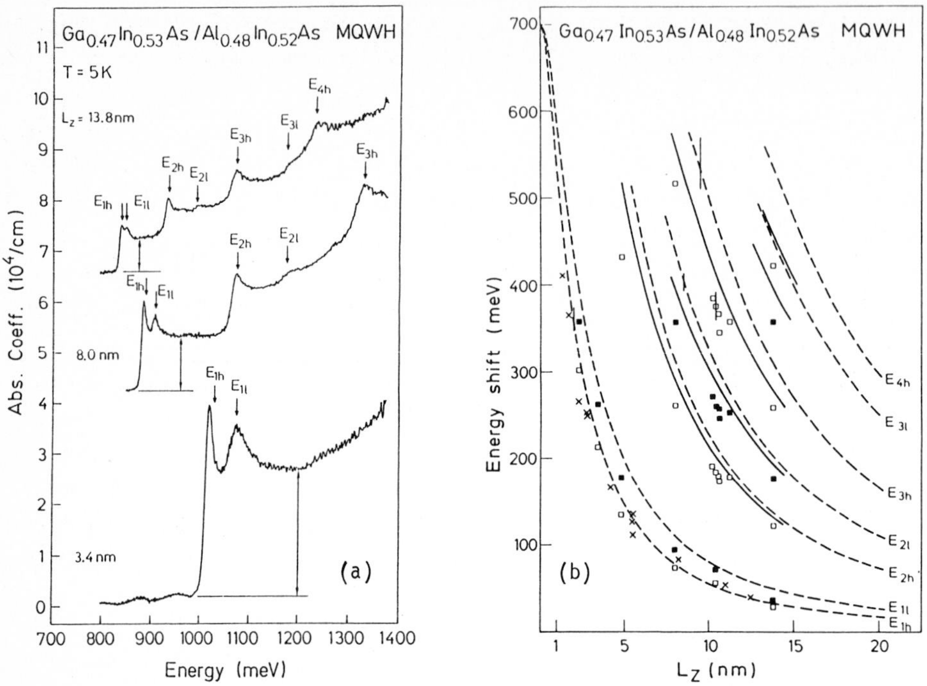

Fig. 9 (a) Absorption spectra of three $Ga_{0.47}In_{0.53}As/Al_{0.48}In_{0.52}As$ MQWH with different well widths L_Z at 5 K. The arrows indicate measured subband transitions, taking into account the exciton binding energy.

(b) Subband transitions with respect to the $Ga_{0.47}In_{0.53}As$ band gap as a function of well width. Open squares (≘ electron-heavy hole) and full squares (≘ electron-light hole) are derived from absorption measurements and crosses from PL measurements. The results of two theoretical models are indicated by the dashed and full lines.

and B_{1h} the electron heavy-hole exciton binding energy. The dependence of the bound and the continuum states on the magnetic field can thus be detected simultaneously. From these data we can directly determine the band edge and the exciton ground state energy and thereby the exciton binding energy B_{1h} of the respective SL / 27 / . For the SL with 8 nm well width we obtain a value of $B_{1h} = 7.5 \pm 2$ meV. From the slope of energy of the continuum transitions with magnetic field we can deduce the reduced effective mass / 26 / . In Fig. 11 we show the observed variation of the exciton binding energy as a function of well width. The data were obtained from the magneto absorption measurements and from a detailed analysis of the absorption measurements using a 2D model / 26 / . Within the error bars these two methods yield consistent results for the exciton binding energies. The full line indicates the theoretical exciton binding energies obtained by a variational calculation in an infinite potential well. This assumption of an infinite barrier height overestimates the value of

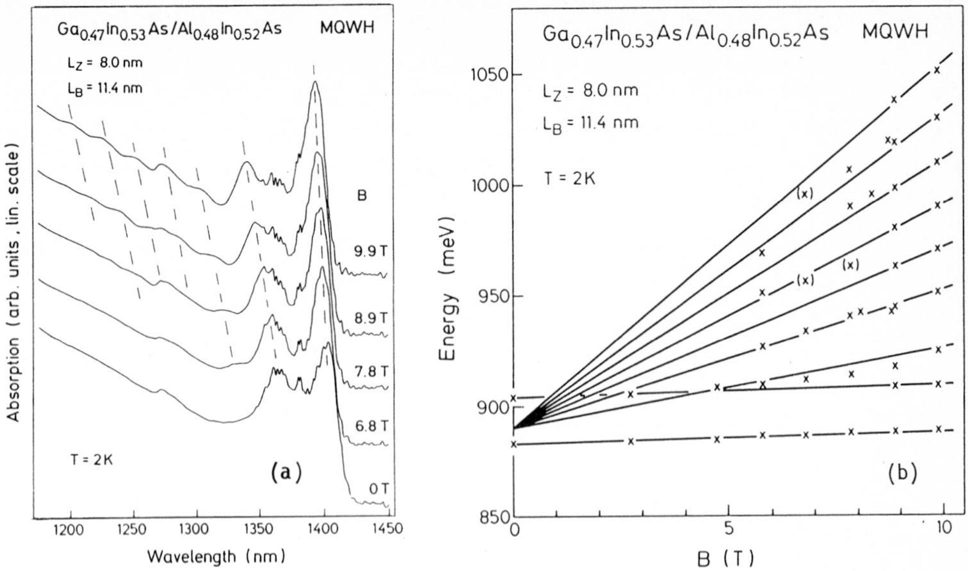

Fig. 10 (a) Absorption spectra of a 8-mm $Ga_{0.47}In_{0.53}As/Al_{0.48}In_{0.52}As$ SL at different magnetic fields. The dashed lines indicate the evolution of the absorption features with magnetic field. (b) Landau level fan for the same SL as a function of the magnetic field.

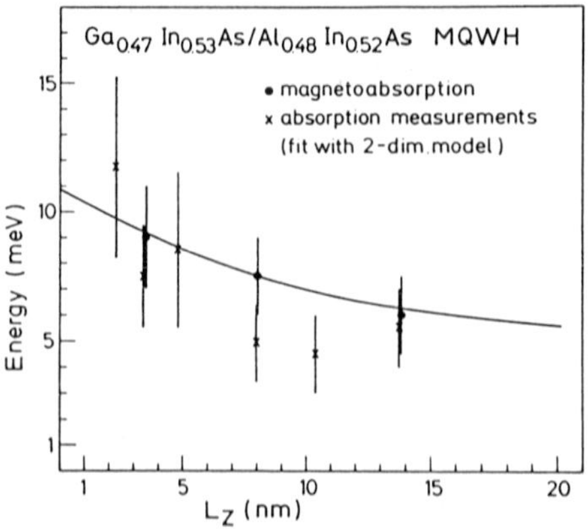

Fig. 11 Electron heavy-hole exciton binding energy B_{1h} in $Ga_{0.47}In_{0.53}As/Al_{0.48}In_{0.52}As$ SL as a function of well width L_Z. The full line represents the result of a variational calculation of B_{1h} assuming an infinite barrier height.

B_{1h}. A value of R_y = 2.7 meV was taken for the 3D Rydberg energy. The large values of the measured exciton binding energies as compared to the overestimated theoretical data manifest an increase of the carrier masses due to nonparabolicity of conduction and valence band, respectively.

Acknowledgements

The active contribution of J. Knecht, J.C. Maan, W. Stolz, L. Tapfer, and J. Wagner to the reviewed results is gratefully acknowledged. This work was sponsored by the Stiftung Volkswagenwerk and by the Bundesministerium für Forschung und Technologie of the Federal Republic of Germany.

References

1. For an extensive survey on MBE see: Proc. 4th Int. Conf. Molecular Beam Epitaxy, J. Cryst. Growth 81 (1987)
2. For an extensive survey on these fields see: Proc. 6th Int. Conf. Electron. Prop. Two-Dimensional Systems (EP 2DS-VI), Surf. Sci. 170 (1986); Proc. 2nd Int. Conf. Modulated Semicond. Struct. (MSS-II), Surf. Sci. 174 (1986); Special issue of IEEE J. Quantum Electron. QE-22, No. 9 (1986)
3. R.C. Miller and D.A. Kleinmann: J. Lumin. 30, 520 (1985)
4. A.C. Gossard: Thin Solid Films 57, 3 (1979)
5. L.L. Chang and K. Ploog (Eds.): Molecular Beam Epitaxy and Heterostructures (Martinus Mijhoff, Dordrecht, 1985) NATO Adv. Sci. Inst. Ser. E 87 (1985)
6. E.H.C. Parker (Ed.): The Technology and Physics of Molecular Beam Epitaxy (Plenum Press, New York, 1985)
7. G.J. Davies, R. Heckingbottom, H. Ohno, C.E.C. Wood, and A.R. Calawa: Appl. Phys. Lett. 37, 290 (1980)
8. H. Fronius, A. Fischer, and K. Ploog: J. Cryst. Growth 81, 169 (1987)
9. T. Sakamoto, H. Funabashi, K. Ohta, T. Nakagawa, N.J. Kawai, T. Kojima, and K. Bando: Superlattices and Microstructures 1, 347 (1985)
10. B.A. Joyce, P.J. Dobson, J.H. Neave, K. Woodbridge, J. Zhang, P.K. Larsen, and B. Bölger: Surf. Sci. 168, 423 (1986)
11. Y. Suzuki and H. Okamoto: J. Appl.Phys. 58, 3456 (1985)
12. M. Tanaka, H. Sakaki, and J. Yoshino: Jpn. J. Appl. Phys. 25, L 155 (1986)
13. F. Voillot, A. Madhukar, J.Y. Kim, P. Chen, N.M. Cho, W.C. Tang, and P.G. Newman: Appl. Phys. Lett. 48, 1009 (1986)
14. L. Tapfer and K. Ploog: Phys. Rev. B 33, 5565 (1986)
15. L. Tapfer, W. Stolz, and K. Ploog: Ext. Abstr. 1986 ICSSDM (Jpn. Soc. Appl. Phys. Tokyo, 1986) 603
16. Y. Kashihara, T. Kase, and J. Harada: Jpn. J. Appl. Phys. 25, 1834 (1986)
17. W. Stolz, K. Fujiwara, L. Tapfer, H. Oppolzer, and K. Ploog: Inst. Phys. Conf. Ser. 74, 139 (1985)
18. W. Stolz, J. Wagner, L. Tapfer, J.L. de Miguel, Y. Ohmori, and K. Ploog: Inst. Phys. Conf. Ser. 79, 457 (1986)
19. J. Wagner, W. Stolz, J. Knecht, and K. Ploog: Solid State Commun. 57, 781 (1986)

20. G. Bastard, C. Delalande, M.H. Meynadier, P.M. Frijlink, and M. Voos: Phys. Rev. B 29, 7042 (1984)
21. E.F. Schubert, E.O. Göbel, Y. Horikoshi, K. Ploog, and H.J. Queisser: Phys. Rev. B 30, 813 (1984)
22. M. Shinohara, T. Ito, and Y. Imamura: J. Appl. Phys. 58, 3449 (1985)
23. K. Fujiwara, J.L. de Miguel, and K. Ploog: Jpn. J. Appl. Phys. 24, L 405 (1985)
24. A. Makamura, K. Fujiwara, Y. Tokuda, T. Nakayama, and M. Hirai: Phys. Rev. B 34, 9019 (1986)
25. Y. Kawamura, K. Wahita, and H. Asahi: Electron. Lett. 21, 372 (1985)
26. W. Stolz, J.C. Maan, M. Altarelli, L. Tapfer, and K. Ploog: Phys. Rev. B 36, 4783 (1987)
27. J.C. Maan, G. Belle, A. Fasolino, M. Altarelli, and K. Ploog: Phys. Rev. B 30, 2253 (1984).

MBE Growth and Characterization of Ternary and Quaternary Semiconductor Alloys and Heterostructures

F. Genova

CSELT, Centro Studi e Laboratori Telecomunicazioni S.p.A.,
Via G. Reiss Romoli, 274, I-10148 Torino, Italy

0. INTRODUCTION

The ability to prepare thin film epitaxial multilayers with high crystallographic quality, high purity and very small thicknesses is opening up new perspectives in the investigation of the fundamental properties of matter and in the design of new optoelectronic devices. Molecular beam epitaxy (MBE) is becoming one of the most promising growth techniques for this purpose.

While highly satisfying results have been obtained in the growth of III/V binary compounds, particularly GaAs, only $Ga_x Al_{1-x}$ As has been intensively studied among the III/V ternary alloys. The material system GaAs/GaAlAs has allowed the realization of quantum structures (MQWs and 2DEG structures) with new optical and electrical properties that have led to new classes of electrical (HEMT, RTBT, ...) and optoelectronic devices (QW lasers, solid state photomultipliers, ...). The growth of other ternary and/or quaternary alloys and related heterojunctions by MBE has been done with greater difficulty, and the demonstrated results are not satisfactory yet, with the partial exception of the InGaAs/InAlAs system grown lattice-matched to InP. The reason for this situation is partly dependent upon the limits of the growth technique, which are being improved. The main limit is the difficulty of growing P-based compounds in the MBE systems, which seems to be overcome by using gaseous sources. Another limit is the reduced purity of the MBE materials compared to other techniques, particularly Al-containing compounds, due to their high reactivity toward the residual oxygen in the growth chamber.

The tendency to treat the ternary and quaternary alloys by linear interpolation of the characteristics of the binary constituents makes it very easy to understand the general properties of these materials and offers a rapid evalutation of their potential for device applications. However, sufficiently sophisticated models to derive all the relevant properties of these alloys have not yet been developed, and the number of effects that may play an important role in the final performances of such compounds is large.

Some of these effects have their origin in the atomic properties of the elements of the compound, while others may depend upon the thermodynamic conditions in which the epitaxial growth takes place. They can be divided into two main categories:

- Strain effects: interface stress due to the difference of the lattice parameters (lattice mismatch, formation of dislocations, tetragonal distortion, modification of the structural properties both electrical and optical);

- Alloy effects: such as miscibility gap, non-uniform distribution of the atoms in the lattice (microphases, alloy clustering), carrier alloy scattering and P.L. alloy broadening.

The aim of this work is to review the state of the art of MBE growth of ternary and quaternary alloys and heterojunctions, the possible developments and to identify the importance of strain and alloy effects by comparing, when possible, the theory with experimental results.

1. GROWTH OF III-V COMPOUNDS AND ALLOYS

1.1 Basic growth process

Current understanding of the growth of binary compounds by MBE can be found in the literature [1-4]. For a simplified view it can be assumed that at low growth temperatures all the incident group III atoms stick on the III-V substrate or growing film and only enough group V atoms adhere in order to satisfy these and give rise to stoichiometric growth. The excess group V species are desorbed so that the control of stoichiometry of the III-V MBE compounds is not a difficult task.

However, it is necessary to note here that the III-V compounds decompose above the congruent evaporation temperature [5], and this has certain consequences when they are grown by MBE, viz:

a. above the congruent sublimation temperature the group V element is preferentially desorbed,
b. at higher temperatures evaporation of the group III element becomes significant.

These two facts produce the effect that in practice an excess of the group V species must be provided during MBE, and that at high growth temperatures the sticking coefficient of group III elements is significantly lower than unity. Although MBE can happen even at very low growth temperatures, the optimal growth temperature for device quality material should be the highest possible. For binary compounds it depends mainly on the congruent evaporation limit and on the surface diffusion coefficient of the group III element. A list of the congruent sublimation temperatures for several binary compounds is given in Table 1.

TABLE I. List of approximate congruent sublimation temperatures (T_C) for Langmuir evaporation

Material	T_C (°C)
AlP	> 700
GaP	670
InP	363
AlAs	~ 850
GaAs	650
InAs	380

The growth of any alloy can be considered as the simultaneous growth of the binary constituents. The main limits to the growth of these films are:

a) the thermal stability of the less stable of the binary compounds;

b) the competition between the group V elements on the surface of the crystal before incorporation. It was found [6] that the element with the lowest vapor pressure incorporates preferentially.

From these two points we can argue that:

- the maximum growth temperature for the alloys is often too low to guarantee the necessary surface mobility of the group III element for the most stable compound (ex:Al). Alloy disorder can arise and films with poor electrical and optical properties are grown;

- Higher group V pressures allow higher growth temperatures. In this case corresponding reevaporation of the group III atoms occurs, as predicted by thermodynamical calculations [7] and measured by layer-by-layer desorption techniques [8]. The calculated normalized growth rates for several alloys are shown in Fig. 1 and measured normalized growth rates for InGaAs are shown in Fig. 2a;

- The growth of compounds with two group V elements can be achieved by keeping constant the ratio between the group III flux and the most stable of the two group V elements, while the other one can be kept in excess. This is very difficult to do with thermal beams and this is the reason for the difficulty in growing such alloys by MBE.
  As an example, the alloy composition versus the mole fraction of P over P+As in the incident beams is reported in Fig. 2b, for the growth of InGaAsP quaternary layers.

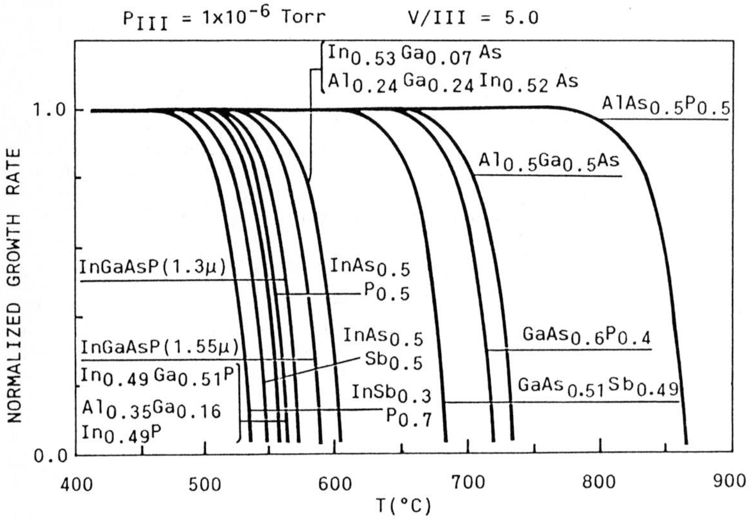

Fig. 1 - The normalized growth rate of ternary and quaternary alloys as a function of temperature

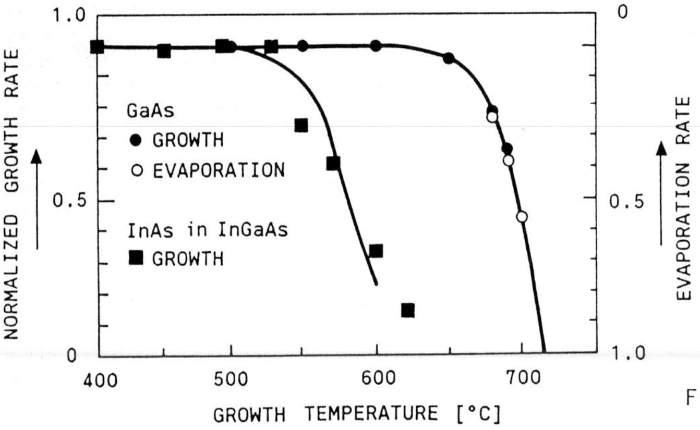

Fig. 2a

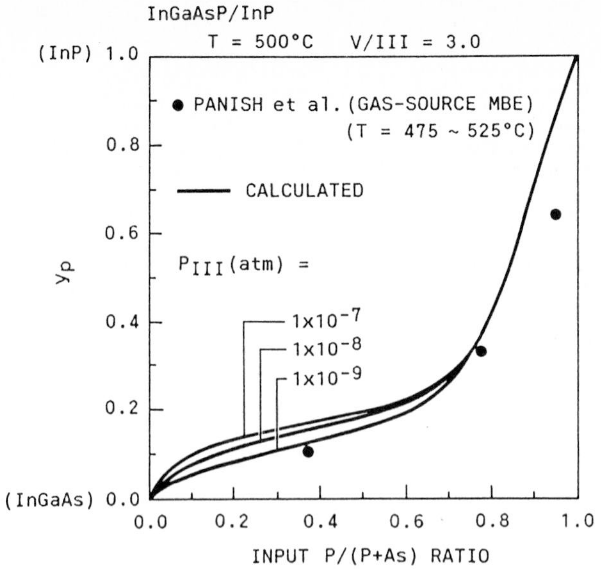

Fig. 2b

Fig. 2 - a) Temperature dependence of the normalized growth rate (filled circles) and the evaporation rate (open circles) for GaAs, and the normalized growth rate for InAs in $In_xGa_{1-x}As$ (squares). b) Comparison between the experimental and calculated compositions of InGaAsP

1.2 Group V element sources

In this chapter arsenic will be considered as representative of most group V elements. The most common source of As in a MBE system is the As_4 produced by the sublimation of elemental arsenic. This is not an ideal situation for at least two reasons:

- the high vapour pressure of As makes it difficult to achieve stable fluxes;
- the tetrameric molecules are less efficient than the dimeric ones for MBE growth and for thermal stabilization of the crystal.

In addition the furnaces are quickly depleted, which results in frequent openings of the MBE system. In order to overcome this problem several improvements have been proposed (high volume cell, fast load cell, ...) but none are yet satisfactory. An interesting source of dimeric molecules is the thermal cracking of As_4. Two-temperature-zone arsenic evaporation cells are now available to produce As_2: in a first low temperature stage, arsenic is thermally evaporated and in a following high temperature (~ 900 °C) zone tetrameric molecules are decomposed into dimers. This results in a reduced consumption of arsenic and more stable fluxes.

By this method it is also possible to grow ternary and quaternary alloys containing Al at high temperatures, because the As_2 flux is more efficient than As_4 in preventing arsenic evaporation from the epilayer.

1.3 Gas sources and metallorganic molecular beam epitaxy

Recently, to overcome some of the limits of the conventional MBE technique, the use of gas sources (decomposed hydrides) has been proposed [9] for the group V beams. Moreover, group III metal alkyls have been used as group III sources [10]. The principal advantages over conventional MBE are:

- an infinite source of group V elements and a good control of their fluxes;
- easy maintenance of the MBE system;
- less expensive furnaces and sources.

This modified MBE is still a molecular beam technique, because the mean free path for the impinging molecules is several centimeters long, but the crystal growth chemisty is different because the pyrolysis of the metallorganic occurs at the surface of the crystal.

1.3.1 Gas sources MBE (GSMBE)

With small modifications a conventional MBE system can host gas sources. The gas sources replace the conventional effusion ovens.

In order to avoid condensation of liquid AsH_3 or PH_3 on the LN_2 walls for safety reasons, the hydrides must be cracked. The ion pumps usually used in conventional MBE must be replaced with faster pumps that do not permanently entrap the H_2: diffusion, turbomolecular or cryopump.

Because of the high pressure due to the H_2 generated by cracking, both arsine and phosphine are expected to decompose according to

$$2MH_3 \rightarrow M_2 + 3H_2. \qquad (1)$$

If sufficient amounts of arsine or phospine are decomposed, the resulting gaseous arsenic and phosphorus species can be a mixture of M_1, M_2 and M_4; the equilibria involved are

$$M_2 \rightarrow 1/2M_4 \qquad (2)$$

and

$$M \rightarrow 1/2M_2. \qquad (3)$$

The ratio of the partial pressures (P) of the dimers to tetramers P_{M2}/P_{M4} for arsenic and phosphorus is given as a function of $P_{M2} + P_{M4}$ at 1000 and 1200K in Fig. 3 [11].

At equilibrium, monomers will predominate only at temperatures in excess of about 1500K for As_1 and 2000K for P_1, for pressures of approximately 10^{-4} to 10^{-6} torr. Dimers are the predominant equilibrium species at low pressures when the temperature is above about 1000K.

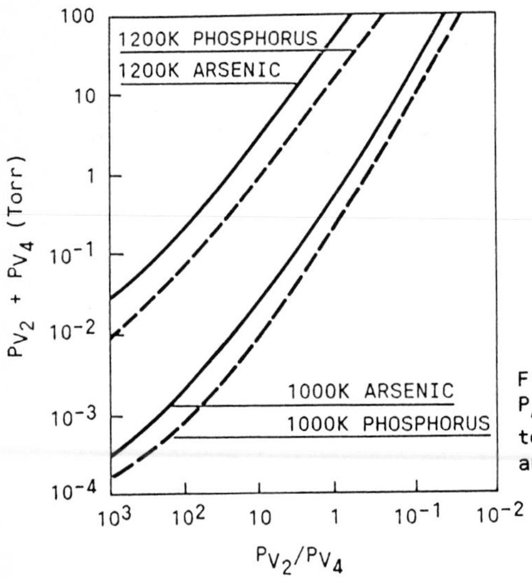

Fig. 3 - Equilibrium partial pressure ratios P_{As_2}/P_{As_4} and P_{P_2}/P_{P_4} as a function of the total pressure of group V species at 1000 and 1200K

The gas sources have been classified as Low Pressure Gas Sources (LPGS) or High Pressure Gas Sources (HPGS) depending upon whether the hydrides are decomposed at pressures near the MBE system pressure or at considerably higher pressure.

The first successful use of a LPGS was obtained by Calawa [12]. In this source the AsH_3 passes through a fused silica tube that incorporates a Ta wire as heater. The resulting epitaxial GaAs had high mobility and low net carrier concentration. In addition there was mass spectroscopic evidence for an appreciable amount of monomeric As in the impinging molecular beam.

Subsequent studies by Chow and Chai [13] and Panish and Sumski [14] showed that the Ta was required for efficient decomposition of both AsH_3 and PH_3 under vacuum conditions. Panish and Sumski found that AsH_3 and PH_3 decomposed essentially quantitatively at 900°- 1000°C when passed through a Ta baffle, but that in the same cracker a significant amount of the AsH_3 or PH_3 remained uncracked when the Ta was replaced by finely divided Al_2O_3 or W.

In the HPGS the species generated by the high pressure source are favoured by thermodynamic equilibrium; in addition, the source can be designed with materials that are inert to all the gaseous components. A HPGS was used for the first reported gas source growth of both epitaxial GaAs and InP [15]. The source consists of "high pressure" cracking regions in alumina tubes that have leaky seals at one end, and a low pressure cracking region. The entire assembly is heated, usually at 900°- 1000°C. Pure AsH_3 and PH_3 are separately introduced into the alumina tubes from a gas handling system having separate manifolds for each leak tube. The gas handling system is illustrated in Fig. 4. The pressures of AsH_3 and PH_3 in the tubes are precisely regulated at selected pressures, usually within the range 200 - 2000 torr. In this pressure range the hydrides decompose completely as measured by the mass spectrometer in the MBE system. Because of the high pressure in the tube, equilibrium between the gas phase species is readily established and the predominant species present in the PH_3 and AsH_3 tubes are P_4 and As_4, respectively. The leaky seal at the end of the decomposition tubes permits M_4 and H_2 to leak from the high pressure region into the low pressure region of the source. In the low pressure region a crucible filled with matted PBN favours M_2 formation, according to (2) and Fig. 3.

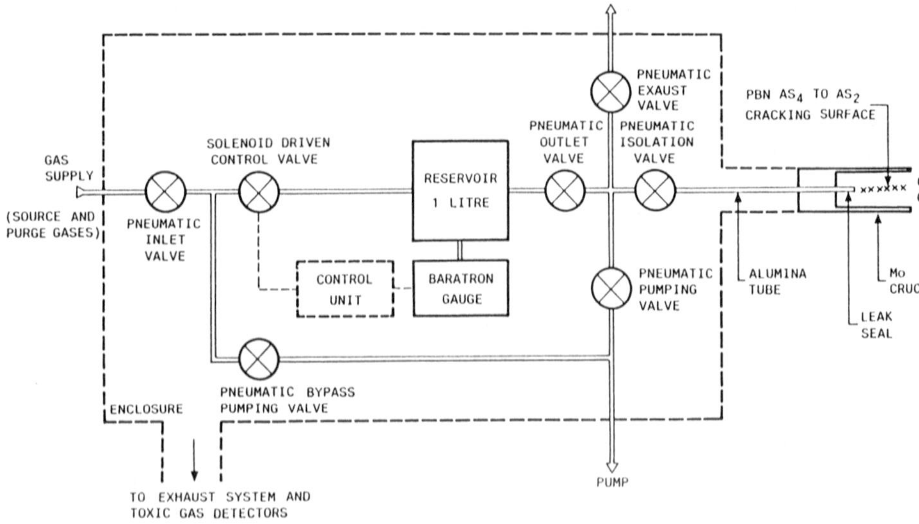

Fig. 4 - Group V gas handling system

1.3.2 MOMBE

In this epitaxial growth technique, the growth kinetics are completely different from that of conventional MBE and in some respects also quite different from MO-CVD. In MOMBE, the beam of group III alkyl molecules impinges directly line-of-sight on the heated substrate surface as in the conventionl MBE process. There is no boundary layer in front of the substrate surface because of the long mean free path of the molecules at the pressure of 10^{-4} Torr. Thus, after a group III alkyl molecule strikes the substrate surface, it can either acquire enough thermal energy from the heated substrate and dissociate all three of its alkyl radicals leaving the elemental group III atom on the surface or reevaporate undissociated or partially dissociated. The probability of each process occurring depends on the substrate temperature. Figure 5 shows the growth rates of InP from TMIn and GaAs from TEGa as a function of substrate temperature for different absolute flow rates of the group III alkyls [16].

In MOMBE two possible kinds of flux control can be used, the first one proposed by Tsang [17] is very similar to an MOCVD control, the other is based on the use of a motorized leak valve, as shown in Fig. 6.

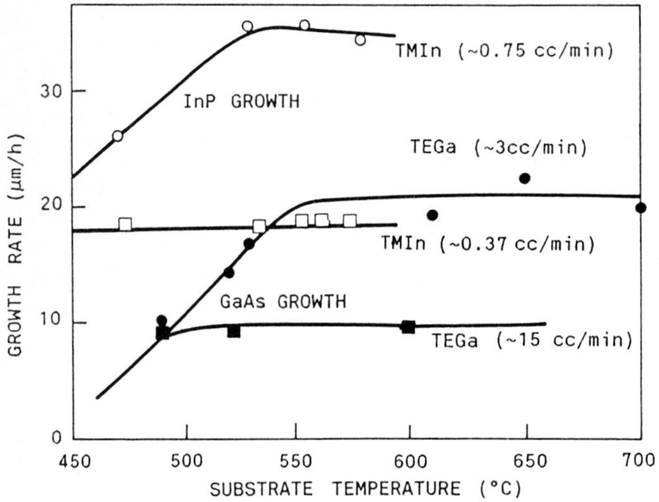

Fig.5 (caption see opposite page)

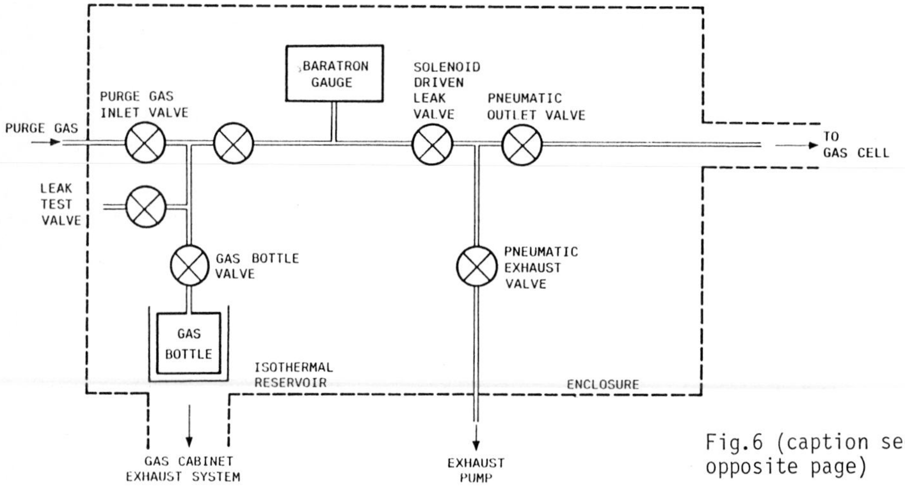

Fig.6 (caption see opposite page)

Using arsine and phosphine and/or metallorganic compounds, high quality ternary and quaternary alloys have been grown. Tsang obtained InGaAs/InP thick layers with 77K Hall mobilities over 55000 cm^2/Vs [16] and Panish [14] obtained InGaAsP quaternary lasers at 1.55 μm with threshold current as low as 1.3 kA/cm^2. Both authors obtained high quality InGaAs/InP quantum well structures.

Modification to the use of gaseous sources has been proposed by several authors, i.e. arsenic transported by H_2 [18] and AsH_3 and PH_3 decomposition at low energy electron impact [19], indicating the great interest in this process.

1.4 Pulsed molecular beam epitaxy (PMBE)

A new technique to grow ternary and quaternary layers, which in principle could be extremely interesting, was developed by Kawabe and coworkers in 1982 [20]. They demonstrated growth of GaAs/GaAlAs heterostructures with good control of the Al profile. With this technique the Al flux is pulsed mechanically with a chopper driven by a rotary solenoid. The Al pulses, detected by a QMS, are integrated and compared with the programmed value in a computer. The computer sends a control signal to eliminate the difference. Alternatively, a photodiode placed so that the inside of the Al cell is visible can be used to monitor pulses, assuming the Al flux is constant.

The final result of this technique might be the growth of a superlattice alloy of alternate monolayer (or very thin layers) of binary compunds (even strained) in the proper proportion in order to obtain a macroscopic alloy with the desired properties and composition. More recently Hiyamitzu and coworkers [21] improved PBME to grow InGaAs/InGaAlAs heterostructures lattice-matched to InP: they adjusted In,Ga,Al cells in order to give InAlAs and InGaAs lattice-matched to InP, then they shuttered alternately the Ga and Al cells for n and m seconds, respectively; the x composition of $In_{.52}(Al_xGa_{1-x})_{.48}As$ was given by

$$x = \frac{m}{n+m} .$$

1.5 Phase Locked Epitaxy (PLE)

Strong intensity oscillations in reflection high energy electron diffraction (RHEED) from several semiconductors during molecular beam epitaxy have been reported in the last few years [22-26]. The measured period of these oscillations was exactly equal to the time necessary for the deposition of one monolayer (one layer of group III and one of group V in the (001) plane) of the semiconductor.

In a "naive" explanation of this phenomenon, the surface roughness caused by island formation in a "layer by layer" growth model was responsible for changes in the surface reflectivity.

Oscillations observed at [110] and [110] azimuth damp rapidly after a few tens of cycles. However, at [100] azimuth it was possible to observe RHEED intensity oscillations for several hundred of cycles, and layer thickness of several hundred nanometres.

By analyzing the phase of these oscillations by computer and operating beam shutters at a particular phase of the oscillations, it is possible to computer control the growth with high precision. This "phase locked epitaxy" (PLE) [27] presents a great advantage over conventional MBE because it allows a precise control of very thin films and superlattice structures, it is independent of any

---

Fig. 5 - Growth rates of InP from TMIn and GaAs from TEGa as a function of substrate temperature at different absolute flow rates of the group III alkyls

Fig. 6 - Group III gas handling system

beam fluctuation and allows interfaces with roughness lower than a monolayer. Unfortunately it is not compatible with substrate rotation, and that limits its practical application for the moment.

2. MISCIBILITY GAP AND CLUSTERING EFFECTS

The properties of III-V alloy semiconductors can be designed to some extent by varying the composition of the binary compound semiconductors. Some material parameters can be obtained rather accurately by linear interpolation from the binary compounds, but others deviate from the linear relation. Moreover, there are some phenomena observed only in alloy semiconductors, such as alloy scattering of carriers [28] and alloy broadening of luminescence linewidth [29]. Most theoretical studies of such phenomena are based on the virtual-crystal approximation or assume that the atom arrangement is completely random. For example, although the probability of alloy scattering depends on the order parameters of the atom arrangement, these parameters are usually assumed to be zero [48]. It is very important to investigate whether deviations from randomness exist in III-V alloy semiconductors, since the deviations may affect many aspects of the material properties.

Miscibility gaps have been reported in several III-V ternary and quaternary alloys [30]. In addition, thermodynamic calculations indicate that nearly all the quaternary III/V alloys are unstable and should undergo phase separation at room temperature [31]. As these materials are used to fabricate devices, we would expect cluster formation during device operation or thermal processing. The fact that no deleterious effects are observed is often attributed to the extremely slow diffusion coefficients expected near room temperature. In any case the growth of metastable alloys by kinetically controlled techniques is possible [32, 33], and no evidence of phase separation is observed for these thermodynamically unstable alloys.

Such observations indicate that these alloys which would be judged to be unstable based on the phase diagram, i.e., considering only chemical free energy, may in fact be stable against clustering or decomposition after they are formed. In any system for which the lattice parameter is a function of composition, clustering or spinodal decomposition would introduce a coherency strain energy term which should be added to the total free energy. Cahn [34] demonstrated that in binary metal systems this coherency strain will indeed stabilize chemically unstable systems against spinodal decomposition.

Alloys like $A_{1-x-y} B_x C_y D$ and $A_{1-x} B_x C_{1-y} D_y$ are stable if they satisfy the following condition on the Helmholtz free energy:

$$\left(\frac{\delta^2 F}{\delta x^2}\right)_{T,P} \cdot \left(\frac{\delta^2 F}{\delta y^2}\right)_{T,P} - \left(\frac{\delta^2 F}{\delta x \delta y}\right)^2_{T,P} \geqslant 0 \ .$$

The value of F can be obtained from a simple thermodynamic model of the solid, such as the regular solution model [35] or the DLP (delta lattice parameter) model [36]. In Fig. 7 are shown the spinodal curves for almost all the III-V ternary and quaternary alloys calculated by Onabe [37] on the basis of the regular solution model.

It is easy to see that almost all the alloys are expected to present spinodal decomposition at the common MBE (or VPE and LPE) growth temperatures. As discussed before, the coherency strain energy acts to stabilize the alloy considered unstable. Adding this energy to the Gibbs free energy and reconsidering the stability condition, Stringfellow [31, 36] found that most alloys were stable even at T=0 K except those with a very large miscibility gap (i.e. some of the Sb compounds). In accordance with these considerations a recent paper of Ichimura and Sasaki [38] predicts that at thermal equilibrium, for most III-V alloys, there is a preference for ordering and not for clustering. In calculating the strain energy

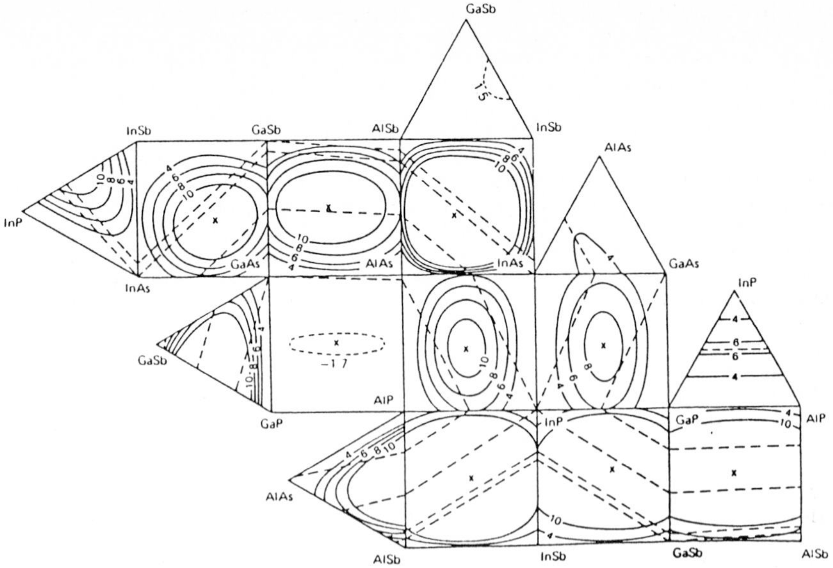

Fig. 7 - Spinodal curves for III-V quaternary solid solutions at 400°-1000°C (solid lines). Temperatures are indicated as 4 for 400°C, etc. Dashed lines represent the compositions for lattice-matching to GaAs, InP, InAs and GaSb

both bond stretching and bond angle distortion were taken in account, and Martin's microscopic elastic constants were used.

On the other hand the poor optical quality of InAlAs grown by MBE suggest the presence of alloy clustering [30,39-41]. A paper based on Monte Carlo simulation [42] predicts the formation of alloy clusterings if the growth temperature is higher than 600 K. In the same paper a formalism is developed for estimating the effect of alloy clustering on the optical and electrical properties of the alloy. Even a small amount of clustering is very detrimental to the material properties.

Alloy clustering has been observed also in $Al_{0.25}Ga_{0.75}As$ layers grown on non-polar, (110) oriented substrates at 600°C [43]. Transmission electron microscope cross sections of these layers revealed the presence of quasi-periodic variations in chemical composition (Al content) in a direction perpendicular to the growth direction, the periods ranging from 15 to 300 Å. Layers grown under identical conditions on (100) oriented surfaces show no evidence of clustering and are uniform within the sensitivity of the analysis technique (composition variations of ±5% of x).

The surface thermodynamics of the system is not well enough understood to establish a surface phase diagram that takes into account the effects of surface reconstruction, lattice strain, and the presence of exchange reactions which depend on the polar nature of the substrate [44]. It is believed, however, that the result of these different effects may be to introduce a surface miscibility gap for growth on the (110) orientation, which would be responsible for the observed composition fluctuations.

3. LATTICE MISMATCHED SYSTEMS

Epitaxial layers with highly uniform surfaces and low dislocation densities can be deposited over a wide range of growth conditions as long as the lattice mismatch stays below a critical value.

In heteroepitaxial systems where a close lattice match is achieved only over a limited range of composition, the quality of the epitaxial layers depends critically on the degree of lattice parameter mismatch and on the composition uniformity over the substrate area. The maximum values of lattice mismatch that can be accommodated by elastic strain in thick layers (> 5 μm) lie in the range $\delta \leqslant 1$ to 2×10^{-3} depending on the alloy system considered. For mismatch values beyond this level extensive arrays of misfit dislocations are generated at the interface and propagate throughout the layer. Surfaces of dislocated layers show a characteristic cross-hatched surface appearance with dislocation lines running parallel to the [110] and [1$\bar{1}$0] directions for growth on the (001) orientation. In addition, surfaces of layers under tensile stress may show evidence of cracking. In all cases the electronic properties of the layers degrade rapidly with increasing lattice mismatch.

3.1 Misfit dislocations

The accommodation of misfit across the interface between an epitaxial film and its substrate has been considered by Frank and Van der Merwe [45, 46]. A review on the subject can be found in [47] and in [48 - 50] for epitaxial multilayers.

A critical thickness exists above which it is energetically favourable to share the misfit between dislocations and strain. Predictions are based on the minimization of the total energy of the crystal: the energy associated with the tensile or compressive stress (elastic strain) and the energy of a grid of misfit dislocations. The evaluation of these terms is based on very simple models, nevertheless the agreement between prediction and experiments has been found to be satisfactory for several bicrystals, even if for some of the III-V compounds and alloys the measured critical thickness seems to be much higher than predicted.

Results on InGaAs grown on GaAs and InP pointed out clearly that misfit dislocations (m.d.s) are parallel to the two [110] directions lying in the (001) plane. They are partially screw and mainly 60° oriented [51]. Their location is at the interface between the film and the substrate and they penetrate into the substrate for several micrometres. Dislocations parallel to the [1$\bar{1}$0], (α type) have higher mobility than those in the [110] (β type), that means that the two directions are not equivalent and their m.d. density can be different.

The ion channeling technique and asymmetric D.C. X-ray diffraction have also been used to determine the tetragonal distortion due to the bidirectional strain parallel to the interface. The so-called tetragonal distortion of the lattice cell, which arises from the lateral strain, in the elasticity theory should follow the well-known Poisson effect governed by the bulk lattice constants of the material. However, at least in the case of Si on GaP [52], there is evidence of a large discrepancy, the results being more consistent with conservation of the covalent bond length than with the Poisson effect. This result is in agreement with EXAFS measurements of ternary alloys [53], where it is observed that Vegard's law is obeyed on the average, but locally the atoms are arranged so that they tend to conserve the bond length.

Moreover, a series of experiments [54-56] indicate that the critical thicknesses are far in excess of the values predicted by the equilibrium theory of Van der Merwe [47].

Recently Drigo et al. measured the strain and the critical thickness of InGaAs/GaAs layers. The observed tetragonal distortion is in agreement with the Poisson effect while the critical thicknesses were about one order of magnitude higher than those predicted by the equilibrium theory [57].

Preliminary structural data exist for many other lattice-mismatched heteroepitaxial systems such as InAs on GaAs [58,59], GaSb and GaAsSb on GaAs or GaSb [60], GaPAs on GaAs or GaP [61-63], GaInSb on GaAs [64], etc., and we refer to the original publications for further details.

3.2 Buffer layers

The heteroepitaxial growth of semiconductor films on lattice-mismatched substrates is currently being investigated for a large number of technological applications. These films may be either single composition epilayers [66] or superlattices combining two or more lattice-mismatched materials. In both cases, the misfit strain has been reported to remove threading dislocations which would otherwise propagate into the film [67]. This filtering effect is thought to arise from the misfit strain which "forces" threading dislocations away from the growth direction. In the superlattice films, adjacent layers typically have opposite strains, so the threading dislocations weave back and forth from layer to layer [68]. To achieve a net deflection away from the growth direction, the "forces" in adjacent layers must be slightly different. This can be accomplished by differences in elastic properties of adjacent layers.

Several kinds of buffer layers have been proposed for this goal:

a. Growth of a continuously graded buffer layer.

b. Step-graded growth, resulting in partial confinement of the misfit dislocations at intermediate interfaces. Reductions in dislocation densities up to 2 - 3 orders of magnitude can be achieved by this method in the case of heteroepitaxy of InAs on GaAs substrates [59].

c. Growth of superlattice dislocation barriers on top of graded buffer layers. This latter technique, applied in the case of growth of $Ga_{0.23}In_{0.77}As$ on InP, was found to lead to substantial improvements compared to simple graded buffer layers [65].

This last kind of buffer layer repeated several times has been proved to be effective in reducing threading dislocations even in the InGaAs/GaAs [69] and in InGaAs/InAlAs/GaAs [70] systems.

3.3 Effect of strain on bandgap

The tetragonal distortion is known to shift bulk energy levels and to split some band degeneracies. The epitaxial layers are subject to a biaxial compressive (tensile) stress along the [100] and the [010] directions in the (001) plane that modifies the lattice parameter. This breaks the symmetry of the zinc blende and splits the degeneracy of valence band into $V_{1/2}$ and $V_{3/2}$ bands, in addition, the hydrostatic component shifts the band gap to a higher (lower) energy [71].

The schematic band structures near the centre of the Brillouin zone under biaxial tensile and compressive stress are shown in Fig. 8. The solid lines in Figs. 8a and c indicate the valence bands along the biaxial stress directions [100] and [010], while the dotted lines indicate the valence bands along the [001] direction [72]. This splitting can be many tens of meV for layer strain in the percent range, and results in a significant change in the valence band dispersion on the layer plane.

Systems which have hole confinement in biaxially compressed layers can exibit two-dimensional hole masses near the top of the valence band maxima which are roughly 4/3 m^*_l where m^*_l is the bulk light hole mass [73]. Such mass values are significantly less than those of the bulk heavy hole valence band which are preferentially populated in unstrained III-V materials. The effect of layer strain on the band gap has been examined, in particular for strained layer superlattices (SLS) [73-82].

There is also particular interest in materials based on the InAsSb system [83]. These structures are predicted to exhibit low temperature absorption at long wavelength values (~ 12 μm) which are not accessible to any of the bulk III-V materials. Such structures are relatively stable (compared to the II-VI HgCdTe

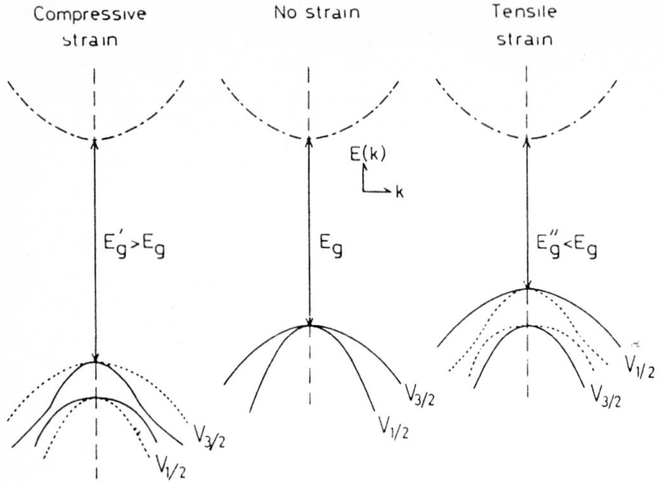

Fig. 8 - Schematic of the band structures in the presence of biaxial tensile and compressive strains near the Brillouin zone centre

alloys) and are of interest for infrared detector materials. Strain-induced reduction of the band gap has been observed experimentally in the GaP/GaAsP, GaSb/AlSb [84, 79], GaAs/InAlAs [82], and Si/SiGe SLS systems [81]. Strain-induced increases have been observed in the GaAs/GaAsP [77] and InGaAs/GaAs [78, 85] SLS systems. Layer strain clearly provides another mechanism for band gap engineering.

4. STRAINED LAYER SUPERLATTICES (SLS)

Semiconductor superlattices have attracted a considerable amount of interest in recent years due to the emergence of new material properties in these thin-layered structures. Recent studies started to explore the possibilities offered by the lattice-mismatched superlattice systems. These superlattices, called strained-layer superlattices (SLS), can be grown from layers with lattice mismatches if the layers are kept sufficiently thin. The flexibility in the choice of SLS layer materials and the interesting effects of the large elastic strains allow SLS materials to exhibit a wide range of tailorable properties which are of interest for scientific and device purposes [75].

A typical SLS structure is shown schematically in Fig. 9a. The thin SLS layers are alternately in compression and tension so that the in-plane lattice constants of the individual strained layers are equal. All the lattice mismatch is accommodated by layer strains without the generation of misfit dislocations if the individual layers are below the critical thickness for dislocation generation, so that the SLS layer can be of high crystalline quality.

The lattice constant of the superlattice which determines its lattice matching to other materials is that parallel to the superlattice interfaces (a"). A superlattice made from closely lattice-matched bulk materials has an a" which is fixed and equal to that of the bulk materials. On the other hand, SLS grown from lattice-mismatched bulk materials have a" values which are a function of the SLS structure. The SLS a" is given by

$$a'' = a_1 \left[1 + \frac{f}{1 + (G_1 h_1 / G_2 h_2)} \right], \tag{4a}$$

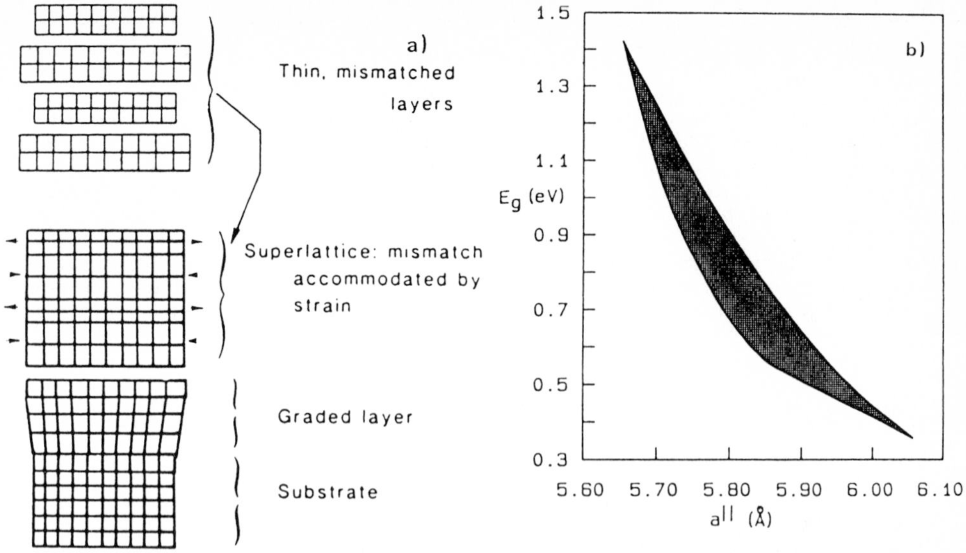

Fig.9- a) Schematic of an SLS. b) Calculated range of band gaps vs lattice constant for $In_xGa_{1-x}As/In_yGa_{1-y}As(100)$SLSs obtained by varying the layer thicknesses and layer compositions. The obtainable gaps are contained in the shaded area

$$G_i = 2\left[C^i_{11} + C^i_{12} - \frac{2\,(C^i_{12})^2}{C^i_{11}}\right], \quad i = 1,2, \qquad (4b)$$

where a_1 is the unstrained lattice constant of layer 1, h_1 and h_2 are the thickness of the individual superlattice layers, f is the lattice mismatch of the unstrained layer materials, G_1 and G_2 are the shear moduli of the two layers, and C^i_{11}, C^i_{12} are the elastic constants of material i. It is possible to vary the layer composition and the layer thickness while keeping a" fixed. This provides an additional degree of freedom in varying the material properties, as shown in Fig. 9b.

A number of prototype devices have been successfully fabricated in the GaAsP, InGaAs, and GaAsSb SLS systems. The motivation of this early work was to demonstrate that good device characteristics could be obtained when the active region of the device contained multiple interfaces with lattice mismatches in the percent range. Devices made in the InGaAs/AlGaAs SLS system include efficient photodetectors, infrared LEDs, pulsed and cw injection lasers and modulation-doped field effect transistors. All these devices contained lattice mismatches of ~ 1.4% [86] except for the pulsed injection laser which contained ~ 2.5% [87]. The photodetector work included both ion-implanted structures and as-grown diode structures. Recent SLS device research focussed on specific SLS devices which are superior to present candidates for a specific application of current interest; for example, Modfet structures in the InGaAs/AlGaAs system with electron transport in single strained InGaAs quantum well structures [88-90].

5. CONCLUSIONS

We have briefly reviewed the growth of III-V alloys by molecular beam epitaxy, pointing out the more recent developments of this technique (i.e. Gas Source MBE) which overcome some traditional limits for the growth of some alloys, or can improve even more the control of the grown layers (i.e. Pulsed MBE and Phase Locked MBE).

In spite of this progress the quality of several alloys is still poor and should be improved if we want to use them for the realization of electronic and photonic devices.

Particular attention has been paid to the problem of cluster formation due to the miscibility gap and to the properties of the mismatched systems, because unwanted strain and cluster in epitaxial films can deeply affect the device properties. Under certain limits, strain can offer an additional degree of freedom by tailoring the electronic and optical properties of new structures.

Since it was not possible to extend this discussion to all the electrical and optical properties that are peculiar to the III-V alloys, such as alloy scattering effects, alloy photoluminescence broadening, etc., the reader is referred to the following publications for more information.

References

1. A.Y. Cho, J.R. Arthur: Progr. Sol. State Chem. 10, 157 (1975)
2. C.T. Foxon and B.A. Joyce: Surf. Sci. 64, 293 (1977)
3. C.T. Foxon and B.A. Joyce: J. Cryst. Growth 44, 75 (1978)
4. C.T. Foxon and B.A. Joyce: and M.T. Norris, J. Cryst. Growth 49, 132 (1980)
5. C.T. Foxon, J.A. Harvey, and B.A. Joyce: J. Phys. Chem. Solids 34, 1693 (1973)
6. J.R. Arthur and J.J. Lepore: J. Vac. Sci. Technol. 6, 545 (1969)
7. H. Seki, A. Koukitu: J. Cryst. Growth 78, 342 (1986)
8. S. Chika et al.: Jpn. J. Appl. Phys. 25, 1441 (1986)
9. M.B. Panish: J. of Electrochem. Soc. 127 27-29, (1980)
10. E. Veuhoff, W. Pletschen, P. Balk and M. Luth: J. Cryst. Growth 55, 30 (1981)
11. M.B. Panish: Progr. Cryst. Growth and Charact. 12, 1 (1986)
12. A.R. Calawa: Appl. Phys. Lett. 38, 701 (1981)
13. R. Chow and Y.G. Chai: J. Vac. Sci. Technol. A 1, 49 (1983)
14. M.B. Panish and S. Sumski: J. Appl. Phys. 55, 3571 (1984)
15. M.B. Panish, H. Temkin and S. Sumski: J. Vac. Sci. Technol. B3, 657 (1985)
16. W. Tsang: Appl. Phys. Lett 45, 1234 (1984)
17. W. Tsang: J. Appl. Phys 58, 1415 (1985)
18. H. Sugiura, M. Kawashima, and Y. Horikoshi: Jpn. J. Appl. Phys. 25, 950 (1986)
19. M. Uematsu and Y. Imamura: Jpn. J. Appl. Phys. 25, L940, (1986)
20. M. Kawabe, N. Matsuura, and M. Inuzuka: Jpn. J. Appl. Phys. 21, L447 (1982)
21. T.Fujii, Y. Nakata, Y. Sugiyama and S. Hiyamitzu: Jpn. J. Appl. Phys. 25, L254 (1986)
22. J.J. Harris and B.A. Joyce: Surf. Sci. Lett.: 103, L90 (1981)
23. C.E.C. Wood: Surf. Sci. Lett.: 108, L441 (1981)
24. J.J. Harris, B.A. Joyce, P.J. Dobson: Surf. Sci. Lett. 108, L444 (1918)
25. J.H. Neave, B.A. Joyce, P.J. Dobson and N. Norton: Appl. Phys. A-31, (1983)
26. J.M. Van Hove, C.S. Lent., P.R. Pukite and P.I.Cohen: J. Vac. Sci. Technol. B-1, 741 (1983)
27. T. Sakamato et al.: Superlattices and Microstructures 1, 347, (1985)
28. Y. Takeda: in GaInAsP Alloy Semiconductors, ed. by T. P. Pearsall (Wiley, Chichester, 1982) Chap. 9.
29. E.F. Schubert, E.O. Gobel, Y. Horikoshi, K. Ploog, and H.J. Queisser: Phys. Rev. B30, 813 (1984)
30. M.F. Gratton, R.G. Goodchild, L.Y. Juravel, and J. C. Woolley: J. Electron. Mater. 8, 25 (1979)
31. G.B. Stringfellow: J. Cryst. Growth 58, 194 (1982)
32. J. Waho, S. Ogawa, and S. Maruyama: Jpn. J. Appl. Phys. 16, 1875 (1977)
33. C.B. Cooper, M.J. Ludowise, V. Aebi, and R.L. Moon: J. Electron. Mater. 9, 299 (1980)
34. J.W. Cahn: Acta Met. 9, 795 (1961)
35. I. Prigogine and R. Defay: in Chemical Thermodynamics, (Longmans, London) (1965) p. 257
36. G.B. Stringfellow: J. Cryst. Growth 27, 21, (1974)

37. K. Onabe: N.E.C. Res. and Develop. 72, 1, (1984)
38. M. Ichimura and A. Sasaki: J. Appl. Phys. 60, 3850, (1986)
39. D.F. Welch, G.W. Wicks, and L.F. Eastaman: Appl. Phys. Lett. 43, 762 (1983)
40. K.Y. Cheng, A.Y. Cho, T.J. Drummond, and H. Morkoç: Appl. Phys. Lett. 40, 147 (1982)
41. K. Najajima, T. Tanahashi, and K. Akita: Appl. Phys. Lett. 41, 194, (1982)
42. J. Sing. S. Dudley, B. Davies and K. Bajaj: J. Appl. Phys. 60, 3167 (1986)
43. P.M. Petroff, A.Y. Cho, F.K. Reinhart, A.C. Gossard, and W. Wiegmann: Phys. Rev. Lett. 48, 170 (1982)
44. J. Singh and A. Madhukar: J. Vac. Sci. Technol. B1, 305 (1983)
45. F.C. Frank and J.H. van der Merwe: Proc. Roy. Soc. A 198, 216 (1949)
46. J.H. van der Merwe: J. Appl. Phys. 34 117 (1963)
47. J.W. Matthews: Misfit Dislocations in Dislocations in Solids, ed. by Nabarro (North-Holland 1979), 461
48. J.W. Matthews and A.E. Blakeslee: J. Cryst. Growth 27, 118, (1974)
49. J.W. Matthews and A.E. Blakeslee: J. Cryst. Growth, 29, 273, (1975)
50. J.W. Matthews and A.E. Blakeslee, J. Cryst. Growth. 32, 265, (1976)
51. P. Franzosi, G. Salviati, F. Genova, A. Stano and F. Taiariol: J. Cryst. Growth 75, 521, (1986)
52. P.M.J. Marée, R.I.J. Althof, J.W.M. Frenken, T.F. van der Veen, Bulle-Lieuwma, M.P.A. Viegers, and P.C. Zalm: J. Appl. Phys. 58, 3097 (1985)
53. J.C. Mikkelsen, Jr. and J.B. Boyce: Phys. Rev. Lett. 49, 1412, (1982)
54. G.C. Osbourn: J. Vac. Sci. Technol. B2, 176 (1984)
55. J.C. Bean, L.C. Feldman, A.T. Fiory, S. Nakaharo and I.K. Robinson: J. Vac. Sci. Technol. A2 436 (1984)
56. K. Kamigaki, H. Sakashita, R. Kato, M. Nakayama, N. Sano, H. Terauchi: Appl. Phys. Lett. 49, 1071 (1986)
57. Drigo et al. to be published (acts of the 1987 MRS Conference, Strasbourg)
58. L. Goldstain, F. Glas, T.Y. Harzin, M.N. Charasse and G. Le Roux: Appl. Phys. Lett. 47, 1099 (1985)
59. C.M. Serrano and C.A. Chang: Appl. Phys. Lett. 39, 808 (1981)
60. C.A. Chang. H. Takaoka, L.L. Chang, and L. Esaki: Appl. Phys. Lett. 40, 983 (1982)
61. J.R. Arthur and J.J. LePore: J. Vac. Sci.Technol. 6, 545 (1969)
62. M. Naganuma and K. Takahashi: Phys. Status Solidi A-31, 187 (1975)
63. Y. Matsushima and S. Gonda: Jpn. J. Appl. Phys 15, 2093 (1976)
64. M. Yano, T. Takase, and M. Kimata: Jpn. J. Appl. Phys. 18, 387 (1979)
65. Y.G. Chai and R. Chow: J. Appl. Phys. 53, 1229 (1982)
66. T.W. James and R.E. Stoller: Appl. Phys. Lett. 44, 56 (1984)
67. J.W. Matthews, A.E. Blakeslee, and S. Mader: Thin Solid Films 33, 253 (1976)
68. J.W. Matthews and A.E. Blakeslee: J. Vac. Sci. Technol. 14, 989 (1977)
69. S.M. Bedair et al.: Appl. Phys. Lett. 49, 942 (1986)
70. P.L. Gourley, T.J. Drummand and B.L. Doyle: Appl. Phys. Lett. 49, 1101, (1986)
71. F.M. Pollack and M. Cardona: Phys. Rev. 172, 816, (1986)
72. H. Kato, N. Iguchi, S. Chika, M. Nakayama, and N. Sano: Jpn. J. Appl. Phys. 25, 1327, (1986)
73. G.C. Osbourn: Superlattices and Microstructures, 1, 223 (1985)
74. G.C. Osbourn: J. Appl. Phys. 53, 1586 (1982)
75. G.C. Osbourn: Mat. Res. Soc. Symp. Proc., 25 455 (1984)
76. N.G. Anderson, W.D. Laidig, G. Lee, Y. Lo, and M. Ozturk: Mat. Res. Soc. Symp. Proc. 37, 223 (1985)
77. P.L. Gourley and R.M. Biefeld: Appl. Phys. Lett., 45, 749 (1984)
78. J.Y. Marzin and E.V.K. Rao: Appl. Phys. Lett., 43, 560 (1983)
79. P. Voison, C. Delalande, M. Voos, L.L. Chang, A. Segmuller, C. A. Chang, and L. Esaki: Phys. Rev. B, 30, 2276 (1984)
80. G. Abstreiter, H. Brugger, T. Wolf, H. Jorke, and H.J. Herzog: Phys. Rev. Lett., 54, 2441 (1985)
81. R. People and J.C. Bean: Appl. Phys. Lett., 4, 538 (1986)
82. H. Kato, N. Iguchi, S. Chika, N. Nakayama, and N. Sano: J. Appl. Phys., 59, 588 (1986)
83. G.C. Osbourn: IEEE J. Quantum Electron., QE22, 1677 (1986)

84. R.M. Biefeld, P.L. Gourley, I.J. Fritz, and G.C. Osbourn: Appl. Phys. Lett., 43, 759 (1983)
85. I.J. Fritz, L.R. Dawson, G.C. Osbourn, P.L. Gourley, and R.M. Beifeld: Proc. 1982, Int. Symp. GaAs and Related Compounds, (1983), p. 241
86. T.E. Zipperian, L.R. Dawson, C.E. Barnes, J.J. Wiczer, and G.C. Osbourn, in: Proc. IEEE Int. Electron Devices Meet., (1983), p. 696
87. W.D. Laidig, P.J. Caldwell, Y.F. Lin, and C.K. Peng: Appl. Phys. Lett., 44, 653 (1984)
88. W.T. Masselink, A. Ketterson, J. Klem, W. Koop, and H. Morkoc: Electron. Lett., 21, 937 (1985)
89. T.E. Zipperian and T.J. Drummond: Electron. Lett., 21, 823 (1985)
90. J.J. Rosenberg, M. Benlamri, P.K. Kirchner, J.J. Woodall, and G.D. Petit: in 1985 IEEE Device Res. Conf. Tech. Dig., EDL-6, (1985), p. 491

Crystallographic and Electronic Properties of the $(GaAs)_1(InAs)_1(001)$ Superlattice

*P. Boguslawski**[1,2] and A. Baldereschi*[2,3]

[1]International School for Advanced Studies, I-34100 Trieste, Italy
[2]Institute of Applied Physics, EPFL, CH-1015 Lausanne, Switzerland
[3]Department of Theoretical Physics and GNSM-CISM, University of Trieste, I-34100 Trieste, Italy

Self-consistent ab initio pseudopotential calculations of the crystallographic and electronic properties of the $(GaAs)_1(InAs)_1$ superlattice oriented along (001) are reported. We find that this system is unstable against phase segregation at zero temperature, due to the excess elastic energy of stretched Ga-As and compressed In-As bonds. The calculated reduction of the optical band gap with respect to the gap obtained in the virtual crystal approximation is larger than that of random GaInAs alloys.

1. Introduction

The thermodynamic stability against phase segregation of several semiconductor superlattices has recently been investigated theoretically [1,5]. For lattice matched systems, BYLANDER and KLEINMAN [1] predicted the instability at $T = 0$ of the $(AlAs)_1(GaAs)_1(001)$ superlattice, a result later confirmed by WOOD et al. [2], who also found an analogous instability of $(CdTe)_1(HgTe)_1(001)$, and by CIRACI and BATRA [3]. Instability results from the electron charge transfer induced by the electronegativity difference between the two cations. Instability has also been predicted [3] for several crystallographic configurations of strained Si/Ge superlattices. The only unstrained lattice-mismatched system studied till now is $(GaP)_1(InP)_1$ (001) for which SRIVASTAVA et al. [4] and MBAYE et al. [5] predict stability due to the large relaxation energy of the anion sublattice. The same authors also predict the stability of the cubic compounds Ga_3InP_4 and $GaIn_3P_4$.

In this work we intend to investigate further the formation energy at $T = 0$ of unstrained lattice-mismatched systems. We show that $(GaAs)_1(InAs)_1(001)$ is unstable against compound segregation due to the large excess elastic energy of stretched (compressed) Ga-As (In-As) bonds. The instability is also found for the Ga_3InAs_4 and $GaIn_3As_4$ cubic ordered phases, which we study here for the sake of completeness. Given the strong similarity between the GaInAs and GaInP systems, we conclude that the claimed stability [4,5] of $(GaP)_1(InP)_1$ (001), Ga_3InP_4, and $GaIn_3P_4$ needs verification.

We also examine the influence of the microscopic crystallographic structure on the energy bands of the GaInAs ordered phases. The long-range order considerably modifies the band structure obtained within the Virtual Crystal Approximation, causing in particular a large reduction of the optical band gap.

2. Equilibrium configuration of $(GaAs)_1(InAs)_1$ (001)

We consider $Ga_{4-n}In_nAs_4$ ($n = 1,2,3$) ordered supercrystals. The case $n = 2$ refers to the $(GaAs)_1(InAs)_1$ (001) superlattice, as grown in Ref. 6, whose tetragonal cell has lattice parameters a, and $\eta = c/a$. The Ga (In) atoms occupy the cell centers (corners). The As positions are (a/2, 0, uc) and (0, -a/2, -uc), where (u-1/4)c measures the anion internal displacement along the z axis. The Ga_3InAs_4 and $GaIn_3As_4$ supercrystals are assumed to have simple cubic structure with lattice constant a. Cations occupy ideal fcc sites with the $AuCu_3$ arrangement, while anions are at a(u, u, u), and tetrahedrally equivalent sites, where again (u -1/4) measures the internal distortion.

Crystal total energies are calculated following Ref. 7 using atomic ab initio pseudopotentials and the exchange correlation functional of Ref. 8. We use a 12 Ry cutoff for the basis set, which corresponds to about 100 plane waves per atom. Brillouin zone integrations for zinc blende (zb) lattice are performed with the 2 mean-value-point method [9]. To allow for comparison between total energies of different structures, equivalent points are used for the supercrystals : $k = (2\pi/a)(1/4, 1/4, 1/4)$ for the cubic cell, and $k_1 = (2\pi/a)(1/4, 0, 1/4\eta)$ and $k_2 = (2\pi/a)(1/2, 1/4, 1/4\eta)$ with equal weights for the tetragonal cell. The accuracy of calculations was tested by evaluating the excess internal energy of $(AlAs)_1(GaAs)_1$ (001) for $a = 5.65/\sqrt{2}$ Å, $\eta = \sqrt{2}$ and $u = 1/4$. The value we obtain is 2.7 meV/atom which is close to the value 2.3 meV/atom calculated in Ref. 1 with the same set of parameters.

Table 1. Structural parameters, excess internal energy ΔE^{tot}, and its decomposition $\Delta E^{tot} = \Delta E^{V} + \Delta E^{ct} + \Delta E^{r}$ for the $Ga_{4-n}In_nAs_4$ supercrystals at equilibrium.

n	1	2	3
a_{eq}(Å)	5.67	4.07[(a)]	5.86
u_{eq}	0.242	0.267	0.259
u_{exp}[(b)]	0.24	0.27	0.26
η_{eq}		1.42	
ΔE^{V}(meV/atom)	39.5	49.0	32.4
ΔE^{ct}(meV/atom)	1.6	2.0	1.6
ΔE^{r}(meV/atom)	-24.1	-30.1	-20.7
ΔE^{tot}(meV/atom)	17.0	20.9	13.3

a) The corresponding zinc blende lattice constant is $\sqrt{2}\ a_{eq} = 5.76$ Å

b) After Ref. 11.

The resulting equilibrium structural parameters are summarized in Table I. The lattice constants of GaAs (5.57 Å) and InAs (5.95 Å) compare well with experimental values (5.65 Å and 6.05 Å, respectively). For intermediate compositions n, the lattice constant follows Vegard's law to within 0.1 %. The tetragonal distortion of the superlattice differs from $\eta_{ideal} = \sqrt{2}$ by less than 1 %, in agreement with experiment [10]. Thus, the 7 % bond-length misfit is accommodated for by the distortion of the anion sublattice, while the cations remain at ideal fcc sites. The equilibrium nearest neighbor distances (Fig. 1) are closer to the model of bond length conservation than to the

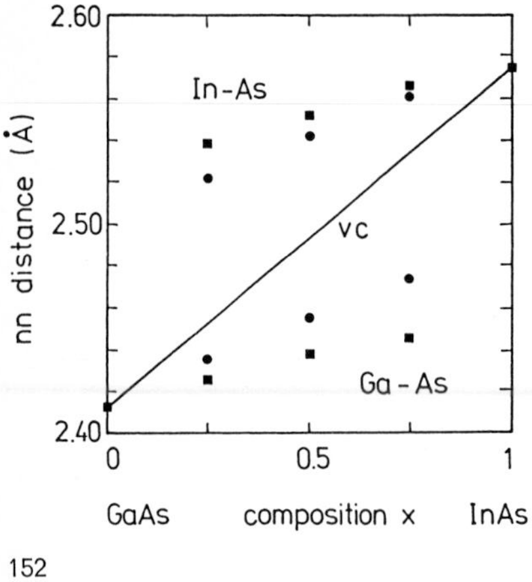

Fig. 1. Calculated nearest neighbor distances as a function of composition for the $Ga_{4-n}In_nAs_4$ supercrystals (squares) and $Ga_{1-x}In_xAs$ random alloys (dots). The straight line gives the linear interpolation between the two end compounds (virtual crystal model).

virtual crystal (VC) assumption that all bond lengths are the same. This result extends to the ordered phases the behavior observed by EXAFS in random $Ga_{1-x}In_xAs$ alloys [11], where values of the internal distortion parameter u similar to those reported in Table I have been measured.

3. Thermodynamic instability of $(GaAs)_1(InAs)_1$

The instability of $Ga_{4-n}In_nAs_4$ supercrystals against phase segregation at T = 0 is demonstrated by the positive values (Table I) of the excess internal energy $\Delta E^{tot}(n)$, i.e. the internal energy of the supercrystal relative to those of the segregated phases at equilibrium. The value $\Delta E^{tot}(2)$ = -20.9 meV/atom that we obtain for the $(GaAs)_1(InAs)_1$ superlattice is one order of magnitude bigger than the values 2.3 meV/atom [1] and 3 meV/atom [2] recently calculated for the lattice-matched systems $(AlAs)_1(GaAs)_1$ and $(HgTe)_1(CdTe)_1$, respectively.

In order to understand the instability of $Ga_{4-n}In_nAs_4$ supercrystals and the quantitative difference between GaInAs and lattice-matched systems, we follow SRIVASTAVA et al. [4], and consider the formation of the supercell at equilibrium as a three-step process. In the first step GaAs is compressed and InAs is dilated to the equilibrium lattice constant of the supercrystal. In the second step the supercell is formed, but all atoms still occupy ideal zb sites. The third step consists of the internal structural distortion.

The first step requires a positive elastic energy ΔE^V, which for $Ga_{4-n}In_nAs_4$ is

$$\Delta E^V(n) = [(4-n)\,\Delta E_{GaAs}(n) + n\,\Delta E_{InAs}(n)] / 4 \qquad (1)$$

where e.g. $\Delta E_{GaAs}(n) = E_{GaAs}[a_{eq}(n)] - E_{GaAs}[a_{eq}(0)]$, $a_{eq}(n)$ being the equilibrium lattice constant of $Ga_{4-n}In_nAs_4$. We shall not analyze in detail pressure-induced effects in this paper, since they were discussed recently by one of the authors [12].

The ordered alloy formed in the next step has all atoms at ideal zb sites, but its valence charge distribution does not have zb symmetry [13]. Differences between the electron densities along Ga-As and In-As bonds (Fig. 2a) are as high as 15 % of the maximum density, which is about 30

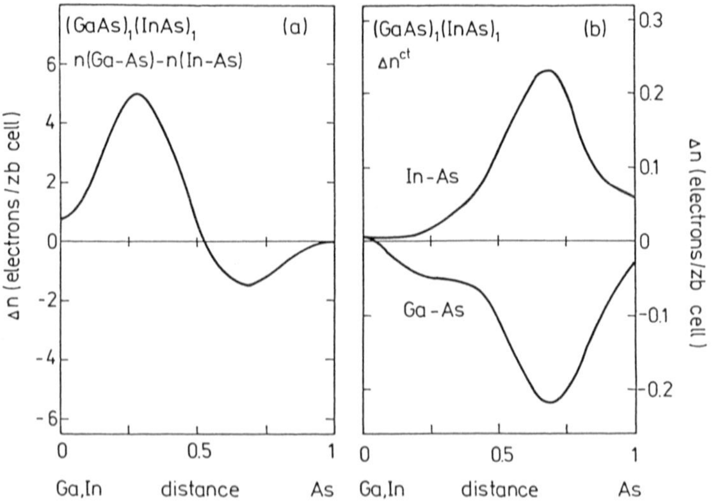

Fig. 2. (a) Difference between the electron densities along Ga-As and In-As bonds in the unrelaxed $(GaAs)_1(InAs)_1$ (001) superlattice.
(b) Difference between the electron density of the unrelaxed $(GaAs)_1(InAs)_1$ (001) superlattice and that of the corresponding pure compounds calculated at the lattice constant of the superlattice. Notice the small values of the density difference.

electrons per zb cell. On the other hand, the supercrystal charge density along the Ga-As (In-As) bonds strongly resembles that of pure GaAs (InAs) at the same lattice constant, deviations being less than 1 % of the maximum electron density (Fig. 2b). One may additionally observe that while the non-zb component of the electron density (Fig. 2a) is large close to the cations, the much smaller charge transfer (Fig. 2b) mainly occurs close to the anions. The internal energy variation in this step

$$\Delta E^{ct}(n) = E_{Ga_{4-n}In_nAs_4}[a_{eq}(n),\ u = 1/4,\ \eta = \sqrt{2}]$$

$$- [(4-n)\, E_{GaAs}[a_{eq}(n)] + n\, E_{InAs}[a_{eq}(n)]] \;/\; 4 \qquad (2)$$

is positive like $\Delta E^V(n)$, but is about 35 times smaller. The magnitude of $\Delta E^{ct}(n)$ for $Ga_{4-n}In_nAs_4$ agrees with the values calculated for $(AlAs)_1(GaAs)_1$ [1,2], and $(CdTe)_1(HgTe)_1$ [2].

As follows from Fig. 3, the anion sublattice distortion (third step) is driven by electron-core (E_{e-c}) attractive forces and is opposed by the core-core (E_M) and electron-electron (E_H) repulsions. Variations of the remaining contributions to the total energy (i.e., the kinetic and the exchange-correlation terms) are one order of magnitude smaller. The non-zb component of the electron density (Fig. 2a) forces the As^{5+} core to move out of the ideal zb site towards the more electronegative cation (Ga). The variation of internal energy in this step

$$\Delta E^r(n) = E_{Ga_{4-n}In_nAs_4}[a_{eq}(n),\ u_{eq}(n),\ \eta_{eq}\ (n)]$$

$$- E_{Ga_{4-n}In_nAs_4}[a_{eq}(n),\ u = 1/4,\ \eta = \sqrt{2}] \qquad (3)$$

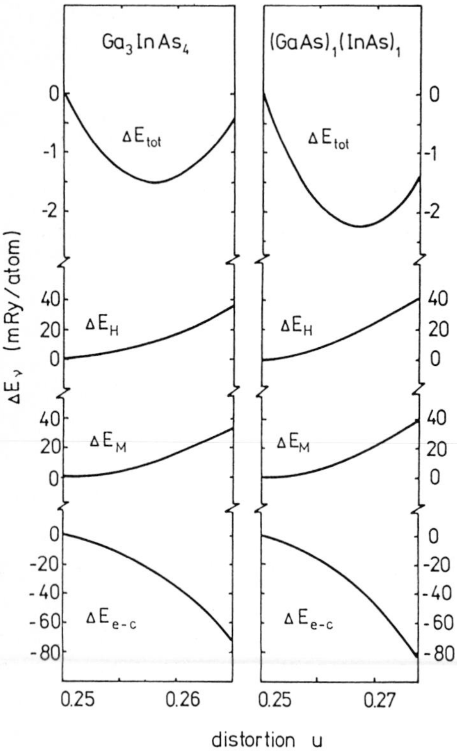

Fig. 3. Variations of ΔE_{tot} and its decomposition for Ga_3InAs_4 and $(GaAs)_1(InAs)_1(001)$ as a function of the internal distortion parameter u.

is comparable in magnitude but opposite in sign to $\Delta E^{v}(n)$, which is easily understood since part of the energy lost in bringing bond lengths to the common virtual crystal value is recovered when the bonds relax back towards the equilibrium lengths of pure compounds. In all cases we find $\Delta E^{r}(n) \approx -0.6\ \Delta E^{v}(n)$, since the bonds remain slightly distorted (Fig. 1). The excess elastic energy $\Delta E^{v} + \Delta E^{r}$ is one order of magnitude higher than $\Delta E^{ct}(n)$, and is the main contribution to the instability of the lattice-mismatched $Ga_{4-n}In_nAs_4$ supercrystals. Although of less importance in the case under study, the charge transfer term further destabilizes the supercrystals.

The above analysis shows that lattice-mismatched systems are controlled by different mechanisms than those valid in the lattice-matched case. The variation of internal energy of strongly lattice-mismatched supercrystals is dominated by the elastic contribution $\Delta E^{v} + \Delta E^{r}$, which is positive and larger than ΔE^{ct}. These systems are therefore unstable against phase segregation at T = 0. On the other hand, in lattice-matched systems one has $\Delta E^{v} + \Delta E^{r} \approx 0$, and the formation energy is dominated by the small charge transfer term. Since the latter can in principle be either positive or negative, one can obtain stable lattice-matched supercrystals at T = 0 if $\Delta E^{ct} < 0$.

Our results for the GaInAs supercrystals are at variance with the recently predicted stability of lattice-mismatched GaInP ordered phases [4,5], a surprising result given the similarity between the two systems. The discrepancy is discussed in more detail in Ref. 14. Further work is therefore necessary in order to establish structural properties of the GaInP supercrystals which, according to our conclusions, should be unstable against phase segregation.

4. Bowing of the band gap in GaInAs

All supercrystals are direct-gap semiconductors at Γ. The computed values $E_g(n)$ of the optical gap in $Ga_{4-n}In_nAs_4$ (n = 0,..,4) at theoretical equilibrium are shown in Fig. 4 by full dots. The considerable bowing, given by the reduction $\Delta E_g(n)$ of the supercrystal band gap with respect to the linearly interpolated value $E_{lim}(a)$ (full line) between the GaAs and InAs band gaps, is well approximated by

$$\Delta E_g(n) = E_g(n) - E_{lim}(a_n) = -b\, n(n-4)/16 \qquad (4)$$

or $\Delta E_g = -bx(1-x)$, with $x = n/4$. The bowing parameter obtained by us $b \approx 1$ eV is larger than the experimental value $b \approx 0.45$ eV for random GaInAs alloys. Similar enhancement of bowing induced by the transition from the random to the ordered phase has recently been observed for GaInAs [6] and GaInP [15]. This is a surprising effect since in random alloys the bowing is usually thought to be due to the presence of the random scattering potential, which is obviously absent in the ordered phases.

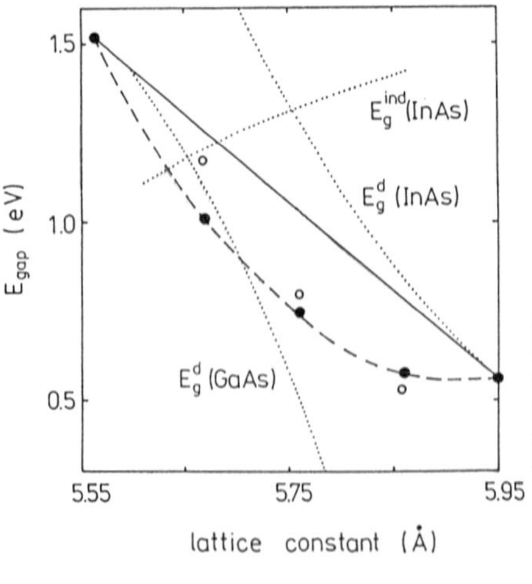

Fig. 4. Band gap energies of $Ga_{4-n}In_nAs_4$ (n = 0,..,4) for the ideal (open circles) and distorted (full dots) configurations. For GaAs and InAs, the lattice-constant dependence of the direct Γ–Γ and the indirect Γ-X band gaps is given by dotted lines. The dashed line is to guide the eye.

We examine the enhancement of bowing by noting first that the virtual crystal approximation renders the same band structure for both supercrystal and a corresponding random alloy with x = n/4. In the random alloy, this band structure is modified by the second order corrections due to the random scattering potential [16]. In a supercrystal, modifications of the VCA band structure are due to a coherent potential ΔV_{coh} (given by the difference between the supercrystal and the VCA potentials), which has the supercrystal periodicity. The presence of ΔV_{coh} leads to the folding of the Brillouin zone, and to the coupling between states which differ by a reciprocal lattice vector of the supercrystal.

We discuss these effects for $(GaAs)_1(InAs)_1$, whose energy bands are shown in Fig. 5. To display the influence of ΔV_{coh}, we also show by dots the energy levels of $Ga_{0.5}In_{0.5}As$ calculated within the VCA. Crosses denote the states at Γ folded back from $Z^* = (2\pi/c)(001)$ of the VC (the zb representations are denoted here by asterisks), according to $Z_1^* \rightarrow \Gamma_1$, $Z_3^* \rightarrow \Gamma_4$, and $Z_5^* \rightarrow \Gamma_5$. One observes that the energies of the valence states at Γ are well approximated by those of the VC. On the other hand, the conduction band energies of the VC and of the superlattice differ considerably. In fact, see Fig. 5, the VC has an indirect gap with the conduction band minimum at X^*. In the superlattice, a reordering of levels occurs, and the lowest conduction band at Γ has Γ_1 symmetry. In order to explain this behavior we have projected the superlattice conduction band wave functions at Γ on s, p, and d-symmetry atomic states. The analysis reveals that the contribution of the s(Ga) orbitals is higher than that of s(In). This corresponds to the closure of the gap since, as follows from Fig. 4, the direct band gap of GaAs is smaller than that of InAs for all values of the lattice constant. The repulsion between $\Gamma_1^{c^*}$- and Z_2^*- derived Γ_1 states further lowers the bottom of the conduction band. Finally, the large calculated bowing is partially due to the tetragonal splitting of the valence band top $\Gamma_{15}^* \rightarrow \Gamma_4 + \Gamma_5$ which amounts to ≈ 100 meV.

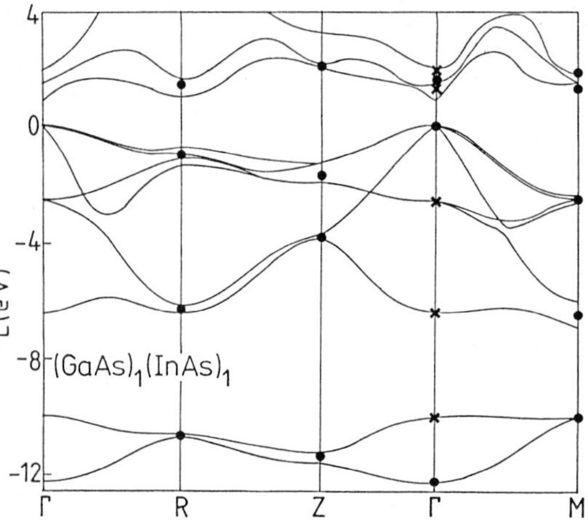

Fig. 5. Energy bands of $(GaAs)_1(InAs)_1(001)$ at theoretical equilibrium. Dots and crosses represent the band energies calculated within the virtual crystal approximation.

Considering other high-symmetry points we notice that the coherent potential ΔV_{coh} affects the folded VCA band structure in the first order, by splitting the degenerate VC levels typically by 0.5 eV. For instance, the conduction band splitting at $L^* = R$, $2L_1^* \rightarrow R_1 + R_4$ is found to be ≈ 0.7 eV, similarly to what is reported by BYLANDER and KLEINMAN [1] for AlGaAs.

5. Effect of internal distortion on the optical energy gap

ZUNGER and JAFFE [17] and BERNARD and ZUNGER [18] have shown that the energy gap in lattice-mismatched superlattices is reduced by the internal lattice distortion. The present computations for $(GaAs)_1(InAs)_1$ confirm this result, but demonstrate also that the effects of the internal distortion are strongly composition dependent. The gaps of the undistorted (u = 1/4) supercrystals are shown in Fig. 4 by open circles. We see that for n = 1 and 2 the distortion closes the gap, contributing ≈ 70 % and ≈ 20 % of the calculated bowing, respectively, while for n = 3 the ≈ 20 % distortion-induced contribution opens the gap. Analysis of the conduction band wave function reveals that for n = 1 the contribution of In orbitals decreases with internal distortion, and the wave function becomes more GaAs-like. As indicated above, this fact corresponds to the closure of the gap. On the contrary, for n = 3, the contribution of In orbitals increases with distortion, thus opening the gap.

6. Conclusions

In conclusion, the present work on the structural stability of the lattice-mismatched systems $(GaAs)_1(InAs)_1$, Ga_3InAs_4 and $GaIn_3As_4$ shows that all systems are unstable against phase segregation at T = 0. Differently from lattice-matched systems as $(GaAs)_1(AlAs)_1$ and $(HgTe)_1(CdTe)_1$, the instability is mostly due to the excess elastic energy $\Delta E^v + \Delta E^r$ of the distorted bonds. Considering the small values of the charge transfer contribution ΔE^{ct}, we propose that instability is a general feature of lattice-mismatched systems, a result which is at variance with recent computations [4,5] on $Ga_{4-n}In_nP_4$. Stability can only occur in nearly lattice-matched cases, when the excess elastic energy is small and can be overcompensated by a negative charge transfer contribution. The calculated bowing of the optical gap of the GaInAs ordered phases is larger than that measured in random alloys of equal composition, in agreement with experimental data. Finally, the effect of the internal distortion on the optical gap depends strongly on composition, being negative for the Ga-rich system and positive for the In-rich one.

Acknowledgements.

One of us (P.B.) takes a real pleasure in thanking Prof. Erio Tosatti, Roberto Car and Raffaele Resta for their hospitality during his stay in Trieste.

References

1. D.M. Bylander and L. Kleinman, Phys. Rev. B 34, 5280 (1986).
2. D.M. Wood, S.H. Wei, and A. Zunger, Phys. Rev. Lett. 58, 1123 (1987).
3. S. Ciraci and I.P. Batra, Phys. Rev. Lett. 58, 2114 (1987).
4. G.P. Srivastava, J.L. Martins, and A. Zunger, Phys. Rev. B 31, 2561 (1985).
5. A.A. Mbaye, L.G. Ferreira, and A. Zunger, Phys. Rev. Lett. 58, 49 (1987).
6. T. Fukui and H. Saito, Japan. J. Appl. Phys. 23, L521 (1984); Inst. Phys. Conf. Ser. 79, 397 (1986).
7. J. Ihm, A. Zunger, and M.L. Cohen, J. Phys. C 12, 4409 (1979).
8. G.B. Bachelet, D.R. Hamann, and M. Schluter, Phys. Rev. B 26 4199 (1982).
9. J.D. Chadi and M.L. Cohen, Phys. Rev. B 8, 5747 (1973).
10. The authors of Ref. 6 do not give explicitly η . It can be deduced, however, from the measured value of c and the fact that a is determined by the InP substrate.
11. J.C. Mikkelsen and J.B. Boyce, Phys. Rev. B 28, 7130 (1983); and cited references.
12. P. Boguslawski, Solid State Commun. 57, 623 (1986).
13. P. Boguslawski and A. Baldereschi, Proc. 17th Int. Conf. Phys. Semicond., ed. W.A. Harrison (Springer-Verlag, New York 1985), p. 939.
14. P. Boguslawski and A. Baldereschi, to be published.
15. A. Gomyo, T. Suzuki, K. Kobayashi, S. Kawata, I. Hino, and T. Yuasa, Appl. Phys. Lett. 50, 673 (1987).

16. A. Baldereschi and K. Maschke, Solid State Commun. 16, 99 (1975)
17. A. Zunger and J.E. Jaffe, Phys. Rev. Lett. 51, 662 (1983).
18. J.E. Bernard and A. Zunger, Phys. Rev. B 34, 5992 (1986).

Electronic States in Heterostructures

Y.C. Chang, G.D. Sanders†*, and D.Z.-Y. Ting*

Department of Physics, University of Illinois at Urbana-Champaign, 1110 West Green Street, Urbana, IL 61801, USA

†Present address: Universal Energy Systems, 4401 Dayton-Xenia Rd., Dayton, OH 45432, USA

Electronic states of semiconductor heterostructures (superlattices and quantum wells) are discussed. Various theoretical techniques and their applicability for different types of heterostructures are reviewed.

1. Introduction

Research interest in the area of semiconductor superlattices[1] has grown tremendously during the past few years. To date, lattice-matched materials such as GaAs-$Al_xGa_{1-x}As$ and InAs-GaSb superlattices can be grown by molecular Beam Epitaxy (MBE) with nearly perfect interface quality[2]. Good quality strained-layer superlattices have recently been successfully grown[3,4]. These include Ga(As,P), Ga(As,Sb), (Al,Ga)Sb, (In,Ga,Al)As, and Si-Ge systems. Other novel superlattices with unique properties such as HgTe-CdTe[5,6], CdTe-CdMnTe[7], ZnSe-ZnMnSe[8] and doping (nipi)[9] superlattices have been proposed and fabricated. Detailed information about the electronic, optical, and transport properties of semiconductor superlattices (or quantum wells) has been accumulated via various experimental techniques such as photoluminescence, photoabsorption, magneto-absorption, Raman scattering, cyclotron resonance, Shubnikov-de Haas, resonant tunneling, Hall, and I-V measurements.

Theoretically, both pseudopotential[10] and tight-binding[11] methods have been used to study the interface properties and the global band structures of thin-layer semiconductor superlattices. Simple effective-mass (Kronig-Penney) [12] and two-band methods[13] have been used to examine the superlattice subband structures near the zone center. SCHULMAN and CHANG have recently developed a new theoretical method within the tight-binding framework[14-16]. This method utilizes the Bloch states associated with complex wave vectors of constituent bulk semiconductors for constructing the total wavefunctions of the heterostructure. The energy eigenvalues of the system can be found in a very efficient way by diagonalizing a "reduced Hamiltonian", whose dimension is twice that of the bulk tight-binding Hamiltonian and is independent of the size of the heterostructure[14]. This method allows us to obtain the electronic structures of semiconductor superlattices with arbitrary sizes efficiently. Furthermore, the method can lead to very accurate superlattice subband structures near the fundamental band gap, provided the empirical tight-binding parameters are chosen to reproduce effective electron and hole masses of constituent bulk semiconductors. An efficient computational technique for obtaining the complex wave vector solutions (complex band structures) has also been developed[17] to make the above method feasible.

Other techniques which are more advantageous for some applications have also been developed recently. These include the full-zone k•p method[18], mini k•p method[19], multi-band effective-mass method[20-22], and the Wannier-orbital method[23]. The multi-band effective mass method is designed for calculating the valence band structures of wide band gap superlattices. The method treats a hole in a quantum well in the same manner as treating a shallow acceptor problem in a bulk semiconductor with the spherical Coulomb potential replaced by a square well potential appropriate for the superlattice. Within the effective mass approximation, the kinetic energy term of the hole is described by the Luttinger-Kohn-

Hamiltonian[24]. The subband energies of the superlattice are obtained by variational method[20,21] or matching method[23]. The Wannier-orbital method is ideal for treating the conduction band states of superlattices made of indirect semiconductors. The method incorporates the full conduction-band structures of the constituent semiconductors in the most efficient way. Only one Wannier orbital per unit cell is used and the interaction parameters between Wannier orbitals up to twentieth neighbors are included in the fit of the bulk conduction-band structure. Effective masses associated with Γ, X, and L valleys along all directions are fitted to the known experimental values. Thus the method is capable of determining the ordering and mixing of states derived from various valleys. We shall classify the superlattices into two categories: the wide-band-gap and narrow-band-gap superlattices.

2. Wide-band-gap Superlattices

Superlattices with fundamental band gap larger than 0.5 eV are considered as wide band gap superlattices. These superlattices include $Al_x Ga_{1-x} As$-$Al_y Ga_{1-y} As$, $In_x Ga_{1-x} As$-GaAs, $GaAs_{1-x} P_x$-GaP, and $Si_{1-x} Ge_x$-Si superlattices. For wide-band-gap superlattices, the conduction and valence band states of interest are almost decoupled, and many theoretical methods which treat electrons and holes independently can be used here. Since the methods used for treating conduction and valence band states are quite different, we shall discuss them separately.

2.1 Conduction-band States

For superlattices made of direct semiconductors such as the well-studied GaAs-$Al_x Ga_{1-x} As$ superlattices or multiple quantum wells, the conduction-band states of interest are mainly derived from the Γ valley. In this case it is often sufficient to use the simple Kronig-Penney method[12] to calculate the energy dispersion and envelope functions. The two-band method[13] is just as easy to implement as the Kronig-Penney method and it includes the nonparabolicity of the conduction band and therefore is more reliable for predicting higher energy superlattice states. It has been shown that[14] when the superlattice energy levels are below the X-minima of GaAs, the results obtained by the two-band model agree well with those obtained by a ten-band tight-binding model, whereas the Kronig-Penney method tends to predict energies that are too high (mainly due to the neglect of nonparabolicity). For superlattice states with energies above the X-minimum of GaAs, a full band-structure calculation is needed. When a tight-binding model is adopted, one may use the reduced Hamiltonian method to obtain the superlattice energies efficiently. However, since the coupling between the conduction band states with the valence-band states is very weak for wide-gap semiconductors, it is sufficient to use the one-band Wannier-orbital approach as recently reported by TING and CHANG[23].

The one-band Wannier-orbital method is most suitable for studying the conduction bands of superlattices made from indirect semiconductors. Using this method TING and CHANG[23] have examined the mixing of Γ and X states in the (001) GaAs-$Al_x Ga_{1-x} As$ and $Al_x Ga_{1-x} As$-AlAs superlattices.

In treating this problem it is convenient to think of the superlattice as consisting of different types of quantum wells: Γ-well for the Γ-valley electrons and X-well for the X-valley electrons. In the $Al_x Ga_{1-x} As$-$Al_y Ga_{1-y} As$ superlattices, the Γ-wells and the X-wells are staggered, with the Γ-well in the $Al_xGa_{1-x}As$ layers and the X-well in the $Al_yGa_{1-y}As$ layers ($x < y$). There are three X-valleys in $Al_x Ga_{1-x} As$ oriented along the [100],

[010], and [001] directions. For a superlattice grown along the [001] direction, we shall refer to these three X-valleys as X_x, X_y, and X_z valleys. If the wave vector parallel to the interface, $\vec{k}_{\parallel}$ is set at (0,0), then the superlattice states will be a mixture of the Γ- and the X_z-valley states. If $\vec{k}_{\parallel}$ is set at (1,0), then the superlattice states will be a mixture of the X_x- and X_y-valley states. The envelope functions of these confined states can be considered as particle-in-the-box wave functions modulated by a spatially varying phase factor. This phase factor has an important effect on the property of the envelope function. For example, at zero superlattice wave vector along the growth direction (q=0), the modulating phase factor associated with the X_z- and the X_y-valleys is either an even- or an odd-parity function, depending on whether the number of atomic layers in the X-well (L_B) is an odd or an even number, respectively. Therefore, in this case the overall parity of the envelope functions depends on the thickness of the X-well. The phase factor is important because it can determine how the confined states interact with one another. By appro-

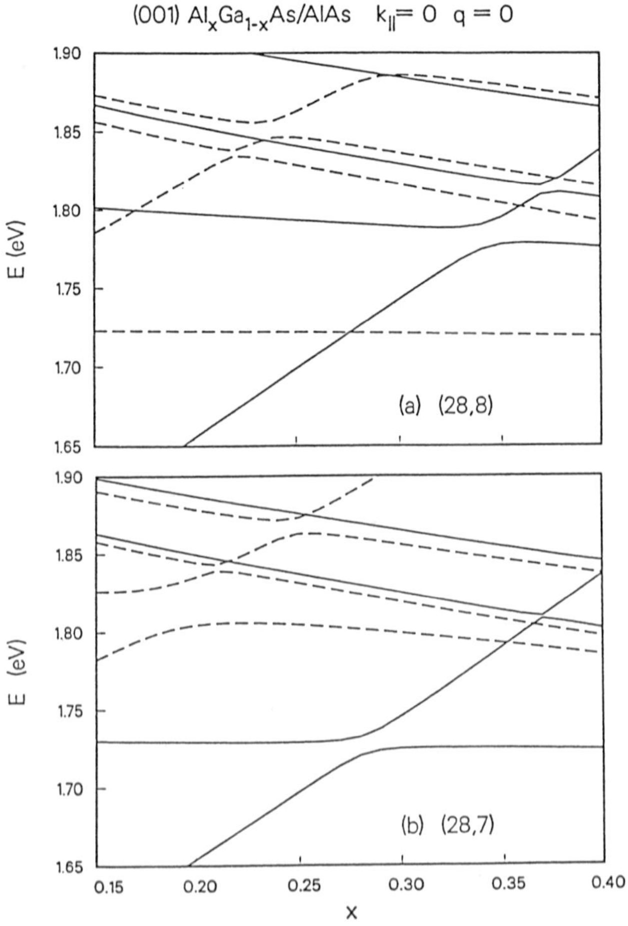

Fig. 1. Energy levels of the (001) $Al_xGa_{1-x}As$-AlAs superlattice as functions of x near the crossover point. The layer thicknesses are (28,8) in 1 (a), and (28,7) in 1 (b). Solid and dashed curves are used to identify the even and odd parity states, respectively. Dotted curves denote bulk conduction minima at various valleys.

priately adjusting the alloy compositions or external hydrostatic pressure, one can bring the confined states associated with the different conduction band valleys close together in energy and they can couple with each other strongly if they have the same parity. The Γ-valley states and the X_z-valley states can couple together at $\vec{k}_\parallel = (0,0)$, and so can the X_x and the X_y-valleys at $\vec{k}_\parallel = (1,0)$. The Γ -X_z mixing and X_x-X_y mixing are found to be similar in nature. The only difference is that the X_x- and the X_y-valleys have the same quantization mass, while the Γ-valley and the X_z-valley have different quantization masses. Since the parity of the zone-folded states is determined by the X-well layer thickness, the mixing between the Γ and X_z states or between the X_x and X_y states depends sensitively on X-well thickness. This phenomenon is illustrated in Fig. 1 for the Γ-X_z mixing and in Fig. 2 for the X_x-X_y mixing with q=0. The layer thicknesses in these figures are labeled by (M,N), where M is the number of AlGaAs atomic layers and N is the number of AlAs atomic layers. The switching of parities of the X-valley states and its effect on the mixing of states are apparent in these figures.

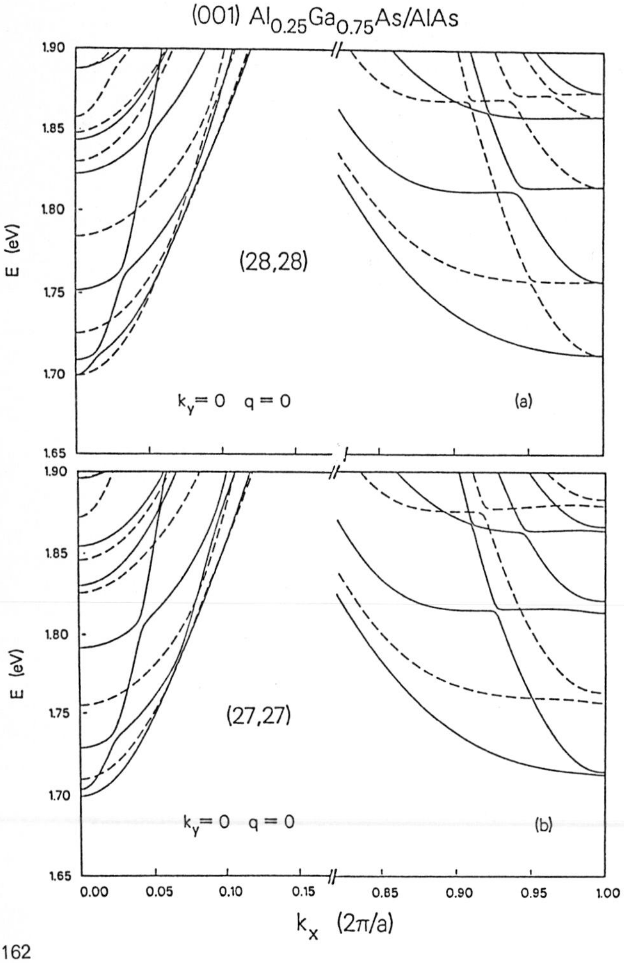

Fig. 2. Energy levels of the (001) $Al_{0.25}Ga_{0.75}As$-AlAs superlattice as functions of k_x, with k_y and q fixed at 0. The solid and dashed curves are used to identify the even and odd parity states, respectively. The layer thicknesses are (28,28) in 2(a), and (27,27) in 2(b).

The Γ-X_z and X_x-X_y mixing also depends on the superlattice wave vector q. For instance if for a particular superlattice Γ-X_z mixing occurs at q=0, then the Γ state and the X-state are non-interacting at $q=q_{max}$. This is because the parities of the phase factor for X-states at q=0 and those at $q=q_{max}$ are exactly opposite.

2.2 Valence-band States

The valence band structures in superlattices are rather intriguing. Because of the mixing of the heavy and light hole bulk states in the superlattice, the valence subband structures are highly nonparabolic even at wave vectors very close to the zone center (within 1% of the superlattice Brillouin zone along directions parallel to the interface) and the valence band states contain substantial admixtures of heavy and light components[16,20,21]. Such a "valence band mixing" phenomenon has nontrivial and very interesting effects on the optical properties. Both the tight-binding method and the multi-band effective mass method (or equivalently the valence-band k • p method) are adequate for studying the valence band states. These methods produce almost identical results, provided that the band parameters used give the same bulk heavy and light hole ef fective masses. However, the multi-band effective mass method is easier to implement and computationally more efficient. Furthermore, it is a trivial matter to include the effect of smoothly-varying external potential such as those due to the electric field and modulation-doping. We shall briefly describe the multi-band effective-mass method for calculating the valence subband structures of quantum wells. The extension to the case of superlattices is simple. We first review some basic knowledge about the valence band structure of semiconductors. Since only the states with energies close to the top of valence bands are of interest, one can describe the valence band structures accurately by the k•p perturbation theory[25]. The valence band states at the zone center (Γ) are described by three p-like states coupled with the electron spin via spin-orbit interaction. The six spin-orbit coupled states are denoted by $|JM\rangle$, with J = 3/2, 1/2 and M = -J, ..., J. For future references we relabel them by u_M for J = 3/2 and v_M for J = 1/2.

Using the k•p perturbation theory, one obtains a 6x6 Hamiltonian matrix, $H^{(o)}(\vec{k})$ within the representation described in (15). Diagonalization of the matrix $H^{(o)}(\vec{k})$ will give rise to valence band structures near the zone center (Γ). Following the notation of Luttinger[26], we write

$$H^{(0)}(\vec{k}) = \begin{pmatrix} P+Q & L & M & o & -L/\sqrt{2} & -\sqrt{2}\,M \\ L^{\star} & P-Q & o & M & \sqrt{2} & \sqrt{3/2}\,L \\ M^{\star} & o & P-Q & -L & \sqrt{3/2}\,L^{\star} & -\sqrt{2}\,Q \\ o & M^{\star} & -L^{\star} & P+Q & \sqrt{2}\,M & -L^{\star}/\sqrt{2} \\ -L^{\star}/\sqrt{2} & \sqrt{2}\,Q & \sqrt{3/2}\,L & \sqrt{2}\,M^{\star} & P+\Lambda & o \\ -\sqrt{2}M^{\star} & \sqrt{3/2}\,L^{\star} & \sqrt{2}\,Q & -L/\sqrt{2} & o & P+\Lambda \end{pmatrix}, \quad (1)$$

where

$$P \equiv -\gamma_1 (k_x^2 + k_y^2 + k_z^2)$$

$$Q \equiv -\gamma_2 (k_x^2 + k_y^2 - 2k_z^2)$$

$$L \equiv 2\sqrt{3}\,\gamma_3 (k_x - ik_y)k_z$$

$$M \equiv -[\sqrt{3}\,\gamma_2(k_y^2 - k_x^2) + 2\sqrt{3}\,\gamma_3 i k_x k_y]$$

and Γ is the spin-orbit splitting.

In the presence of the quantum well potential, we write the eigenstates of the system as linear combinations of bulk valence band basis states, viz.

$$\psi(\vec{k}_\parallel, \vec{r}) = \sum_{\nu k_z} \tilde{F}_\nu(\vec{k}_\parallel, k_z)\, e^{i\vec{k}\cdot\vec{r}}\, |\nu; \vec{k}_\parallel, k_z\rangle , \qquad (2)$$

where $|\nu;\vec{k}_\parallel,k_z\rangle$ are basis states associated with wave vector $|k_\parallel,k_z\rangle$, correct to first order in the k•p perturbation; $\nu = (J,M)$. $\tilde{F}_\nu\,(\vec{k}_\parallel,k_z)$ are Fourier transforms of envelope functions, $F_{\nu'}(\vec{k}_\parallel,z)$. Within effective-mass approximation, it can be shown that $F_\nu(\vec{k}_\parallel,z)$ satisfy the effective-mass equation

$$\sum_{\nu'} \{-H^{(0)}_{\nu\nu'}(\vec{k}_\parallel,-i\partial_z) + V(z) + \delta_{\nu\nu'}\}F_{\nu'}(\vec{k}_\parallel,z) = E\, F_\nu(\vec{k}_\parallel,z), \qquad (3)$$

where $H^{(o)}_{\nu\nu'}\,(\vec{k}_\parallel,-i\partial_z)$ are matrix elements of $H^{(o)}(\vec{k})$ with k_z replaced by the operator $-i\partial_z$. Note that the Luttinger parameters γ_1, γ_2 and γ_3 take on two different values for z inside and outside the well. V(z) is the quantum well potential for holes.

For semiconductors with spin-orbit splitting (Λ) much larger than the quantization energy of the quantum well states, the split-off states ($v_{1/2}$ and $v_{-1/2}$) may be ignored. This is true for GaAs-$Al_xGa_{1-x}A_s$ and GaAs - $Ga_{1-x}In_xAs$ systems. For Si-$Si_{i-x}Ge_x$ systems, which are attracting a great deal of interest recently, the split-off states must be included. For GaAs-$Al_xGa_{1-x}As$ quantum wells, one only needs the upper-left 4x4 black of matrix $H^{(o)}(\vec{k})$ as described in [1].

The envelope functions and energy eigenvalues for GaAs-$Al_xGa_{1-x}As$ quantum wells can be solved by either variational method or matching method. As an illustration we show in Fig. 3 the valence-subband structure calculated with the multi-band effective mass method for a number of GaAs-$Al_{0.25}$ $Ga_{0.75}$ As quantum wells.

For strained-layer quantum wells, some modifications on $H^{(o)}(\vec{k})$, are needed to incorporate the effect of large built-in strain. The built-in bi-axial strain in (001) strained-layer quantum wells due to lattice mismatch can be decomposed into a hydrostatic component plus a [001] uniaxial component[27]. The hydrostatic component will shift all valence band energies uniformly. The uniaxial component has a substantial effect on the quantum well states because it causes additional splittings between the heavy-hole and light-hole states. Such a splitting is given by [27] (according to the theory of PIKUS and BIR [28])

$$\varepsilon = \frac{2}{3} D_u\, X/(C_{11} - C_{12}) = \frac{2}{3} D_u\, [2(C_{12}/C_{11}) + 1]\, (\frac{\Delta a}{a}), \qquad (4)$$

where D_u is the deformation potential, C_{11} and C_{12} are elastic constants, X is the effective unaxial stress. $(\frac{\Delta a}{a})$ is the fractional lattice mismatch, $\Delta a = (a - a_s)$ where a is the lattice constant of the material before it's strained and a_s is the lattice constant of the material after it's strained.

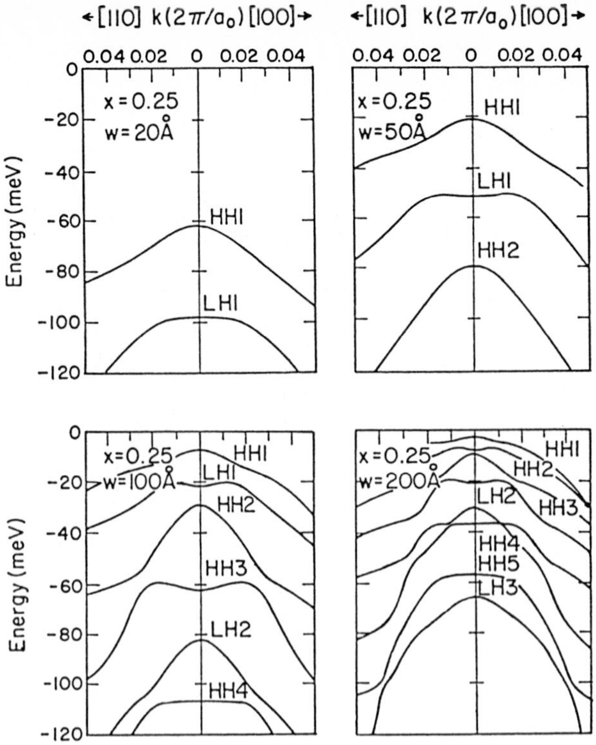

Fig. 3. Valence-subband structures of a number of GaAs-$Al_{0.25}Ga_{0.75}As$ quantum wells.

The correction to the effective-mass Hamiltonian due to [001] uniaxial stress in the (J,M) representation is given by H_S, which together with the bulk Hamiltonian $H^{(o)}$ at $\vec{k} = 0$ is written as[29]

$$\tilde{H}^{(o)}(o) \equiv H^{(o)}(o) + H_S = \begin{pmatrix} \varepsilon & 0 & 0 & 0 & 0 & 0 \\ 0 & -\varepsilon & 0 & 0 & -\sqrt{2}\,\varepsilon & 0 \\ 0 & 0 & -\varepsilon & 0 & 0 & 0 \\ 0 & 0 & 0 & \varepsilon & 0 & \sqrt{2}\,\varepsilon \\ 0 & 0 & -\sqrt{2}\,\varepsilon & 0 & -\Lambda & 0 \\ 0 & 0 & 0 & \sqrt{2}\,\varepsilon & 0 & -\Lambda \end{pmatrix} . \qquad (5)$$

Here $\tilde{H}^{(o)}(\vec{k}) \equiv H^{(o)}(\vec{k}) + H_S$ denote the modified bulk Hamiltonian. Note that in (5), the only contributions from $H^{(o)}(o)$ are the spin-orbit splitting terms.

Although Λ is much larger than the quantization energies in this system, the effect due to the split-off states can still be substantial in the presence of large strain. We can circumvent this difficulty by defining $\tilde{H}^{(o)}(\vec{k})$ in a new representation (u'_m, v'_m) which diagonalize $\tilde{H}^{(o)}$ at $\vec{k} = 0$. We choose

$$\begin{aligned} u'_{\pm 3/2} &= u_{\pm 3/2} , \\ u'_{\pm 1/2} &= C_1\, u_{\pm 1/2} \pm C_2\, v_{\pm 1/2} , \\ v'_{\pm 1/2} &= C_2\, v_{\pm 1/2} \mp C_1\, v_{\pm 1/2} , \end{aligned} \qquad (6)$$

where

$$C_2 = \frac{1}{\sqrt{2}} (1 - \lambda/\varepsilon)\ C_1,$$

$$C_1^2 + C_2^2 = 1,$$

and

$$\lambda \equiv \frac{1}{2} (\varepsilon + \Lambda) - \frac{1}{2}\sqrt{\Lambda^2 - 2\varepsilon\Lambda + 9\,\varepsilon^2}\ .$$

In the new representation, the upper-left 4x4 block of the modified bulk Hamiltonian is given by

$$\overline{H}^{(o)}(\vec{k}) = \begin{bmatrix} P + Q & f_1 L & f_2 M & o \\ f_1 L^* & P - f_o Q - \lambda & o & f_2 M \\ f_2^* & o & P - f\ Q - \lambda & - f_1 L \\ o & f_2 M^* & -f_1 L^* & P + Q + \varepsilon \end{bmatrix},$$

where (7)

$$f_o \equiv \frac{1}{2} \left[1 + \frac{\Lambda - 9\varepsilon}{(\Lambda^2 - 2\varepsilon\Lambda + 9\,\varepsilon^2)^{1/2}} \right],$$

$$f_1 \equiv C_1 - C_2/\sqrt{2} = \frac{1}{2} (1 + \lambda/_{\varepsilon})\ C_1\,,$$

$$f_2 \equiv C_1 + \sqrt{2}\ C_2 = (2 - \lambda/_{\varepsilon})\ C_1\,. \qquad (8)$$

In deriving (7) we have dropped higher-order terms in k^2. Because $\overline{H}^{(o)}$ is diagonal at $\vec{k} = 0$, we can now ignore the new split-off states, $v'_{1/2}$ and $v'_{-1/2}$. Replacing $H^{(o)}_{\nu\nu'}$ in (3) by $\overline{H}^{(o)}_{\nu\nu'}$, we obtain the effective mass equation for holes in quantum wells with [001] uniaxial stress,

$$\sum_{\nu'} \{- \overline{H}^{(o)}_{\nu\nu'} (\vec{k}_{\parallel},-i\partial_z) + V(z)\ \delta_{\nu\nu'}\}\ F_{\nu'}(\vec{k}_{\parallel},z) = E(k_{\parallel})\ F_{\nu}(\vec{k}_{\parallel},z). \qquad (9)$$

3. Narrow-band-gap Superlattices

Both InAs-GaSb and HgTe-CdTe superlattices have very small band gaps (substantially less than 0.5 eV), and are classified as narrow-band-gap superlattices. Both the multi-band effective mass method and one-band Wannier-orbital method described above are not appropriate for treating such superlattices, because the superlattice states are likely to consist of both the valence and conduction band bulk states. The two-band model[13] may be used to study the the superlattice states with wave vectors along the growth direction, but it is not appropriate in general directions. To study the superlattice states for general wave vectors, one needs to use the tight-binding method or a more elaborate k • p method[18].

The reduced Hamiltonian method (within a ten-band tight-binding model) has been used to study the electronic and optical properties of these superlattices [16,30,31]. For the GaSb-InAs superlattices the major findings are :
(a) Valence band mixing effect is strong even at the zone center for superlattices with certain number of InAs and GaSb layers where the second heavy hole and the first light hole state energies nearly coincide. This is a result of

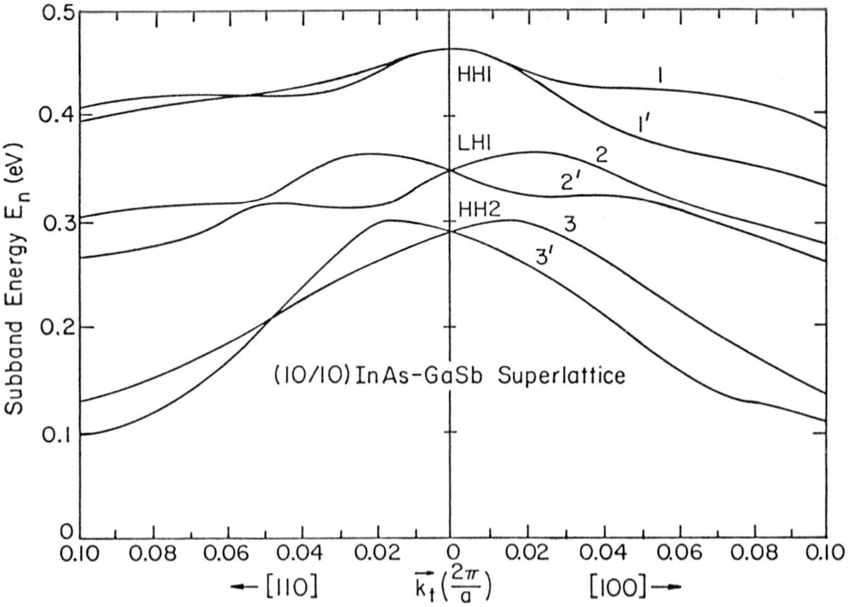

Fig. 4. Valence-subband structure of a (10,10) GaSb-InAs superlattice.

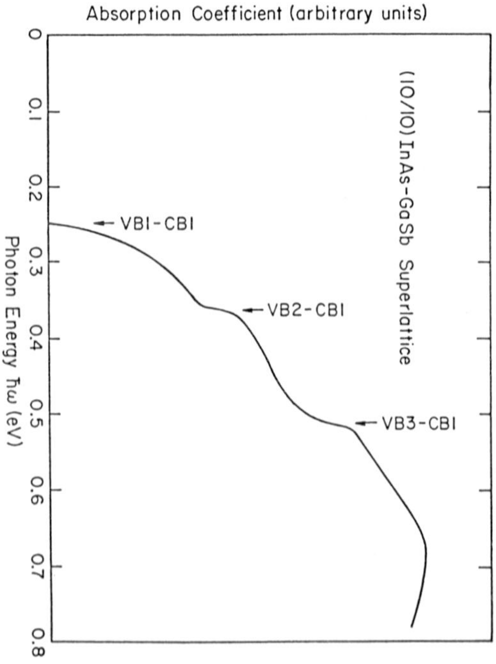

Fig. 5. Absorption spectrum of a (10,10) GaSb-InAs superlattice.

large dissimilarity in the InAs and GaSb tight-binding interactions. Such an effect is not expected in any effective-mass or k.p approach in which the artificial quantum well or superlattice potential used is diagonal in the hole spin indices. This effect is demonstrated in Fig. 4 in which the valence-subband structure of a GaSb-InAs superlattice is shown.

(b) The squared optical matrix elements decrease monotonically with increasing number of either InAs or GaSb layers.
(c) The absorption spectrum is featureless, because the formation of spatially separated excitons is unfavorable and the conduction band typically has large dispersion in the growth direction. The calculated absorption spectrum for a (10/10) GaSb-InAs superlattice is shown in Fig. 5.

For the HgTe-CdTe superlattice, the subband structures in both the semiconductor and semimetallic regimes are studied. The major findings are [30,31]:
(a) A new type of quasi-interface state exists in such a system. The quasi-interface state is a consequence of matching up of bulk states belonging to the conduction band in HgTe and the light-hole valence band in CdTe. Such a matching is possible because of the inverted band structure in HgTe (the bulk states to be matched are made of atomic orbitals of the same symmetry type and the effective masses on either side of the interface have the opposite sign). Energies of the quasi-interface states of a HgTe-CdTe superlattice as functions of the HgTe layer thickness are plotted in Fig. 6.
(b) For HgTe-CdTe superlattices in the semiconductor regime, the lowest conduction band is a quasi-interface state, thus having an unusual wavefunction which contains a p-like component peaked at the interfaces and an s-like component peaked at the center of HgTe material.
(c) The optical matrix element for the fundamental transition in HgTe-CdTe superlattices is found to be comparable (although somewhat smaller) to that in bulk HgCdTe alloy of the same band gap. However the absorption coefficient of the superlattice can become stronger than the alloy, because the conduction band effective mass in the superlattice is found to be substantially larger.
(d) Calculation including the effect of internal strain shows that the subband structures of the semimetallic superlattice are strongly influenced by the strain, implying that a reappraisal of the 40 meV valence band offset determination[31] is needed.

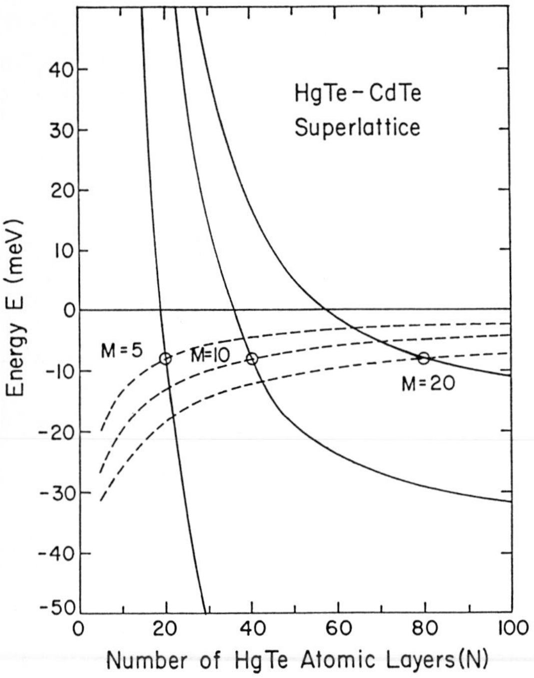

Fig. 6. Energies of the quasi-interface states of a HgTe-CdTe superlattice, plotted as functions of the number of HgTe atomic layers (N). M is the number of CdTe atomic layers. Solid curves (odd parity), Dashed curves (even parity).

Acknowledgements

The authors have benefitted from fruitful discussions with J. N. Schulman. This work was supported in part by the U. S. Office of Naval Research (ONR) under Contract No. N00014-81-K-0430.

References

1. L. Esaki, R. Tsu, IBM J. Res. Develop. 14, 61 (1970)
2. A. C. Gossard, P. M. Petroff, W. Wiegman, R. Dingle, A. Savage, Appl. Phys. Lett. 29, 323 (1976); E. E. Mendez, L. L. Chang, C. A. Chang, L. F. Alexander, L. Esaki, Surf. Sci. 142, 215 (1984)
3. G. C. Osbourn, J. Appl. Phys. 53, 1586 (1982); G. C. Osbourn, R. M. Riefield, P. L. Gourley, Appl. Phys. Lett. 41, 172 (1982)
4. T. P. Pearsall, F. H. Pollak, J. C. Bean, Bull. Am. Phys. Soc. 30, 266 (1985)
5. J. N. Schulman, T. C. McGill, Appl. Phys. Lett. 34, 663 (1979)
6. J. P. Faurie, A. Millon, J. Piaguet, Appl. Phys. Lett. 41, 713 (1982)
7. X. C. Zhang, S. K. Chang, A. V. Nurmikko, L. A. Kolodziejski, R. L. Gunshor, S. Datta, Phys. Rev. 31, 4056 (1985)
8. L. A. Kolodziejski, R. L. Gunshor, T. C. Bonsett, R. Venkatasubramanian, S. Datta, R. B. Bylsma, W. M. Becker, N. Otsuka, Appl. Phys. Lett. 47, 169 (1985)
9. G. H. Dohler, H. Kunzel, D. Oiego, K. Ploog, P. Ruden, H. J. Stolz, G. Abstriter, Phys. Rev. Lett. 47, 864, (1981)
10. E. Caruthes, P. J. Lin-Chung, Phys. Rev. Lett. 39, 1543 (1977)
11. J. N. Schulman, T. C. McGill, Phys. Rev. Lett. 39, 1680 (1977)
12. L. L. Chang, L. Esaki, Surf. Sci. 98, 70 (1980)
13. G. A. Sai-Halasz, L. Esaki, W. A. Harrison, Phys. Rev. B 18, 2812 (1978); G. Bastard, Phys. Rev. 25, 7584 (1982)
14. J. N. Schulman, Y. C. Chang, Phys. Rev. 44, 4445 (1981); Phys. Rev. B 27, 2346 (1983); Phys. Rev. B 31 2056 (1985)
15. Y. C. Chang, J. N. Schulman, J. Vac. Sci. Technol. 21, 540 (1982)
16. Y. C. Chang and J. N. Schulman, Appl. Phys. Lett 43, 536 (1983); Phys. Rev. B. 31, 2069 (1985)
17. Y. C. Chang, Phys. Rev. B 25, 3297 (1982); Y. C. Chang, J. N. Schulman, Phys. Rev. B 25, 3975 (1982)
18. C. Mailhiot, D. L. Smith, T. C. McGill, J. Vac. Sci. Technol. B2 (3), 371 (1984)
19. L. J. Sham, Bull. Am. Phys. Soc., 30, 395 (1985)
20. A. Fasolino, M. Altarelli, In Two-Dimensional Systems, Heterostructures, and Superlattices, edited by G. Bauer, F. Kucher, and H. Heinrich (Springer-Verlag, New York, 1984); M. Altarelli, U. Ekenberg, A. Fasolino, Phys. Rev. B 32, 5138 (1985)
21. G. D. Sanders, Y. C. Chang, Phys. Rev. B 31, 6892 (1985); Phys. Rev. B, 35 1300 (1987)
22. H. Y. Chu, G. D. Sanders, Y. C. Chang, unpublished
23. D. Z.-Y. Ting, Y.-C. Chang, Bull. Am. Phys. Soc., 32, 760 (1987); Phys Rev. B (in press)
24. J. M. Luttinger, W. Kohn, Phys. Rev. 97, 869 (1956)
25. E. O. Kane, J. Phys. Chem. Solids, 1, 82 (1956)
26. J. M. Luttinger, Phys. Rev. 102, 1030 (1956)
27. G. C. Osbourn, Phys. Rev. B 27, 5126 (1983)
28. G. E. Pikus, G. L. Bir, Fiz. Tverd. Tela 1, 1642 (1959); [Sov. Phys. Solid State 1, 1502 (1960)]
29. H. Hasegawa, Phys. Rev. 129, 1029 (1961)
30. Y. C. Chang, J. N. Schulman, G. Bastard, Y. Guldner, M. Voos, Phys. Rev. B 31, 2557 (1985)
31. J. N. Schulman, Y. C. Chang, Phys. Rev. B 33, 2594 (1986)

Electronic States of Heterostructures in the Envelope-Function Approximation

M. Altarelli

European Synchrotron Radiation Facility (ESRF),
B.P. 220, F-38043 Grenoble Cedex, France and
Max-Planck-Institut für Festkörperforschung,
Hochfeld-Magnetlabor, B.P. 166X, F-38042 Grenoble Cedex, France

1. INTRODUCTION

In this paper, calculations of electronic states in heterostructures are reviewed, with emphasis on the envelope-function method and on comparison with experiments. Energy levels in the presence of an external magnetic field are discussed and compared to experiments. Finally, envelope-function calculations of excitons in quantum wells are briefly discussed.

Semiconductor heterostructures are obtained by alternating thin layers of different materials with interfaces of high perfection on an atomic scale [1]. Typical values of layer thicknesses are $\sim 10^2$ Å. Of particular interest are quantum well systems, in which a thin layer is sandwiched between two thicker layers of a different material, and superlattices, in which the layers are alternated in a periodic fashion. The constituent semiconductors may differ in chemical composition e.g. as in GaAs-AlGaAs superlattices, or simply in the doping, as in the GaAs n-i-p-i systems [2], which are termed doping superlattices.

From the point of view of electronic states, a crucial quantity characterizing an A-B compositional superlattice or quantum well system is the band offset of materials A and B. The concept of band offset is based on the fact that the one-electron effective potential near an A-B heterojunction is characterized by two length scales [3]. One is the width of the space charge or band-bending region, which is typically of order $\sim 10^3$ Å, and depends on the doping of A and B. The other is of the order of a few interplanar distances and therefore corresponds to rather abrupt potential steps, or offsets, independent of the A and B doping in the relevant ranges, and reflecting the rearrangement of the bonding charges in the immediate neighbourhood of the interfaces. It must be kept in mind that, even if the band gap difference ΔE_g between materials A and B is accurately known, its breakdown into conduction, ΔE_c, and valence, ΔE_v, offsets is often known with unsatisfactory precision.

To show the relevance of the band offsets, one can proceed to a qualitative classification of compositional superlattice and quantum well systems in terms of the possible types of band line-ups and the resulting potentials. In one type of structure, the smaller of the two energy gaps is entirely contained in the larger one. This happens in GaAs-AlGaAs, InGaAs-InP, etc. It is then easy to realize that the lowest electron or hole states correspond to confinement in the low-gap material (e.g. GaAs in GaAs-AlGaAs). Another interesting type of system (e.g. InAs-GaSb or InAs-AlSb) is characterized by having energy gaps that are only partly overlapping or not overlapping at all. In this case electrons and holes will localize in different layers of the superlattice. In the case of InAs-GaSb, electrons tend to flow to InAs, leaving holes behind in GaSb. A third qualitatively different kind of superlattice is CdTe-HgTe, characterized by the Fermi level of the zero-gap semiconductor HgTe lying in the lower part of the CdTe gap.

Another very important parameter for the electronic structure is the lattice mismatch of the two semiconductors. For the thicknesses characteristic of superlattices, the mismatch is accommodated by elastic strain [4]. In practice, this often means that, in the planes parallel to the interfaces, the whole superlattice assumes the lattice parameter of the substrate on which it is grown. The resulting strains can be described as a superposition of a hydrostatic and a uniaxial component, both of which affect the band structure of the constituent semiconductors considerably. A lattice mismatch inferior to 1% corresponds indeed to several kbars of pressure. The ability to grow high-quality superlattices with strained layers leads to very interesting systems, such as the $Si-Si_xGe_{1-x}$ superlattices, where strain is the dominant factor affecting band offsets and the character of the electronic states [5].

A short review of electronic states in quantum wells and superlattices is attempted, with emphasis on theoretical aspects but in close relation to the interpretation of experiments. In Section 2 a short introduction to the advantages and limitations of the envelope-function description is given, while Section 3 discusses the results obtained by this method in GaAs-AlGaAs systems. A short discussion of electronic states in an external magnetic field is given, including a comparison with magneto-optical experiments. Finally, Section 4 is devoted to a brief discussion of exciton binding energies in quantum wells.

2. ENVELOPE-FUNCTION APPROXIMATION FOR THE CALCULATION OF THE SUBBAND STRUCTURE

The calculation of the band structure of a superlattice can be viewed, in principle, as any other problem of band structure theory. Its difficulty, or at least what makes it different, is that there are typically ~100 atoms per unit cell, so that the computational complexity of ab initio electronic structure methods is prohibitive, except for very short periods [6]. Furthermore, the features of the electronic structure which are relevant for the interpretation of experiments are on the 1-10 meV scale, far beyond the reach of band structure methods which are reliable on the 0.1 eV scale at the very best. Therefore, almost all approaches to the band structure of superlattices are semi-empirical in nature, in that they assume: (1) that the band structure of the two bulk semiconductors is known, and parametrized in some convenient form; (2) that the band offset between the two materials is also known. Examples of the parametrization chosen for the bulk band structures are the empirical tight-binding method [7], the pseudopotential method [8], or the k.p method which is adopted in the envelope-function approximation [9]. In this method, the wavefunction in each layer of the superlattice is written in terms of products of a k = 0 Bloch function of the corresponding bulk semiconductor with a slowly-varying envelope-function (with wavelengths of the order of the layer thicknesses). The envelope functions satisfy effective-mass-like equations, and are joined at the interfaces by appropriate boundary conditions which are compatible, under the assumptions specified below, with the smoothness of the total wavefunction. Suppose for example that the two materials have a parabolic conduction band edge at the Γ point, with effective masses m_A, m_B and energies $E_{\Gamma A}$, $E_{\Gamma B}$ respectively. Then we write the wavefunction of the conduction subbands as

$$\begin{aligned} \psi^A(\underline{r}) &\simeq F^A(\underline{r})\, u_{\Gamma A}(\underline{r}) \text{ in the A layers,} \\ \psi^B(\underline{r}) &\simeq F^B(\underline{r})\, u_{\Gamma B}(\underline{r}) \text{ in the B layers,} \end{aligned} \tag{1}$$

with F^A, F^B, satisfying the equations

$$-\frac{\hbar^2}{2m_{A,B}} \nabla^2 F^{A,B} = (E - E_{\Gamma A,B})\, F^{A,B} \tag{2}$$

and joined at an A-B interface, say the z = 0 plane, by the boundary conditions

$$F^A(x, y, z=0^-) = F^B(x, y, z=0^+),$$

$$\frac{1}{m_A}\frac{\partial}{\partial z} F^A(x, y, z=0^-) = \frac{1}{m_B}\frac{\partial}{\partial z} F^B(x, y, z=0^+). \qquad (3)$$

It can be argued [9,10] that if the assumption $u_{\Gamma A}(\underline{r}) \approx u_{\Gamma B}(\underline{r})$ is tenable then the boundary conditions (3) ensure the conservation of the total probability current at the interface and are therefore physically justified. Notice that the difference in effective mass between the two materials is reflected in a discontinuity in the derivatives of the envelope functions.

The assumption concerning the Bloch functions, $u_{\Gamma A}(\underline{r}) \approx u_{\Gamma B}(\underline{r})$, restricts the applicability to pairs of materials with very similar chemical nature and band structure; in practice, it has proven applicable to III - V compounds and alloys with direct band-gap, such as GaAs-$Al_xGa_{1-x}As$, for $x < 0.4$. If $x > 0.4$, the ternary compound has an indirect gap, with conduction band minima at the X points. Superlattice band states resulting from the admixture of Bloch functions from different points of the Brillouin zone are beyond the reach of the envelope-function method described here. In these situations, one must resort to other methods such as the empirical tight-binding [7] or the pseudopotential scheme [8]. In either case, some of the attractive features of the envelope-function method are lost. Alternatively, one can devise an extension of the envelope-function method, but at the price of introducing some additional parameters in the theory. Ando [11] has recently proposed such an approach for GaAs-indirect AlGaAs heterostructures.

The interest of Eqs (1)-(3) is that they are susceptible of far-reaching generalizations. First, any potential varying slowly on the scale of the lattice parameter of the constituents can be included in Eq. (2):

$$\frac{\hbar^2}{2m_{A,B}} (-\nabla^2 + V^{A,B}(\underline{r}))\, F^{A,B}(\underline{r}) = (E - E_{\Gamma\, A,B})F^{A,B} \qquad (4)$$

thus allowing consideration of space charge effects in doped superlattices, of charge transfer across the interfaces, of external electric fields (which, however, break the lattice periodicity), etc. A second very important generalization is the inclusion of band coupling, whenever it is impossible to construct the subband wavefunctions from a single non-degenerate, nearly parabolic band edge as implied by Eq. (1). This happens frequently, because of various factors:

(i) band degeneracy near an extremum, as in the case of the valence band maximum at Γ in all cubic semiconductors.

(ii) Deviations from parabolicity, as in the conduction band of direct-gap semiconductors. For narrow-gap materials, like InAs or InSb the non-parabolicity is quite large for energies near the band minimum, but even in GaAs it has a sizeable effect on levels with energy larger than 0.1 eV above the band edge.

(iii) There are situations specific to heterostructures in which the single-band approach fails; if the two materials have a band line-up without overlap of the energy gaps, then, in a large and interesting energy region, the wavefunction has conduction band character on one side of the heterojunction, and valence band character on the other. InAs-GaSb superlattices provide an example of this situation.

In all these cases it is necessary to treat more than one band at a time, say n bands, on the same footing. This is accomplished via the k.p formalism [12], in which the n band energies $E_l(\underline{k})$, $l = 1,2...,n$ are obtained as eigenvalues of the nxn matrix $H_{lm}(\underline{k})$, expanded, for k near the origin, up to 2nd order in k:

$$H_{lm}(\underline{k}) = E_l(O)\ \delta_{lm} + \sum_{\alpha=1}^{3} P_{lm}^{\alpha}\, k_\alpha + \sum_{\alpha,\beta=1}^{3} D_{lm}^{\alpha,\beta}\, k_\alpha k_\beta . \qquad (5)$$

Here α,β run over the x,y and z directions. The P and D coefficient matrices are written in terms of momentum matrix elements between the Bloch functions at k = 0 of the n bands in question, and they are, for each material, a set of parameters playing the role of the effective masses.

Following the general ideas of the effective-mass theory in its many-band formulation (as adopted e.g. in the theory of acceptor impurities), we write, instead of Eq. (1)

$$\begin{aligned} \psi^A(\underline{r}) &\simeq \sum_{l=1}^{n} F_l^A(\underline{r})\, u_{\Gamma lA}(\underline{r}) + \dots , \\ \psi^B(\underline{r}) &\simeq \sum_{l=1}^{n} F_l^B(\underline{r})\, u_{\Gamma lB}(\underline{r}) + \dots , \end{aligned} \qquad (6)$$

where the dots replace higher order corrections proportional to ∇F_l which need not be detailed here [13]. The effective-mass equation,

Eq. (4), is now replaced by a set of n differential equations

$$\sum_{m=1}^{n} \left[H_{lm} (-i\hbar \underline{\nabla}) + V(\underline{r}) \; \delta_{lm} \right] F_m(\underline{r}) = EF_l(\underline{r}) \tag{7}$$

for each material, A or B, where, as in Eq. (4), $V(\underline{r})$ is a slowly varying potential. The boundary conditions which must complement Eq. (7) are

$$F_l^A(x, y, z = 0^-) = F_l^B(x, y, z = 0^+),$$
$$\sum_{m=1}^{n} \left[\sum_{\alpha=x,y} (D_{lm}^{z\alpha} + D_{ml}^{\alpha z})k_\alpha - 2iD_{lm}^{zz} \frac{\partial}{\partial z} \right] F_m \quad \text{continuous at } z = 0 \; . \tag{8}$$

The boundary conditions Eq. (8), in analogy to Eq. (3) ensure the continuity of the probability current, provided that one can assume

$$u_{\Gamma lA}(r) \simeq u_{\Gamma lB}(r), \quad l = 1,2,\ldots n \; . \tag{9}$$

They are therefore valid under the condition that the two materials have a set of k = 0 band edges which can be grouped in pairs with similar symmetry and chemical origin. Given the success of the envelope function method, it appears that Eq. (9) is reasonably appropriate for pairs of lattice-matched III-V (or II-VI) compounds, as far as the lowest conduction band or the upper valence-band edge is concerned. It is important to notice that the equality Eq. (9) implies [13] also the equality of the P coefficient matrix for the two materials A and B.

The envelope-function equations for the superlattice eigenstates, in their various forms, can easily be solved numerically, also in the presence of external fields. They provide therefore a simple, accurate and versatile tool for the investigation of the electronic structure. We must however stress that their range of applicability is restricted to energies in the neighborhood of a band edge, where the second order expansion of $E(\underline{k})$ is valid, and to pairs of semiconductors having the relevant band edges at the same k-point and fulfilling Eq. (9). It was already mentioned that the GaAs-indirect AlGaAs system has conduction subbands inaccessible to the method; it is important to remember that the Si-Si_xGe_{1-x} system also has conduction band valleys centered at different k-points and therefore outside the scope of the method.

3. RESULTS FOR GaAs-AlGaAs SYSTEMS

The most extensively investigated system is GaAs-$Al_xGa_{1-x}As$, generally in the low-concentration region, $x \leq 0.4$ so that the envelope-function method is applicable to all conduction states related to the Γ point of the Brillouin zone. This excludes the conduction subbands high enough in energy to approach the secondary minima at X and L. The valence bands, on the other hand, are accessible for all alloy compositions. The band parameters of the two materials are reasonably well known [14]; the band offset value is believed to be well described by the rule that the conduction band discontinuity is between 60 and 70% of the difference between the gaps of AlGaAs and GaAs [15,16]. A description of the conduction and valence band edges for direct-gap III-V compounds is given by an 8 x 8 scheme, including the Γ_6 conduction band, and the Γ_8 and Γ_7 valence band edges, split by the spin-orbit interaction. In selected energy ranges, however, the calculation can be simplified by neglecting some of the bands. Thus, for example, a 6 x 6 model is obtained by neglecting the split-off band, as long as we are not interested in valence subbands lying lower than the J = 3/2 edge by an energy comparable to the spin-orbit splitting Δ ($\Delta \sim 340$ meV for GaAs). For conduction subbands lying close to the band edge, so that non-parabolicity is not important, a simple one-band model is applicable, which produces an analytically soluble dispersion relation [9]:

$$\cos(k_z d) = \cos(k_{zA} d_A) \cos(k_{zB} d_B) - \frac{1}{2}\left(\frac{m_A\, k_{zB}}{m_B\, k_{zA}} + \frac{m_B\, k_{zA}}{m_A\, k_{zB}}\right) \sin(k_{zA}\, d_A)\, \sin(k_{zB}\, d_B), \tag{10}$$

where k_z is the wavevector in the superlattice growth direction, d_A and d_B are the thicknesses of the individual layers and d is the period $d_A + d_B$, k_{zA} and k_{zB} are functions of the energy E and of the momenta k_x and k_y parallel to the interfaces, defined by the relations

$$k^2_{zA,B} = \frac{2m_{A,B}}{\hbar^2} (E - E_{CA,B}) - k_x^2 - k_y^2 . \tag{11}$$

Eqs (10), (11) give the energy dispersion $E(k_x,k_y,k_z)$ in implicit form.

For the valence subbands, the simplest appoach in the neighbourhood of the J = 3/2 valence band top is to neglect the conduction and the split-off bands, and to consider the 4 x 4 k.p matrix known as the Luttinger Hamiltonian [17]. For given k_x, k_y, a suitable canonical transformation can further reduce this problem into two 2 x 2 differential equation systems [18]. This is a good illustration of the complicated subband dispersion which arises from band coupling [19]. In Fig. 1, results are shown for a superlattice composed of 140 Å thick GaAs layers separated by 200 Å thick barriers of $Al_{0.21}Ga_{0.79}As$. The barriers are thick enough to prevent sizeable coupling of different GaAs wells and the subbands can be regarded as those of an isolated quantum well. Notice the anisotropy of the in-plane dispersion, that increases with k. Notice also that the scale of momenta over which the pronounced non-parabolicities take place is of order 10^6 cm^{-1}, i.e. of the inverse of the well width.

At $k_x = k_y = 0$ there is no mixing of $J_z = \pm 3/2$ (heavy holes), with $J_z = \pm 1/2$ (light holes) components. As we move out of the k_z axis, the mixing grows rapidly and produces the non-parabolicities first pointed out by Nedorezov [20], who solved the problem exactly in the limit of infinitely high barriers. Experimentally, the hole subband structure and the mixing effects are best revealed by optical and magneto-optical experiments. Raman experiments [21] on p-type quantum wells give a particularly strong indication of the upward curvature of the light-hole subband. They can be quantitatively compared with theory [22], and good agreement is also found in presence of external stress [23] or magnetic field [24] (see below).

In the example of GaAs-AlGaAs, one has to deal with a system in which no significant charge rearrangement across or near the inter-

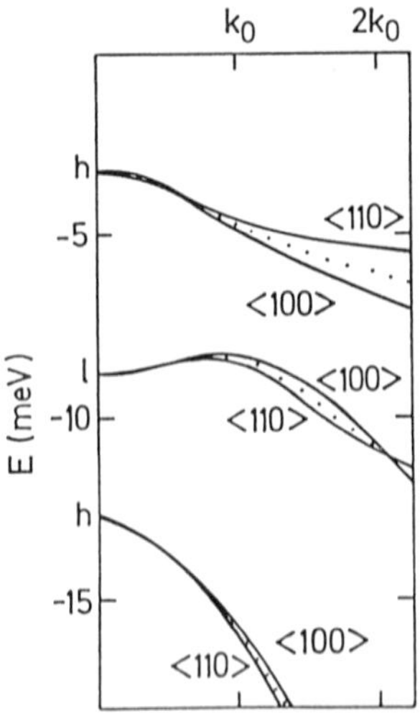

Fig. 1. Valence subband dispersion vs k in two directions in the <001> plane, for a 140 Å GaAs-200 Å $Al_xGa_{1-x}As$ superlattice with x = 0.21. The labels h and l identify heavy-hole and light-hole character at k = 0. The dots indicate an average over the in-plane directions, the axial approximation (Ref. [19]). k_0 denotes $\pi/d = 9.24 \times 10^{-5}$ cm^{-1}

faces takes place, if we consider superlattices involving intrinsic semiconductors. In contrast, in modulation doped superlattices, filling of the quantum wells takes place, with carriers released from impurities located in different layers. In InAs-GaSb superlattices, on the other hand, the band line-up produces the possibility of charge transfer from the GaSb to the InAs layers, even in perfectly intrinsic materials.

Whenever such transfers take place, they contribute to the definition of the one-electron potential $V(\underline{r})$ appearing in the envelope-function equations (4) or (7). The potential depends on the charge density of the states filled by the transferred carriers; and, because the corresponding wavefunctions depend in turn on the potential $V(\underline{r})$, the problem must be solved self-consistently, within some scheme of approximation, e.g. the Hartree approximation, the local density approximation, etc. These procedures are familiar from the treatment of Si MOS systems [25].

The envelope-function method, like the effective-mass method for impurity calculations, is readily extended to include external fields. Examples are pressure and strain fields [26,23], the Coulomb fields of shallow impurities [27,28] or magnetic fields. The last one is particularly important, because a magnetic field perpendicular to the layers of a quasi two-dimensional system has a profound effect on the electronic levels. It quantizes both available degrees of freedom, producing an entirely discrete spectrum of Landau levels. This leads to an enrichment and sharpening of optical structures and to such striking transport phenomena as the Quantum Hall Effect.

The basic transformation to include a magnetic field, described by a vector potential A, into the effective-mass formalism is the replacement of

$$k \quad \text{with} \quad k + \frac{e}{c} A \qquad (12)$$

in the k.p Hamiltonian, Eq. (5). We must then add the direct coupling of the electron and/or hole spin [17,29] to the field B. The corresponding equations for superlattices with appropriate boundary conditions were solved in a simplified 2 x 2 band model [30], in the 4 x 4 Luttinger model [22,31] and in the 6 x 6 model [32]. It is very convenient in the latter two cases to perform the axial approximation, which consists in neglecting the warping of the valence bands in the plane of the interfaces, usually the <001> plane.

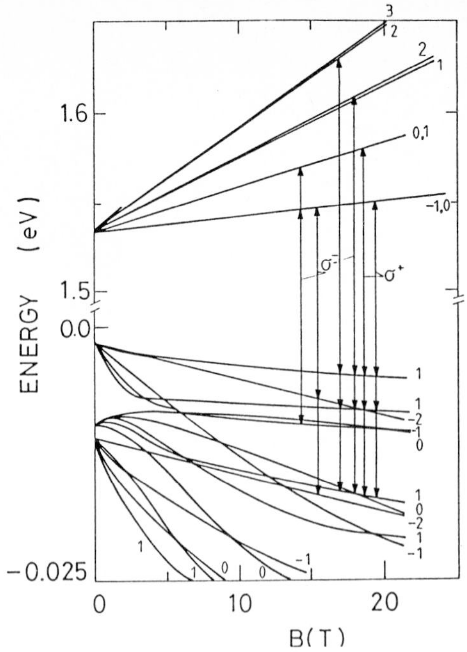

Fig. 2 Landau level diagram for a GaAs quantum well 125 Å thick between $Al_xGa_{1-x}As$ barriers with $x = 0.21$. Allowed transitions for σ^+ and σ^- polarizations in the Faraday configuration are also shown

In GaAs-AlGaAs superlattices, we have a relatively simple dependence of the conduction subbands' Landau levels on the field B, and a very complex one for the valence subbands (Fig. 2). A careful analysis of interband magnetoabsorption data [33] shows however that the conduction band non-parabolicity is larger than predicted by the 8 x 8 model. Ekenberg [34] has shown that a more accurate description of the bulk GaAs conduction band [35] produces the observed mass enhancement. As to the valence bands, the complex dispersion of the sub-bands, shown in Fig. 1, is reflected in a very non-linear field dependence of the Landau levels, shown in Fig. 2.

Intraband magneto-optical experiments [24,36,37] on p-type samples indicate that the theory [24,36,38], although reproducing the main qualitative features, has not been able to reach a detailed quantitative agreement.

4. EXCITONS IN GaAs-AlGaAs QUANTUM WELLS

The calculation of the one-particle energy levels of electrons and holes, as presented in the previous Sections is not sufficient for an accurate interpretation of interband optical experiments in $GaAs\text{-}Al_xGa_{1-x}As$ quantum wells. The optical spectra are often dominated by excitonic transitions from hole subbands below the bulk valence-band to electron subbands above the conduction band edge. As discussed in many contributions to these Proceedings, a knowledge of the exciton binding energies is therefore essential. Recently, variational calculations [39-42] were performed, which showed the importance of the light-heavy hole coupling, together with the confinement effect, in determining the enhancement of the binding energies with respect to the bulk GaAs values.

Here we shall briefly describe a perturbation approach to the problem [43]. The total effective-mass Hamiltonian for the electron and the hole in the quantum well is divided into an unperturbed part H_0, and two perturbation terms, $H_1 + H_2$:

$$H = H_o + H_1 + H_2 \tag{13}$$

where

$$H_o = H_o^e(z_e) + H_o^h(z_h) + H_o^{exc}(\rho). \tag{14}$$

Here z_e, z_h are the coordinates in the confined direction z, while ρ is the radial relative coordinate in the $x = x_e - x_h$, $y = y_e - y_h$ plane, i.e. $\rho = (x^2 + y^2)^{1/2}$. More specifically, H_o^e describes the z-motion of the electron in the quantum well one-dimensional potential, while H_o^h retains only the z_h-dependent diagonal terms of the Luttinger Hamiltonian and the confining potential for holes. It is therefore a diagonal 4 x 4 matrix. These two terms contain therefore the k = 0 quantization of the z-motion into electron, light-hole or heavy-hole subbands. $H_o^{exc}(\rho)$ is a purely two-dimensional hydrogenic Hamiltonian, with interaction strength $\lambda e^2/\varepsilon\rho$, controlled by an adjustable parameter λ. The perturbation terms are

$$H_1(z_e, z_h, \rho) = \frac{\lambda e^2}{\varepsilon\rho} - \frac{e^2}{\varepsilon[\rho^2 + (z_e - z_h)^2]^{1/2}}, \tag{15}$$

which restores the three-dimensional character of the Coulomb interaction, and $H_2(z_h, x, y)$ which restores the off-diagonal components of the Luttinger Hamiltonian. The parameter λ can now be chosen [44-45] to make the effect of Eq. (15) vanishing in first order perturbation theory. As far as H_2 is concerned, it is treated in second order perturbation theory, all necessary matrix elements and energy denominators between unperturbed wavefunctions being determined exactly [43]. This is possible because the unperturbed eigenfunctions are separable and analytically expressed as

$$\Psi_o = \psi_o^e(z_e)\,\psi_o^h(z_h)\,\psi_o^{exc}(\rho,\phi), \tag{16}$$

where the first two factors are one-dimensional particle-in-a-box functions and the third term is a two-dimensional hydrogenic wavefunction. For details of the calculations, the reader is referred to Ref. [43].

A comparison of the results for the binding energies of excitons associated to the first electron subband and to various hole subbands is shown in Table I; the experimental values are obtained by direct observation of the 1s - 2s exciton difference, or by (less reliable) zero-field extrapolations of magneto-optical data. In the former case an estimate of 1 to 2 meV for the 2s exciton binding energies produces the reported spread of 1 meV for the value of the 1s binding energy.

Table I - Comparison with experimental results for the exciton binding energies. x denotes the composition of the $Al_xGa_{1-x}As$ barriers

a: From 1s-2s difference

b: Extrapolated from high-field magneto-optics

c: Extrapolated from low-field magneto-optics

Well width (Å)	x	Hole level	Reference	E_B (meV)	Theory (meV)
75	0.4	H1	46	10.5-11.5(a)	9.4
		L1		11.3-12.3(a)	11.4
92	0.35	H1		9.5-10.5(a)	8.8
		L1		11.2-12.2(a)	10.9
100	0.29	H1	47	13(b)	8.5
		L1		10(b)	10.2
112	0.3	H1	48	12(b)	8.2
75	0.35	H1	33	12(b)	9.3
				10(c)	9.3
		L1		11(c)	11.0
110	0.35	H1		9.5(b)	8.4
				8(c)	8.4
		L1		9(c)	9.9

REFERENCES

1. See e.g. L. Esaki: In Heterojunctions and Semiconductor Superlattices, ed. by G. Allan, G. Bastard, N. Boccara, M. Lannoo, M. Voos (Springer, Berlin, 1986) and references therein
2. K. Ploog, G.H. Döhler: Adv. Phys. 32, 286 (1983)

3. A.G. Milnes, D.L. Feucht: Heterojunctions and Metal Semiconductor Junctions (Academic Press, New York, 1972) Ch. 1-2
4. G.C. Osbourn: J. Appl. Phys. 53, 1586 (1982)
5. G. Abstreiter, H. Brugger, T. Wolf, R. Zachai, C. Zeller: In Two-Dimensional Systems: Physics and New Devices, edited by G. Bauer, F. Kuchar, H. Heinrich (Springer, Berlin, 1986)
6. C.G. Van de Walle, R.M. Martin: Phys. Rev. B34, 5621 (1986)
7. J.N. Schulman, Y.C. Chang: Phys. Rev. B24, 4445 (1981); ibid B27, 2346 (1983)
8. M. Jaros, K.B. Wong, M.A. Gell: Phys. Rev. B31, 1205 (1985); a mixed k.p-pseudopotential scheme has been proposed by C. Mailhiot, T.C. McGill, D.L. Smith: J. Vac. Sci. Tech. B2, 371 (1984)
9. G. Bastard: Phys. Rev. B24, 5693 (1981)
10. D.J. Ben Daniel, C.B. Duke: Phys. Rev. 152, 683 (1966)
11. T. Ando: In Proceedings of the 3rd Brazilian School of Semiconductor Physics, Ed. by C.E.T. Gonçalves da Silva, L.E. Oliveira, J.R. Leite (World Scientific, Singapore, 1987) p. 23
12. E.O. Kane: In Handbook of Semiconductors, edited by W. Paul (North-Holland, Amsterdam, 1982) Vol. 1, p.193
13. M. Altarelli: In Heterojunctions and Semiconductor Superlattices, Ref. 1
14. O. Madelung, editor, Landolt-Börnstein: Numerical Data and Functional Relationships in Science and Technology, group III, Vol. 17 (Springer, Berlin 1982)
15. G. Duggan: J. Vac. Sci. Tech. B3, 1224 (1985)
16. T.W. Hickmott: In Two-Dimensional Systems: Physics and New Devices, Ref. 5
17. J.M. Luttinger: Phys. Rev. 102, 1030 (1956)
18. D.A. Broido, L.J. Sham: Phys. Rev. B31, 888 (1985)
19. M. Altarelli, U. Ekenberg, A. Fasolino: Phys. Rev. B32, 5138 (1985)
20. S.S. Nedorezov: Soviet Phys. Sol. State 12, 1814 (1971) (Fiz. Tverd. Tela 12, 2269 (1970))
21. A. Pinczuk, D. Heiman, R. Sooryakumar, A.C. Gossard, W. Wiegmann: Surf. Sci. 170, 573 (1986)
22. T. Ando: J. Phys. Soc. Japan, 54, 1528 (1985)
23. G. Platero, M. Altarelli: In Proceedings of the 18th International Conference on the Physics of Semiconductors, Stockholm, 1986, edited by E. Engström (World Scientific Publishers, Singapore, 1987) p. 633

24. D. Heiman, A. Pinczuk, A.C. Gossard, A. Fasolino, M. Altarelli: In Proceedings of the 18th International Conference on the Physics of Semiconductors, Stockholm, 1986, Ref. 23, p.617
25. See e.g. T. Ando, A.B. Fowler, F. Stern: Rev. Mod. Physics 54, 437 (1982)
26. G.D. Sanders, Y.C. Chang: Phys. Rev. B32, 4282 (1985)
27. C. Mailhiot, Y.C. Chang, T.C. McGill: Phys. Rev. B26, 4449 (1982)
28. W.T. Masselink, Y.C. Chang, H. Morkoc: Phys. Rev. B28, 7373 (1984); J. Vac. Sci. Tech. B2, 376 (1984)
29. C.R. Pidgeon, R.N. Brown: Phys. Rev. 146, 575 (1966)
30. G. Bastard: Phys. Rev. B25, 7584 (1982)
31. A. Fasolino, M. Altarelli: In Two Dimensional Systems, Heterostructures and Superlattices, ed. by G. Bauer, F. Kuchar, H. Heinrich (Springer, Berlin, 1984)
32. A. Fasolino, M. Altarelli: Surf. Sci. 140, 322 (1984)
33. D.C. Rogers, J. Singleton, R.J. Nicholas, C.T. Foxon, K. Woodbridge: Phys. Rev. B34, 4002 (1987); F. Ancilotto, A. Fasolino, J.C. Maan: Superlattices and Microstructures, 3, 187 (1987)
34. U. Ekenberg: In Proceedings of the MSS-3 Conference, Montpellier, 1987 (Les Editions de Physique, Les Ulis) in press
35. M. Braun, U. Rössler: J. Phys. C: Solid State Phys. 18, 3365 (1985)
36 Y. Iwasa, N. Miura, S. Tarucha, H. Okamoto, T. Ando: Surf. Sci. 170, 587 (1986)
37. H.L. Störmer, Z. Schlesinger, A. Chang, D.C. Tsui, A.C. Gossard, W. Wiegmann: Phys. Rev. Lett. 51, 126 (1983)
38. See a short review by L.J. Sham: In High Magnetic Fields in Semiconductor Physics, edited by G. Landwehr (Springer, Berlin, 1987) p.288 and references therein
39. G.D. Sanders, Y.C. Chang: Phys. Rev. B31, 6892 (1985); 32, 5517 (1985); J. Vac. Sci. Tech. B3, 1285 (1985)
40. K.S. Chan, J. Phys. C19, L125 (1986)
41. D.A. Broido, L.J. Sham: Phys. Rev. B34, 3917 (1986)
42. G.E.W. Bauer, T. Ando: In Proceedings of the 18th Internationalnal Conference on the Physics of Semiconductors, Ref. 23, p. 537
43. U. Ekenberg, M. Altarelli: Phys. Rev. B35, 7585 (1987)
44. Y.C. Lee, W.N. Mei, K.C. Lin: J. Phys. C15, L469 (1982)
45. T.-F. Jiang: Solid State Commun. 50, 589 (1984)

46. P. Dawson, K.J. Moore, G. Duggan, H.I. Ralph, C.T. Foxon: Phys. Rev. B34, 6007 (1986)
47. J.C. Maan, G. Belle, A. Fasolino, M. Altarelli, K. Ploog: Phys. Rev. B30, 2253 (1984)
48. N. Miura, Y. Iwasa, S. Tarucha, H. Okamoto: In Proceedings of the 17th International Conference on the Physics of Semiconductors, San Francisco, 1984, ed. by J.D. Chadi, W.A. Harrison (Springer, New York, 1985) p.359

Near Band Gap Photoemission in $Al_{0.27}Ga_{0.63}As/GaAs$ Quantum Wells

F. Ciccacci[1], *H.J. Drouhin*[2], *C. Hermann*[2], *R. Houdré*[2], *G. Lampel*[2], *and F. Alexandre*[3]

[1]Dipartimento di Fisica, Università di Roma "Tor Vergata", I-00173 Roma, Italy

[2]Laboratoire de Physique de la Matière Condensée, Ecole Polytechnique, F-91128 Palaiseau Cedex, France

[3]C.N.E.T., 196 avenue Henri Ravera, F-92220 Bagneux Cedex, France

1. Introduction

Coadsorption of a monolayer of cesium and oxygen on a clean surface of GaAs in Ultra High Vacuum (UHV) lowers the work function ϕ to a value close to 1 eV. A p-type sample with band gap E_g = 1.50 eV is then activated to a Negative Electron Affinity (NEA) situation in which the vacuum level lies below the bottom of the conduction band in the bulk solid /1/.

We have applied this technique to a Molecular Beam Epitaxy (MBE) grown sample containing ten 53 Å thick GaAs Quantum Wells (QW's), separated by 53 Å thick $Al_{0.27}Ga_{0.73}As$ barriers. The QW's are covered by a 530 Å thick GaAs overlayer. The top layer is heavily p-doped in order to make the activation to NEA conditions possible /1/, whereas the QW's region is undoped in order to insure that the Debye length is much larger than the width of the QW's region itself, so that the energy scheme is as shown in Fig. 1.

Since the light penetration depth is larger than the width d of the entire structure including the GaAs overlayer (d ≃ 1500 Å), absorption of photons with energy $h\nu$ larger than the GaAs gap may induce transitions between the valence and conduction bands (VB and CB) in GaAs, the discrete levels in the QW's and the VB and CB in $Al_{0.27}Ga_{0.73}As$. Moreover the diffusion length being larger than d, electrons excited in the various conduction states may reach the surface before recombination and be emitted into vacuum giving rise to a photoemission current /2/, the study of which is the purpose of the present paper.

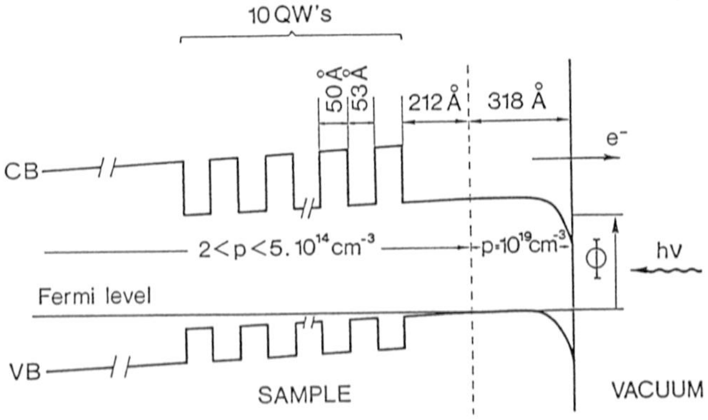

Figure 1. Description in real space of the sample with GaAs wells separated by $Al_{0.27}Ga_{0.73}As$ barriers. The space charge region electric field in the undoped material has been exaggerated for the sake of clarity.

2. Experimental

Samples mounted on a specially designed sample holder are heated up to 600 °C in UHV to obtain a clean surface before activation to NEA by Cs and O_2 coadsorption. Measurements at low temperature are performed by pressing a cold finger against the sample /3/. Light from a quartz halogen lamp through a monochromator (resolution 8 meV) or from a He-Ne laser is focused onto the sample. The total photoemitted current can either be measured as a function of $h\nu$, or energy analyzed by a high resolution (20 meV) electron monochromator at a given photon energy /3/.

3. Results and discussion

Figure 2 shows the number of photoemitted electrons per incident photons versus photon energy $Y(h\nu)$ at 120 K. As discussed in ref. 1, $Y(h\nu)$ is equivalent to an absorption spectrum and reveals the various electronic transitions occurring in the sample. After the low-energy onset corresponding to band gap excitation in the GaAs overlayer, several structures (indicated by arrows in Fig. 2) are observed. They are related to transitions between localized states in the QW's, while the last one E_g^{AlGaAs} occurring at 1.84 eV is due to band gap excitation in $Al_{0.27}Ga_{0.73}As$. The relative amplitudes of the QW's related structures to the GaAs band gap absorption ($h\nu < 1.56$ eV) indicate that all the ten wells contribute to the photoemission current, thus confirming the high efficiency of electron tunneling through the barriers towards the surface /4/.

In order to compare these results with theoretical predictions, we have calculated the transitions energies between valence and conduction states of a single GaAs well embedded in $Al_{0.27}Ga_{0.73}As$, in the parabolic approximation with effective masses as given in Adachi's review /5/. We have also neglected excitonic effects in the optical transitions. This very simple model is well

	$\Delta E_c = 0.85\,\Delta E_g$	$\Delta E_c = 0.6\,\Delta E_g$	experimental
E_{1h}	1.58	1.58	1.57
E_{1l}	1.60	1.61	1.59
E_{3h1c}	-	1.68	1.69
E_{2h}	1.80	1.76	1.77
E_g^{AlGaAs}	1.84	1.84	1.84

Table 1. Experimental energy position of the structures present in the yield of Fig. 2 and calculated transition energies for two values of the conduction band offset in a single well. All values are in eV.

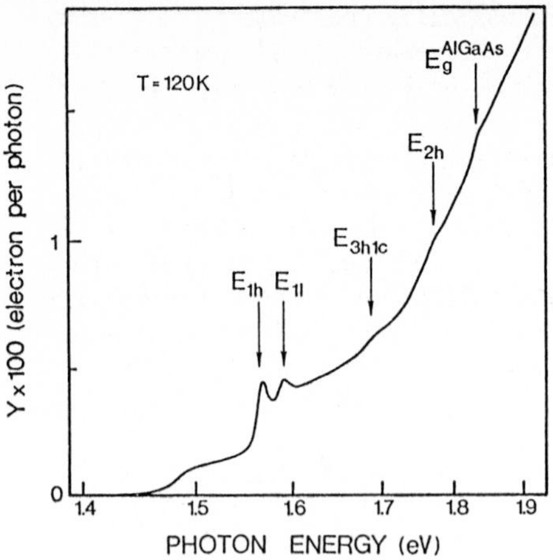

Figure 2. Photoemission yield curve vs photon energy at 120 K. E_{nh} (E_{nl}) refer to transitions from heavy (light) hole levels to conduction levels with $\Delta n = 0$. $E_{nhn'c}$ refer to transitions from heavy hole n levels to conduction n' levels. E_g^{AlGaAs} refers to band gap transitions in $Al_{0.27}Ga_{0.73}As$.

justified in view of the weak coupling between the wells in our sample and the relatively low precision of the present measurements.

The comparison is presented in Table I for two values of conduction band offset ΔE_c, namely $\Delta E_c = 0.6\,\Delta E_g$ and $\Delta E_c = 0.85\,\Delta E_g$ ($\Delta E_g = 0.34$ eV being the difference between the $Al_{0.27}Ga_{0.73}As$ and GaAs gaps at 120 K, as seen in Fig. 2 and according to the gap variation in the alloy /5/).

As can be noted, a good agreement for all the transitions is observed only for $\Delta E_c = 0.6\,\Delta E_g$. In particular the n=3 heavy hole level is absent in the case of $\Delta E_c = 0.85\Delta E_g$. Moreover the high energy transition E_{2h} is more sensitive to the exact value of the offset than the lower ones E_{1h} and E_{1l}.

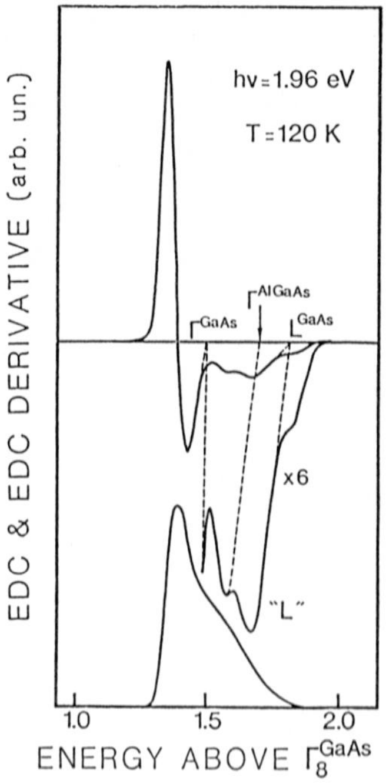

Figure 3.
EDC and EDC derivative at 120 K of electrons emitted from the NEA sample for 1.96 eV excitation energy. Electron energies are referred to the GaAs VB maximum Γ^{GaAs}. The broad "L" peak in the EDC derivative is due to electrons thermalized in the GaAs L minimum.

The total yield measurement is spectroscopically equivalent to usual absorption or excitation luminescence spectra /6/. However, in a photoemission experiment additional information is obtained from the Energy Distribution Curves (EDC's) /7/.

Figure 3 shows such an EDC and its derivative obtained with $h\nu$ = 1.96 eV at 120 K. As compared to bulk GaAs samples an extra peak (indicated by the arrow) appears in the EDC derivative. Such peak is present at the same energy in EDC's taken at other photon energies larger than 1.84 eV, i.e. the $Al_{0.27}Ga_{0.73}As$ gap. Therefore, it is attributed to ballistic electrons injected in the GaAs overlayer from the bottom of the $Al_{0.27}Ga_{0.73}As$ conduction band. By using the extrapolation procedure discussed in ref. 7, we find the energy position of $Al_{0.27}Ga_{0.73}As$ conduction band relative to the GaAs one. In this way E_c is directly measured. We get ΔE_c = (0.20±0.02) eV = (0.60±0.05) ΔE_g in very good agreement with the value obtained on basis of the energy positions of the transitions, as discussed above.

We note, however, that the determination of Γ^{AlGaAs} in Fig. 3 is somewhat unprecise because of the presence of the broad "L" peak. This peak in the EDC derivative is due to electrons thermalized in the GaAs L minimum.
Further measurements on samples with different alloy compositions are in progress, in order to better separate the peak related to the Γ^{AlGaAs} minimum from the "L" structure.

4. References

1. R.L.Bell, Negative Electron Affinity Devices, Clarendon Press (Oxford, 1973)
2. R.Houdré, C.Hermann, G.Lampel, P.M.Frijlink, and A.C.Gossard, Phys. Rev. Lett. 55, 734 (1985)
3. R.Houdré, Thèse de Doctorat, Université de Paris-Sud Orsay (1985)
4. R.Houdré, C.Hermann, G.Lampel, and A.C.Gossard, Physica Scripta T13, 241 (1986)
5. S.Adachi, J. Appl. Phys. 58(3), R1 (1985)
6. C.Weisbuch, R.C.Miller, R.Dingle, A.C.Gossard, and W.Wiegmann, Sol. St. Commun. 37, 219 (1981)
7. H.J.Drouhin, C.Hermann, and G.Lampel, Phys. Rev. B31, 3859 (1985)

Excitons in Heterostructures

Y.C. Chang, G.D. Sanders†*, and H.Y. Chu*

Department of Physics, University of Illinois at Urbana-Champaign,
110 West Green Street, Urbana, IL 61801, USA
†Present address: Universal Energy Systems, 4401 Dayton-Xenia Road,
Dayton, OH 45432, USA

1. Introduction

Semiconductor quantum wells and superlattices[1] have received growing interest in recent years. Optical measurements including photoabsorption, photoluminescence, and Raman scattering are widely adopted for probing the electronic states of these heterostructures. Excitons play a dominant role in determining the optical properties of semiconductor heterostructures. For example, in absorption spectra of quantum wells, they contribute to a series of well-identified peak structures, providing a means for studying the electronic subband energies. By comparing the energy position of these exciton peaks with simple model calculations, much information about the electronic states can be revealed. Using a simple quantum mechanical model which contains essentially a particle in a one-dimensional square-well potential (particle-in-a-box model)[2], one can obtain a fairly accurate description of the energy levels in a GaAs-$Al_xGa_{1-x}As$ quantum well with well size between 50 Å and 300 Å. This model predicts a selection rule for the inter-band optical transitions which requires the difference in principal quantum numbers of the initial hole state and the final electron state in a quantum well to be zero, i.e. $\Delta n = 0$. Indeed, most experimental data indicate that $\Delta n = 0$ transitions are at least an order of magnitude stronger than the other transitions which violate this selection rule.

Recent studies[3,4] of the electronic and optical properties of semiconductor quantum wells (or superlattices) have shown that the mixing of heavy and light hole components (valence band mixing) in the quantum well (or superlattice) states can lead to $\Delta n \neq 0$ (forbidden) interband transitions with strengths much larger than those expected from the simple particle-in-the-box model. The most pronounced $\Delta n \neq 0$ transition is associated with the exciton involving the first light-hole subband and the second conduction subband. This is because the first light-hole subband interacts strongly with the second heavy hole subband. Excellent agreement between theoretical predictions and experimental data for the adsorption spectra of GaAs-$Al_xGa_{1-x}As$ quantum wells has recently been reported[5].

Excitonic effects on the absorption spectra of modulation-doped GaAs-$Al_xGa_{1-x}As$ quantum wells have also been studied within the multi-band effective-mass approximation[6]. The effects of carrier screening on the band structures, excitonic states, and band-to-band transitions are investigated. It is found that the shapes of subband structures are nearly unaffected for doping concentrations less than $10^{12}cm^{-2}$. The binding energies of excitons decrease quickly with increasing doping concentration, but the oscillator strengths of exciton absorptions are found to be much less sensitive to the screening. Peak structures in the absorption spectra can still be observed even when oscillator strengths of exciton states become negligible due to the screening effect. These peak structures are due to the modification of the continuum states via final-state interaction.

There are added complications for the excitonic effect in semiconductor superlattices. At the mini-zone boundary ($q = q_{max}$), the lowest subband energy is a maximum along z, but a minimum along x or y. Thus, we have an M_1 saddle point there. In the past, excitonic effects associated with the M_1 saddle point in bulk semiconductors have attracted a great deal of interest both theoretically and experimentally[7-13]. Although qualitative understanding of this phenomenon can be obtained via a contact-potential model[9,10], quantitative calculations for the Coulombic potential are desired. Chu and Chang[14] have recently performed quantitative calculations for the line shapes of photoabsorption associated with saddle-point excitons in a tight-binding model for bulk semiconductors and in a Kronig-Penney model for semiconductor superlattices, including the electron-hole Coulomb interaction. It is found that when the width of superlattice band dispersion is comparable to the exciton binding energy, some prominent structures appear which can be interpreted as saddle-point exciton resonances.

2. Excitons in GaAs-Al_xGa_{1-x}As Quantum Wells

Binding energies of excitons in GaAs-Al_xGa_{1-x}As quantum wells were first studied by Greene and Bajaj[15] in an effective-mass model which ignores the coupling between the heavy-hole and light-hole bands. In this model the exciton states $\psi(\vec{r})$ satisfy an effective mass equation

$$-\frac{\hbar^2}{2}[m_e^{-1}\nabla_e^2+m_t^{-1}(\partial^2/\partial x_h^2+\partial^2/\partial y_h^2)+m_l^{-1}\partial^2/\partial z_h^2]\psi(\vec{r}_e,\vec{r}_h)$$
$$+[v(\vec{r}_e-\vec{r}_h)+V_e(z_e)+V_h(z_h)]\psi(\vec{r}_e,\vec{r}_h)=E\psi(\vec{r}_e,\vec{r}_h), \qquad (1)$$

where $V_e(z_e)$ and $V_h(z_h)$ are quantum well potentials for the electron and hole, respectively. $v(\vec{r}_e - \vec{r}_h)$ is the electron-hole Coulomb interaction. m_e is the electron effective mass, and m_t and m_l are the transverse (in-plane) and longitudinal (along the growth direction) hole effective masses, respectively. The effective-mass equation for excitons was solved by Greene and Bajaj[15] with a variational method. They found that the binding energies of both the heavy-hole and light-hole excitons as functions of the well width have a maximum somewhere between 30 Å and 70 Å. The maximum exciton binding energy in a quantum well is about twice as large as the bulk value.

Sanders and Chang[4] have considered the effect of band hybridization on the excitonic states and their contributions to the photoabsorption. The band hybridization comes from the mixing of the heavy and light hole bands in the valence subband states[3,4,16]. The n-th conduction subband state of a quantum well can be written as

$$\psi^e_{n,\vec{k}_\parallel}=\sum_{k_z} e^{i\vec{k}\cdot\vec{r}}\, f_{n,\vec{k}_\parallel}(k_z)|c,\vec{k}\rangle, \qquad (2)$$

where $|c,\vec{k}\rangle$ is the cell-periodic function of the conduction band state at $\vec{k}$. The Fourier transform of $f_{n,\vec{k}_\parallel}(k_z)$ satisfies the simple effective-mass equation

$$[\frac{\hbar^2}{2m_e}(k_\parallel^2-\partial^2/\partial z^2)+V_e(z)]f_{n,\vec{k}_\parallel}(z)=E^e_n(\vec{k}_\parallel)f_{n,\vec{k}_\parallel}(z). \qquad (3)$$

It is easy to see that in this approximation, the envelope function f is independent of $\vec{k}_\parallel$, and we shall drop the $\vec{k}_\parallel$ index for f from now on.

The m-th valence subband state can be written as

$$\psi^{h}_{m,\vec{k}_{\parallel}} = \sum_{\nu,k_z} e^{i\vec{k}\cdot\vec{r}}\, g^{\nu}_{m,\vec{k}_{\parallel}}(k_z)\,|\nu,\vec{k}>, \tag{4}$$

where $|\nu,\vec{k}>$'s are the cell-periodic functions of the well-material valence-band states correct to the first order in the $\vec{k}\cdot\vec{p}$ perturbation[17]. The envelope function $g^{\nu}_{m,\vec{k}_{\parallel}}(k_z)$ satisfies a multi-band effective mass equation in $\vec{k}$-space viz.

$$\sum_{\nu'} [H^{(o)}_{\nu,\nu'}(\vec{k}_{\parallel},k_z) - E^{h}_{m}(\vec{k}_{\parallel})\delta_{\nu,\nu'}]g^{\nu'}_{m,\vec{k}_{\parallel}}(k_z) + \sum_{k'_z} <k_z|V_h(z)|k'_z>g^{\nu}_{m,\vec{k}_{\parallel}}(k'_z) = 0,$$

where $H^{(o)}_{\nu,\nu'}(\vec{k}_{\parallel},k_z)$ are matrix elements of the Luttinger-Kohn Hamiltonian[18] for describing the bulk valence band structure. Here we have ignored the mismatch between the valence band parameters for the well and barrier material. Because the wave functions of interest are mostly confined in the well material, the approximation used is justified. Both Eq. (1) and Eq.(2) can be solved by a variational method in which the real-space envelope functions are written in terms of linear combinations of Gaussian-type orbitals[4]. Detailed results for the valence subband structures can be found in Refs. 4,6,16.

The excitonic states associated with the n-th conduction subband and the m-th valence subband can be witten as

$$\Psi^{nm}_{X} = \sum_{n,m} \sum_{\vec{k}_{\parallel}} G_{nm}(\vec{k}_{\parallel})\Psi^{e}_{n,\vec{k}_{\parallel}}\Psi^{h}_{m,-\vec{k}_{\parallel}}. \tag{6}$$

Here we have ignored the interaction between excitonic states associated with different pairs of electron and hole subbands. This approximation is valid when the energy separation between excitons derived from different pairs of subbands is large compared to the exciton binding energy. This is the case for quantum wells with well width between 50 Å and 200 Å. Substituting the expansion into the Schrödinger equation yields an effective-mass equation for the exciton envelope function:

$$[E^{e}_{n}(\vec{k}_{\parallel}) - E^{h}_{m}(\vec{k}_{\parallel}) - E^{ex}_{nm}]G_{nm}(\vec{k}_{\parallel}) + \sum_{\vec{k}'_{\parallel}} V_{nm}(\vec{k}_{\parallel} - \vec{k}'_{\parallel})G_{nm}(\vec{k}'_{\parallel}) = 0, \tag{7}$$

with

$$V_{nm}(\vec{k}_{\parallel} - \vec{k}'_{\parallel}) = -\frac{4\pi e^2}{\varepsilon}\sum_{q}\frac{F_n(q)G_m(\vec{k}'_{\parallel},\vec{k}_{\parallel},q)}{|\vec{k}_{\parallel} - \vec{k}'_{\parallel}|^2 + q^2}, \tag{8}$$

where

$$F_n(q) = \sum_{k_z} f^{*}_{n}(k_z)f_n(k_z - q),$$

$$G_m(\vec{k}'_{\parallel},\vec{k}_{\parallel},q) = \sum_{k_z\nu}(g^{\nu}_{m,\vec{k}'_{\parallel}}(k_z))^{*}g^{\nu}_{m,\vec{k}_{\parallel}}(k_z - q),$$

and ε is the static dielectric constant. We shall further approximate $G_m(\vec{k}'_\parallel, \vec{k}_\parallel, q)$ by its value at the zone center, $G_m(0,0,q)$, and V_{nm} becomes a function of $|\vec{k}_\parallel - \vec{k}'_\parallel|$ only. It was estimated that this approximation introduced an error of about 5 %[6]. Numerical analysis[6] indicates that $V_{nm}(q)$ can be well approximated by a simple analytic expression first introduced by Price[19], viz. $V_{nm}(q) = -4\pi e^2/[\varepsilon q(1 + \rho_{nm} q)]$, where ρ_{nm} is a fitting parameter. The strict two-dimensional (2-d) limit is obtained by letting $\rho_{nm} \to 0$. The variation of ρ_{nm} with well width for several prominent excitons in GaAs-$Al_xGa_{1-x}As$ quantum wells (x=0.25 and 0.4) are shown in Fig. 1. The excitonic states Ψ_X^{nm} are now completely determined by the subband dispersion $E_n^e(\vec{k}_\parallel)$ and $E_m^h(\vec{k}_\parallel)$ and the parameter ρ_{nm}.

To proceed we shall further ignore the angular dependence of the expression $E_n^e(\vec{k}_\parallel) - E_m^h(\vec{k}_\parallel)$. This is a very good approximation since $E_n^e(\vec{k}_\parallel)$ has a circular symmetry and it dominates $E_m^h(\vec{k}_\parallel)$ for all cases of interest. The angular deviation in $E_m^h(\vec{k}_\parallel)$ is found to be around 20 % for GaAs-$Al_xGa_{1-x}As$ quantum wells[3,4]. We have now converted the problem to that of a simple quasi 2-d exciton with circular symmetry. Eq.(10) is now reduced to a one-dimensional integral equation involving the radial component of $\vec{k}_\parallel$. Numerical solution to such an integral equation can be obtained by various techniques. We have used variational method to find the binding energies of several prominent excitons in

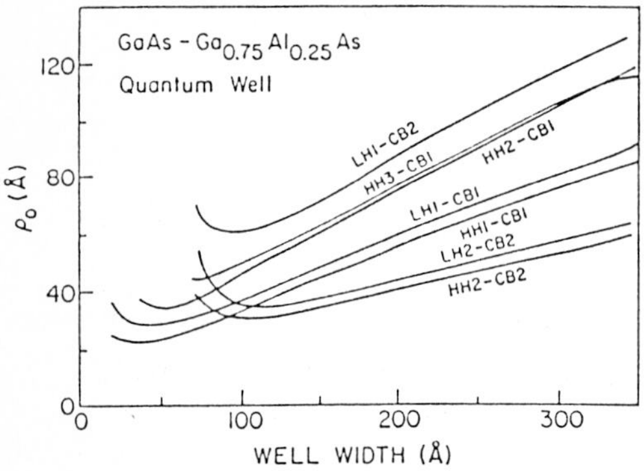

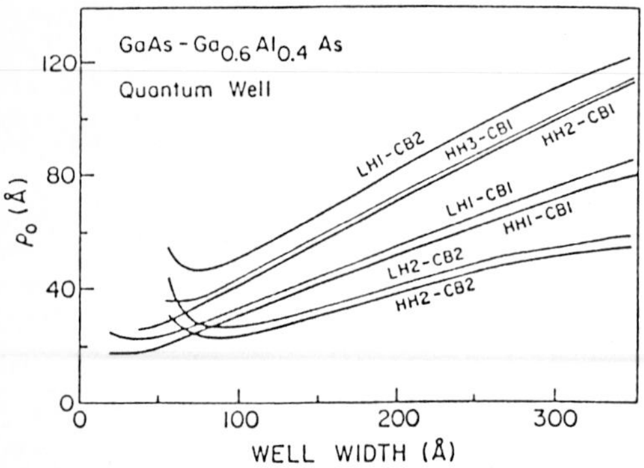

Fig. 1. Potential parameter, ρ_{nm} for several prominent excitons in GaAs-$Al_xGa_{1-x}As$ quantum wells (x = 0.25 and 0.4 plotted as functions of the well width.

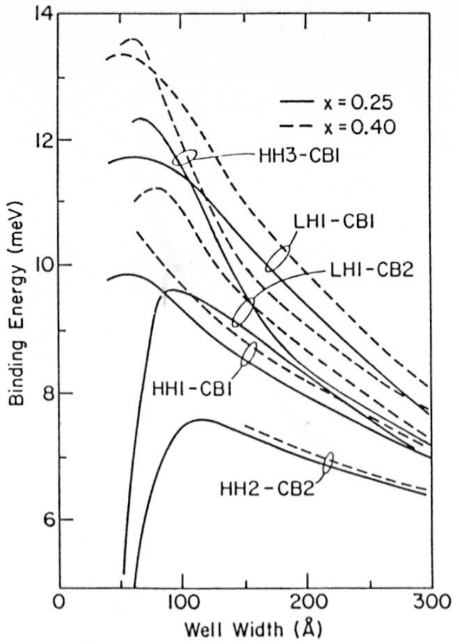

Fig. 2. Binding energies of several prominent excitons in GaAs-$Al_xGa_{1-x}As$ quantum wells (x = 0.25 and 0.4) plotted as functions of the well width.

GaAs-$Al_xGa_{1-x}As$ quantum wells (x=0.25 and 0.4). The results as functions of GaAs well width are shown in Fig. 2.

We have also studied the excitonic effects on the absorption spectra of GaAs-$Al_xGa_{1-x}As$ quantum wells[6]. The absorption coefficent is given by[6]

$$\alpha(\hbar\omega) = C \sum_{nm} [g_{nm}\Delta_{nm}(\hbar\omega - E^{ex}_{nm}) + \sum_{\vec{k}_\parallel} h_{nm}(\vec{k}_\parallel)|\hat{\varepsilon}\cdot\vec{P}_{nm}(\vec{k}_\parallel)|^2\delta(E^e_n(\vec{k}_\parallel) - E^h_m(\vec{k}_\parallel) - \hbar\omega)], \tag{9}$$

where $C = 4\pi^2e^2/n_0cm_0^2V\omega$ and $g^{nm} = (2/m_0)|\sum_{\vec{k}_\parallel}|\hat{\varepsilon}\cdot\vec{P}_{nm}(\vec{k}_\parallel)|^2$. m_0 is the free electron mass, V is the volume of the crystal, and $\hat{\varepsilon}$ indicates the direction of polarization of the incident radiation. $h_{nm}(\vec{k}_\parallel)$ is the Coulomb enhancement factor due to the final-state electron-hole interaction. This factor is proportional to the probability of finding the electron and hole at the same point in space, i.e. $h_{nm}(\vec{k}_\parallel) = |\Psi^{nm}_{X,\vec{k}_\parallel}(o)|^2$, where $\Psi^{nm}_{X,\vec{k}_\parallel}(\vec{r})$ is the $\vec{k}_\parallel$-th exciton continuum state associated with subbands (n,m). Note that in the presence of the electron-hole Coulomb interaction, $\vec{k}_\parallel$ is no longer a good quantum number, but merely a label of the continuum states. $\Delta_{nm}(E) = (\Gamma_{nm}/\pi)(E^2+\Gamma^2_{nm})^{-1}$ is a Lorentzian function of half-width Γ_{nm}. We use the empirical rule $\Gamma_{nm} = \Gamma_0 nn'$, where n and n' are the principal quantum numbers for the electron and heavy- or light-hole quantum well states. Γ_0 is an empirical parameter selcted to be 1 meV. The theoretically predicted absorption spectrum for a 102-Å GaAs-$Al_{0.27}Ga_{0.73}As$ quantum well for unpolarized light propagating along the growth direction (z) is shown in Fig. 3. The experimental data of Miller et. al.[5] is superimposed for comparison. The agreement between the theoretical and

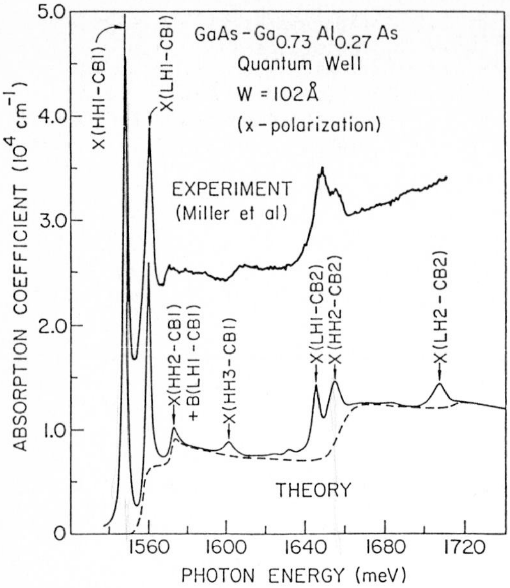

Fig. 3. Theoretical absorption spectrum and experimental excitation spectrum (from Ref. 5) for a 102Å GaAs-$Al_{0.27}Ga_{0.63}As$ quantum well.

experimental results is fairly good. The excitonic effect in conjunction with the valence-band mixing accounted for the strong $\Delta n \neq 0$ forbidden transitions identified as HH2-CB1, HH3-CB1 and LH1-CB2.

3. Effects of Modulation Doping

Modulation doping in the barrier layers of a superlattice (or multiple quantum well) provides free carriers in the well layers and at the same time avoids the scattering of carriers from the ionized impurities. Ultra-high mobilities can thus be achieved[20]. Furthermore, the lack of scattering between carriers and impurities allows us to ignore the disorder effect even in the case of high carrier concentration. The existence of free carriers affects the properties of the superlattice in several ways:

(a) The electrostatic potential introduced by the free carriers modifies the subband energies. This potential can be obtained by solving the Poisson equation and the Schrödinger equation in a self-consistent way[21]. Exchange and correlation effects should also be included to get a more accurate solution[21]. The resulting modulation doping potential is a function of z, which we denote as $V_d(z)$. For electrons in a quantum well treated within the effective-mass approximation, the presence of $V_d(z)$ will not change the subband dispersion, but simply introduce an energy shift to each subband. For holes, because the amount of band mixing depends on the relative subband energies, there is a slight change in the subband dispersion due to the modulation doping, but the change is insignificant for doping concentrations less than 10^{12} cm^{-2}[6].

(b) The Coulomb interaction between carriers (many-body effect) leads to a band renormalization that modifies the energies of the subbands which are partially filled with carriers. It was shown that within the random-phase approximation (RPA), the change of subband dispersion is small, but the energy shift can be substantial for high carrier concentrations[22].

(c) Free carriers block the excitation to subband states which are occupied with free carriers, thus reducing the exciton binding energy or prohibiting the formation of excitons. The blocking also causes a cut-off in the band-to-band absorption spectrum, an effect first realized by Burstein[23] and by Moss[24].

(d) The free carriers screen the electron-hole Coulomb interaction, thus reducing the binding energies of excitons associated with any pair of subbands. The same screening effect also modifies the final-state interaction in the band-to-band transitions, causing excitonic resonance structures in the absorption spectrum[25].

Sanders and Chang[6] have studied the effect of modulation-doping on the excitonic states in GaAs-$Al_xGa_{1-x}As$ quantum wells. They have considered the effects (a), (c), and (d) described above, but neglected the many-body effect (b). The many-body effect is important for the excitons associated with the partially filled subbands, but not for those associated with the unfilled subbands. Furthermore, in their approach, the modulation potential is approximated by a simple parabolic function with a height V_0 determined by the carrier density. This approximation is justified for weak doping potentials ($V_0 < 10$ meV). In Fig. 4, we show calculated results for the binding energies of several prominent excitons versus the hole concentration in a 100-Å p-type modulation-doped GaAs-$Al_{0.25}Ga_{0.75}As$ quantum wells. It is seen that all exciton binding energies except for the HH1-CB1 exciton approach zero asymptotically. This is because in two dimensions an arbitrarily weak potential has at least one bound state. The HH1-CB1 exciton is seen to become unbound at a finite hole concentration due to the carrier blocking effect described in (c). The oscillator strengths of these excitons have also been calculated, and it is found that they are much less sensitive to the doping concentration. The oscillator strength depends on the amplitude squared of the exciton wave function at the origin (i.e. where the electron and hole coincide in real space). The screening effect reduces the range of the Coulomb potential substantially, but not the strength near the origin. Thus, the amplitude of the wave function at the origin is much less sensitive to the screening than the binding energy, and so is the oscillator strength.

The absorption spectra of modulation-doped quantum wells at zero temperature have also been studied. The results for a 115 Å p-type modulation-doped GaAs-$Al_{0.44}Ga_{0.56}As$ quantum well are shown in Fig. 5 along with the excitation spectrum measured by Miller et al. [26] In this figure the dashed curves correspond to the results for band-to-band transitions. The difference between the solid and the dashed curve shows the contribution from the exciton bound state. HH1-CB1 exciton is absent in the theoretical spectrum due to the blocking effect. A weak HH1-CB1 peak appears in the experimental data taken at 5 K, but it is suppressed at lower temperatures. The contribution from the HH2-CB2 exciton is seen to be rather weak, whereas the band-to-band transition shows a resonance structure there. Thus even when the exciton oscillator strength is diminishing due to carrier screening, a peak structure can still show up due to the resonance effect.

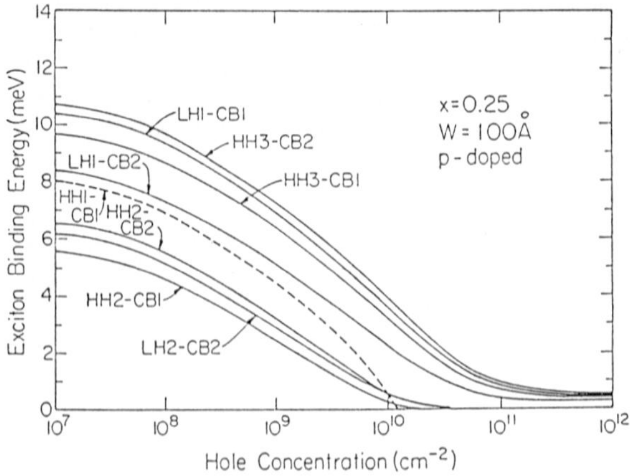

Fig. 4. Binding energies of several prominent excitons in a 100-Å p-type modulation-doped GaAs-$Al_{0.25}Ga_{0.75}As$ quantum well plotted as functions of the hole concentration.

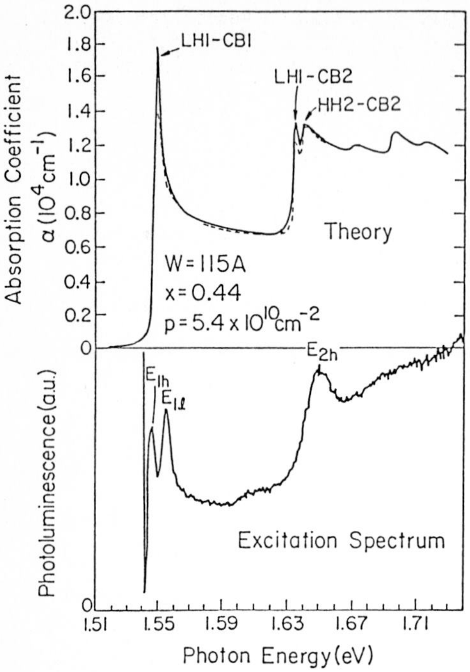

Fig. 5. Theoretical absorption spectrum and experimental excitation spectrum (from Ref. 20) for a 115 Å modulation-doped GaAs-$Al_{0.44}Ga_{0.56}As$ quantum well with a hole concentration n = 5.4 x 10^{10} cm^{-2}.

4. Excitons in Superlattices

Superlattices can be viewed as multiple quantum wells if the barrier material in each superlattice unit cell is sufficiently wide so that interactions between electronic states associated with one well and those with adjacent wells are negligible. For superlattices with narrow barrier width, the energy dispersion for wave vectors along the growth direction (z) is large and its effect on the excitonic states is quite interesting. One expects the binding energy and oscillator strength of the exciton (in the ground state) to be monotonically increasing functions of the barrier width with the lower and upper limits being the corresponding values of the bulk (3-d case) and the quantum well (quasi 2-d case), respectively. More interestingly, the line shape of the absorption spectra changes gradually from a 3-d character in the ultra-thin barrier case to a quasi 2-d character in the wide-barrier case, and for intermediate barrier widths it is difficult to predict what the absorption spectrum is like, unless a computer simulation is done.

For simplicity, we consider the excitonic states for the first pair of subbands and ignore the valence band mixing effect. The Schrödinger equation for an excitonic state Ψ_X can be written in $\vec{k}$-space as

$$\sum_{\vec{k}'} [(E_{cv}(\vec{k}) - E)\delta_{\vec{k},\vec{k}'} - v(\vec{k},\vec{k}')]\phi(\vec{k}') = 0, \quad (10)$$

where $\Psi_X = \sum_{\vec{k}} \phi(\vec{k})|\vec{k}>$ and $|\vec{k}>$ denotes the electron-hole product state at $\vec{k}$. Here $E_{cv}(\vec{k})$ is the energy difference between the conduction and valence subband states at $\vec{k}$ and v denotes the electron-hole Coulomb interaction. For a superlattice, we write $\vec{k} = (\vec{k}_\parallel, q)$, where $\vec{k}_\parallel$ is the projection of $\vec{k}$ in the plane normal to the growth direction and q is the projection of $\vec{k}$ in the growth direction. The Coulomb matrix element for superlattice states is given by

$$v(\vec{k},\vec{k}') = \frac{2\pi e^2}{\varepsilon} \sum_l F_e(q,q' - lK)F_h(q,q' - lK)\frac{1}{|\vec{k}_\parallel - \vec{k}'_\parallel|^2 + (q - q' + lK)^2} , \tag{11}$$

where $F_e(q,q') = \sum_n f(q+nK)f(q'+nK)$ and $f(q+nK)$ is the projection of the superlattice wave function $\psi_q(z)$ for the electron in the plane-wave basis $e^{i(q+nK)z}$. $K = \pi/d$, where d is the length of the superlattice unit cell. $F_h(q,q')$ is similarly defined for the hole.

To solve the Schrödinger equation (10), we expand the wave function Ψ_X in terms of linear combinations of a set of basis states defined by

$$\beta_j = \sum_{\vec{k}\varepsilon\Delta_j} |\vec{k}> / \sqrt{\Delta_j} \quad , \tag{12}$$

i.e. $\Psi_X = \sum_j \psi(\vec{k}_j)\beta_j$, where Δ_j denotes a small volume in $\vec{k}$-space centered at $\vec{k}_j$. Substituting this expansion into the Schrödinger equation for Ψ_X immediately leads to a simple eigen-value problem:

$$\sum_{j'} \overline{H}_{j,j'}\psi(\vec{k}_{j'}) - E\psi(\vec{k}_j) = 0, \tag{13}$$

with

$$\overline{H}_{j,j'} = \frac{\sum_{\vec{k}\varepsilon\Delta_j} \sum_{\vec{k}'\varepsilon\Delta_j} H(\vec{k},\vec{k}')}{\Delta} \quad . \tag{14}$$

After some mathematical manipulations, the multi-dimensional integral in (14) can be reduced to a one-dimensional integral involving special functions, which is then performed by numerical methods. Eq. (13) can then be diagonalized to give the energies and corresponding wave functions for a few low-lying discrete exciton states and a good sampling of the continuum states. The absorption coefficient for interband transitions can be written as

$$\alpha(E) = \sum_i |\Psi_{X,i}(0)|^2 \delta(E_i - E), \tag{15}$$

where $\Psi_{X,i}(\vec{r})$ is the wave function of the i-th excitonic states with an energy eigenvalue E_i. Here the label i runs through discrete states as well as the continuum states. Within the effective-mass approximation, we have $\Psi_X(0) = \sum_{\vec{k}} \phi(\vec{k})$.

To minimize the size of the matrix to be diagonalized, while maintaining high precison, the symmetry of the system must be fully exploited. Note that the absorption spectrum only depends on $\Psi_X(0)$, which is nonzero only for states with full symmetry of the system. For superlattices, we are interested in states with q in the entire mini zone and with $\vec{k}_\parallel$ near the zone center. We approximate the band structure in the parallel direction by a parabolic expression. Thus the system has a circular symmetry in parallel directions and a reflection symmetry in the growth direction. We then use symmetrized basis states labeled by the radial component of $\vec{k}_{\parallel}$ and the absolute value of q. A cut-off Λ is introduced for the sampling of $k_{\parallel}$. The final results for energies near the saddle point are insensitive to the choice of the cut-off, as long as Λ is large enough. In order to obtain a smooth absorption spectrum, we replace the delta function in Eq.(15) by a Lorentzian function with a half-width Γ, viz. the magnitude of Γ must match the size of Δ_j.

Fig. 6 shows the calculated absorption spectra associated with the saddle-point exciton in GaAs-$Al_{0.25}Ga_{0.75}As$ superlattices for a number of $Al_{0.25}Ga_{0.75}As$ layer thicknesses (L_B). The width of the GaAs layer (L_W) is kept at 80 Å. All spectra are broadened by a Lorentzian function with a width of 1 meV. The dashed curves are the absorption spectra in the absence of the electron-hole Coulomb interaction. Prominent structures associated with saddle-point exciton resonances can be seen in this figure for intermediate values of L_B. At large and small L_B's the spectra are similar to the corresponding quantum well and bulk results. High resolution excitation spectroscopy measurements for a large number of GaAs-$Al_xGa_{1-x}As$ superlattices (including part of the series shown in Fig. 6) have recently been performed[27]. Variation of the line shape of the absorption spectra due to the change of barrier thickness is apparent. We find qualitative agreement between our theoretical predictions and the experimental data. However, because of the difficulty in preparing precise superlattice structures, the inhomogeneous broadening tends to smear out the structures associated with the saddle-point excitons.

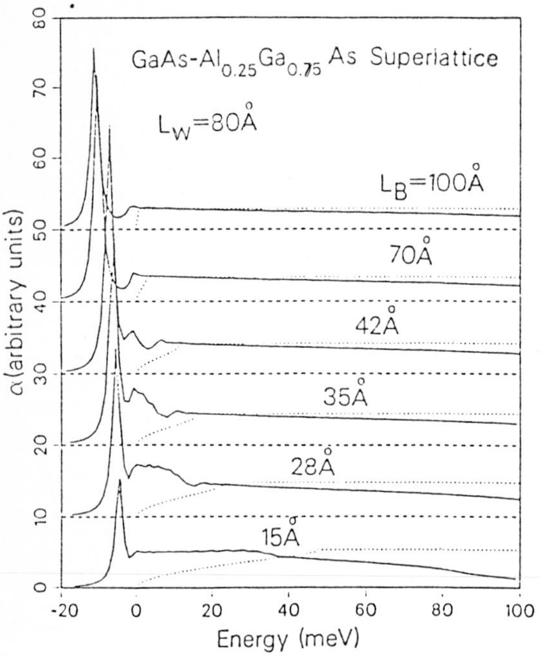

Fig. 6. Theoretical absorption spectra for $GaAs_{0.25}Ga_{0.75}As$ superlattices with well width L_B = 80 Å and barrier widths L_B = 15 Å, 28 Å, 35 Å, 42 Å, 70 Å, and 100 Å.

Acknowledgments

This work was supported by the U. S. Office of Naval Research (ONR) under Contract No. N00014-81-K-0430.

References

1. L. Esaki, R. Tsu, IBM J. Res. Develop. 14, 61 (1970)
2. L. L. Chang, L. Esaki, Surf. Sci. 98, 70 (1980)
3. Y.-C. Chang, J. N. Schulman, Appl. Phys. Lett 43, 536 (1983); Phys. Rev. B. 31, 2069 (1985)
4. G. D. Sanders, Y. C. Chang, Phys. Rev. B 31, 6892 (1985); Phys. Rev. B 32, 4282 (1985)
5. R. C. Miller, A. C. Gosard, G. D. Sanders, Y. C. Chang, J. N. Schulman, Phys. Rev. B 32, 8452 (1985)
6. G. D. Sanders and Y. C. Chang, J. Vac. Sci. Technol. 3, 1285 (1985); Phys. Rev. B 35, 1300 (1987)
7. see for example, F. Bassani and C. P. Parrasvacini, Electronic States and Optical Properties in Solids (Pergammon, New York, 1975).
8. J. C. Phillips, Phys. Rev. 136A, 1705 (1964)
9. B. Velicky, J. Sak, Phys. Status Solidi 16, 147 (1966)
10. H. Kamimura, K. Nakao, J. Phys. Soc. of Japan, V.24, No.6, 1313(1968)
11. E. O. Kane, Phys. Rev. 180, 852 (1969)
12. J. E. Rowe, F. H. Pollak, M. Cardona, Phys. Rev. Lett. 22, 933 (1969)
13. S. Antoci, E. Reguzzoni, G. Samoggia, Phys. Rev. Lett. 24, 1304 (1970)
14. H. Y. Chu, Y. C. Chang, Phys. Rev. B (submitted)
15. R. L. Greene, K. K. Bajaj, Solid State Commun. 45, 831 (1983)
16. A. Fasolino, M. Altarelli, in Two-Dimensional Systems, Heterostructures, and Superlattices, edited by G. Bauer, F. Kucher, and H. Heinrich (Springer-Verlag, New York, 1984); M. Altarelli, U. Ekenberg, A. Fasolino Phys. Rev. B 32, 5138 (1985)
17. E. O. Kane, J. Phys. Chem. Solids, 1, 82 (1956)
18. J. M. Luttinger, W. Kohn, Phys. Rev. 97, 869 (1956)
19. P. J. Price, J. Vac. Sci. Technol. 19, 599 (1981)
20. H. L. Stömer, A. C. Gossard, W. Wiegmann, K. Baldwin, Appl. Phys. Lett. 39, 912 (1981)
21. See for example, T. Ando, A. B. Fowler, F. Stern, Rev. Mod. Phys. 54, 437 (1982)
22. S. Schmitt-Rink, C. Ell, H. Huag, Phys. Rev. B 33, 1183 (1986); D. A. Kleinman, Phys. Rev. B 32, 3766 (1985)
23. E. Burstein, Phys. Rev. 93, 632 (1954)
24. T. S. Moss, Proc. Phys. Soc. London, Sect. B 67, 755 (1954)
25. Y. C. Chang, J. Appl. Phys., 59, 2173 (1986)
26. R. C. Miller, A. C. Gossard, D. A. Kleinman, D. Muntaneu, Phys. Rev, B 29, 3470 (1984)
27. J. J. Song, private communications

Electron-Hole Exchange Interaction in Quantum Wells

Y. Chen[1], *B. Gil*[2], *P. Lefebvre*[2], *H. Mathieu*[2], *T. Fukunaga*[3], and *H. Nakashima*[3]

[1]Scuola Normale Superiore, Piazza dei Cavalieri 7, I-56100 Pisa, Italy
[2]Groupe d'Etudes des Semiconducteurs, Université des Sciences et Techniques du Languedoc, Place E. Bataillon, F-34060 Montpellier, France
[3]Optoelectronics Joint Research Laboratory, Nakahara-Ku, Kawasaki 211, Japan

We study the exchange splitting of quasi-2D excitons confined in quantum wells. The experimental results obtained from luminescence, wave length modulated reflectivity are compared with the theoretical calculations. A giant enhancement of the exchange splitting was found when the 2D confinement of the exciton wave function increased.

Recently a number of works have been devoted to the problem of the exciton in quantum well structures(QWs); it is well known that the confinement of the wave function increases both the binding energy and the oscillator strength of the exciton[1,2]. The intrinsic splitting or the fine structure of the exciton in QWs due to the electron-hole exchange interaction(EHEXI), however, has not yet received enough attention. In this contribution, we study the behavior of the EHEXI with increasing the 2D confinement.

The electron-hole exchange interaction closely depends on the spatial extension of the exciton wave function[3~6]. In 2D materials like QWs, the size of the exciton wave function decreases rapidly and the overlap of the electron and hole increases with the decreasing of the well thickness. This leads to an enhancement of the exchange effect which can be much larger than in 3D cases[7]. Within the effective mass approximation, the wave function of a nondegenerate exciton in a QW, can be expressed as(the spin multiplicity will be included in the next part)

$$\Phi = \frac{1}{\sqrt{A}} e^{iK.R} \varphi_\xi^{2D}(r_e - r_h)\, \chi_e(z_e)\, \chi_h(z_h)\, U_{c,o}(r_e) U_{v,o}(r_h)\,, \tag{1}$$

where A is the area, K and R the in-plane wave vector and coordinate of the center of mass. φ_ξ^{2D} represents a quasi-2D exciton envelope function,specified by the quantum number ξ ; χ_e and χ_h are solutions of the finite square well problem for the electron and the hole respectively. $U_{c,o}$ and $U_{v,o}$ are the zone center Bloch functions of the conduction and the valence bands. The envelope functions, φ_ξ^{2D}, χ_e and χ_h are slowly varying in comparison with the modulation on atomic scale of the Bloch functions. The dominant contribution to the EHEXI arises when the electron and the hole are on the same site. Taking into account the difference in the probability of finding the electron and the hole in different unit cells, a simple algebra gives the exchange energy(short range of the EHEXI) as follows:

$$\varepsilon_{ex}^{2D} = \Omega J_o\, W_{eh}\, |\varphi_\xi^{2D}(0)|^2 \tag{2}$$

with

$$W_{eh} = \int_{-\infty}^{\infty} dz\, [\, \chi_e(z)\, \chi_h(z)]^2, \tag{3}$$

where Ω is the volume of the unit cell and J_o is the exchange integral[3]. Making use of the values of the 1s ground state of a 3D exciton(with the effective Bohr radius a_o^*), (2) becomes

$$\varepsilon_{ex}^{2D} = (\pi a_o^{*3})\, \varepsilon_{ex}^{3D}\, W_{eh}\, |\varphi_\xi^{2D}(0)|^2 . \tag{4}$$

Similarly, the longitudinal-transverse splitting(long-range part of the EHEXI) is given by

$$E_{LT}^{2D} = (\pi a_o^{*3})\, E_{LT}^{3D}\, W_{eh}\, |\varphi_\xi^{2D}(0)|^2 \tag{5}$$

where ε_{ex}^{3D} and E_{LT}^{3D} denote the corresponding values of the 3D exciton.

Now let us consider realistic structures such as GaAs-$Ga_{1-x}Al_xAs$ QWs. Since the light and the heavy hole subbands are energetically split, two types of Wannier excitons can be referred to: the heavy-hole exciton(HHE) and the light-hole exciton(LHE). Consequently, and different from the 3D exciton, two values are expected for $W_{eh}\, |\varphi_\xi^{2D}(0)|^2$; each one with its own well thickness dependence. First, we suppose that in the range of our investigation, the exchange interaction has a small influence on the confinement of the exciton wave functions and the perturbation treatment is still valid. Next, we adopt the approximation of decoupled heavy and light hole subbands[1], so the transition energies without the EHEXI contribution are: $E_h = \varepsilon_g + E_e + E_{HH} - R_H$ and $E_l = \varepsilon_g + E_e + E_{LH} - R_L$ where ε_g is the bulk GaAs band gap. E_e, E_{HH} and E_{LH} are the first subband energies for electron, heavy and light holes respectively, including the effect of effective mass mismatches between GaAs and $Ga_{1-x}Al_xAs$ and the variation of the effective masses with the subband gaps. R_H and R_L are respectively the binding energy of the heavy-hole and the light-hole excitons. In the exciton basis $\{|J,m_h\rangle|S,m_s\rangle\}$ hereafter labeled $\{|m_h,m_s\rangle\}$, and in a spherical approximation, the exchange Hamiltonian of 3D excitons is written as[4]: $H = \Delta_o + \Delta_1 \mathbf{J}.\sigma$; where $\Delta_o = -3\Delta_1/2 = 3\Delta/8$ and $\Delta = 4\,\varepsilon_{ex}^{3D}/3$. In the case of a QW, we have to take into account the difference between the envelope functions of the HHE and the LHE. Let $\Delta_h = 4\varepsilon_{ex}^{2D}(HHE)/3; \Delta_l = 4\varepsilon_{ex}^{2D}(LHE)/3$ and $\Delta_{hl} = W_{eh}^{hl} \times\sqrt{\Delta_h\Delta_l/W_{eh}^{h}W_{eh}^{l}}$, where W_{eh}^{h} and W_{eh}^{l} are given by (3) respectively for the HHE and the LHE; $W_{eh}^{hl} = \int_{-\infty}^{\infty} dz\, \chi_e^2(z)\chi_h^h(z)\chi_h^l(z)$ where $\chi_h^h(z)$ and $\chi_h^l(z)$ are the solutions for the HHE and the LHE respectively. Since the long-range part of the EHEXI only gives rise to an additional splitting of the exciton states which involve the spin singlet and we deal with a delocalized Wannier exciton, the longitudinal-transverse splitting corresponds simply to a shift toward high energy of the longitudinal exciton[9,10]. The total Hamiltonian including the EHEXI contributions takes the following form:

$$\begin{array}{ccccc} |\pm3/2,\pm1/2\rangle & |\pm3/2,\mp1/2\rangle & |\pm1/2,\pm1/2\rangle & |+1/2,-1/2\rangle & |-1/2,+1/2\rangle \end{array}$$

$$\begin{bmatrix} E_h & 0 & 0 & 0 & 0 \\ 0 & E_h+\frac{3}{4}\Delta_h & \mp\frac{\sqrt{3}}{4}\Delta_{hl} & 0 & 0 \\ 0 & \mp\frac{\sqrt{3}}{4}\Delta_{hl} & E_l+\frac{3}{4}\Delta_l & 0 & 0 \\ 0 & 0 & 0 & E_l+\frac{1}{2}(\Delta_l+E_{LT}^{2D}) & -\frac{1}{2}(\Delta_l+E_{LT}^{2D}) \\ 0 & 0 & 0 & -\frac{1}{2}(\Delta_l+E_{LT}^{2D}) & E_l+\frac{1}{2}(\Delta_l+E_{LT}^{2D}) \end{bmatrix}$$

Hence the fine structure of the HHE and the LHE consists respectively of two and three levels; the first 2×2 block diagonal matrix gives dipole-allowed transitions in σ polarization.

Our numerical investigation has been done for GaAs-$Ga_{0.5}Al_{0.5}As$ QWs. Fig.1 displays the well width dependence of the exchange energy of the HHE(solid lines) and the LHE(dashed lines) for two different values of aluminum content. For x=0.5 QWs, the first excitonic transition in the barriers is indirect, but this seems not to influence significantly the resulting EHEXI. It is worth noting that the behavior of the EHEXI in QWs is similar to those of the exciton binding energy[1], however the relative increase in the exchange effect is much more important than the effective Rydberg one when the well width decreases.

In Fig.2, we show a typical luminescence spectrum(bottom) of our samples. Superimposed to the splitting due to well fluctuations, each recombination line reveals a doublet substructure (A_i and B_i transitions). The possibility to observe such a splitting is closely related to growth interruption at the heterointerfaces[8]. All B_i lines arise from dipole allowed transitions. These transitions are observed in the wave length-modulated reflectivity spectrum at top of the figure while the A_i lines are not. The 2K temperature of the sample favors the population rate in the first order forbidden state A_i which can only be observed in luminescence. The σ-allowed transitions involving light hole excitons(C_i) have also been measured.

The well thickness dependence of the splitting between transition energies, calculated for GaAs-$Ga_{0.5}Al_{0.5}As$ QWs is given in Fig.3. The bulk values of GaAs have been used as : $\Delta = 0.3$ meV to obtained the best fit of the HHE exchange splittings; and $E_{LT}^{3D} = 0.08$ meV according to its experimental measurement[11]. All energies are given with respect to the average value $(E_h + E_l)/2$. The fine structure of the HHE exhibits two doublets from which one is dipole-allowed(solid line) and the other dipole forbidden. The fine structure of the LHE exhibits three levels from which only one(the doublet) is dipole-allowed(solid line) in the experimental configuration(σ-polarization). The measured splittings have been included in the same figure; the lowest transition energies have been slightly shifted in order to coincide with $E_{1,2}$.

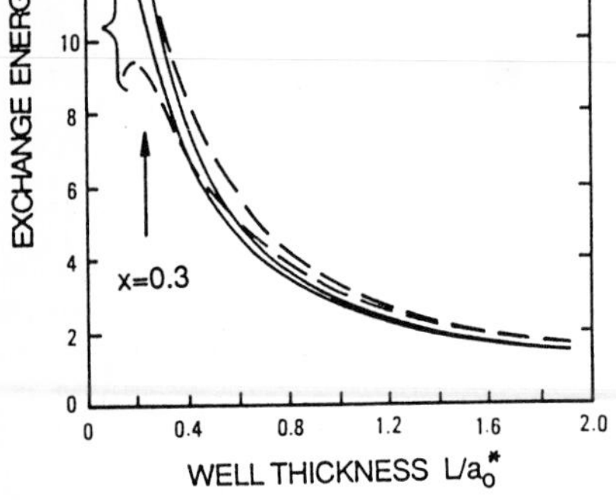

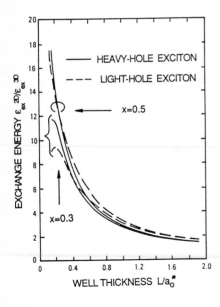

Fig.1. Well width dependence of the exchange energy for GaAs-$Ga_{1-x}Al_xAs$ quantum wells with aluminum contents x=0.3 and x=0.5. The solid lines correspond to the heavy-hole excitons, the dashed lines correspond to the light-hole excitons.

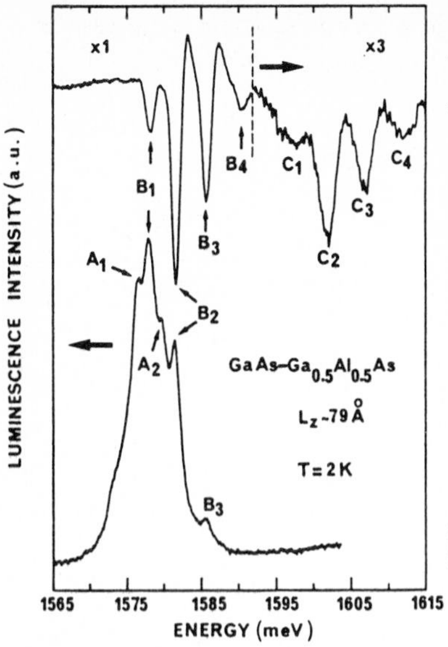

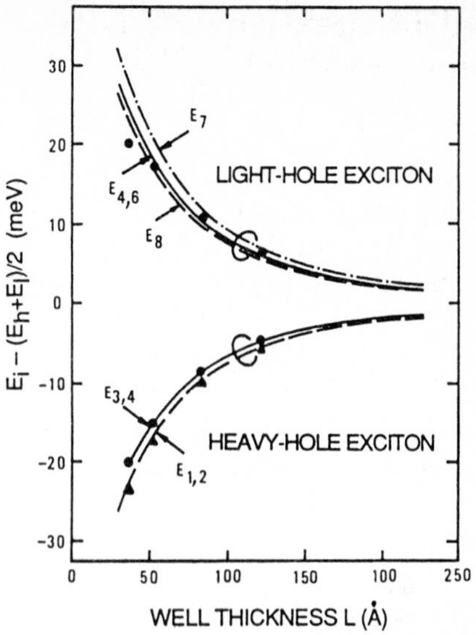

Fig.2. Luminescence spectrum(bottom) and the wave length modulated reflectivity(top). The dipole-allowed states B_i are observed in both spectra; the thermalization effects favor the population rate of low-energy levels and permit to observe luminescence from the dipole forbidden transitions A_i. At 2K, C_i states are easily obtained in reflectance.

Fig.3. Well width dependence of the splitting between the transition energies for GaAs-$Ga_{0.5}Al_{0.5}As$ quantum wells. The $E_{3,5}(B_i)$ and $E_{4,6}(C_i)$ eigenstates are dipole-allowed in σ-polarization. The E_7 state is dipole-allowed in π-polarization and the others are forbidden; A_i lines correspond to the $E_{1,2}$ doublet.

It has been shown that both the short- and the long-range parts of the electron-hole exchange interaction exhibit an important enhancement in QWs. The intrinsic splitting or the fine structure of the heavy-hole and the light-hole excitons are closely related to the 2D confinement of the exciton wave functions.

Acknowledgement : One of the authors(Y.C.) is greatly indebted to Professors F.Bassani and Y.C.Chang for useful discussion and support.

REFERENCES :

1. L. Greene, K.K. Bajaj and D.E. Phelps. Phys. Rev. B**29**,1807(1984); L. Greene and K.K. Bajaj, Phys. Rev. B**31**,913(1985);
2. G.D. Sanders and Y.C. Chang, Phys. Rev. B**32**,5517(1985); Phys. Rev. B**35**,1300(1987)
3. R.J. Elliott: In Polarons and Excitons, ed. by C.G. Kuper and G.D. Whitefield, (Edinburgh, Oliver and Boyd,1961)p.269
4. R.S. Knox, Solid Stat. Phys. (Suppl. Vol. 5), (New York, Academic Press,1963)
5. F. Bassani and G. Pastori Parravicini, Electronic States and Optical Transitions in Solids, (Pergamon press, Oxford,1975)
6. K. Cho: In Excitons, ed. by K. Cho, Top. Curr. Phys. Vol.**14**, (Springer, Berlin, 1979)
7. R. Bauer, D. Bimberg, J. Christen, D. Oertel, D. Mars, J.N. Miller, T.Fukunaka and H. Nakashima, In Proceedings of 18th Int. Conf. on Semicond. Phys., Stockholm, 1986, ed. by O. Engstrom, (World Scientific, Singapore,1986) p.525
8. T. Fukunaga, K.L.I. Kobayashi and H. Nakashima, Surf. Sci. **174**,71(1986)
9. R. Bonneville and G. Fishman, Phys. Rev. B**22**, 2008(1980)
10. M. Suffczynski, L. Smierkowski and W. Wardzynski, J. Phys. C**8**, L52(1975)
11. R. Ulbrich and C. Weisbuch, Phys. Rev. Lett. **38**, 865(1977)

Optical Properties of Excitons in Quantum Wells

E.O. Göbel

Philipps-Universität, Fachbereich Physik,
Renthof 5, D-3550 Marburg, Fed. Rep. of Germany

Some results on the linear optical properties of excitons in GaAs/AlGaAs quantum wells are discussed. Special attention is paid to the recombination dynamics of quasi-two-dimensional excitons and its dependence on external electric fields applied perpendicular to the quantum well layers.

I. Introduction

The linear as well as nonlinear optical properties of direct gap III-V semiconductors are strongly affected by excitonic effects |1| at least at sufficiently low temperatures. The Coulomb interaction between electrons and holes results in a hydrogen-like series for the optical transitions and in an enhancement of the "band to band" transition strength. Coupling of the excitonic polarisation of a direct gap bulk semiconductor to the electromagnetic field of light requires a polariton picture for the description of the optical properties |2|. In addition, binding of free excitons to neutral or charged impurities plays a major role even in relatively pure bulk semiconductors at low temperatures |3|. The strong contribution of bound excitons in optical transitions has been attributed to the "giant oscillator strength" of bound excitons |4|.

Excitons in quantum wells (QW) are different from bulk (3D) excitons in many respects |5,6,7| as has been discussed in detail by G.Bastard |8|. First of all, the exciton motion is "free" only within the layer formed by the substituent with the smaller band gap, i.e. the GaAs layers in a GaAs/AlGaAs QW. The 3D-Polariton model consequently cannot be applied. Nevertheless, polariton-like excitations exist in QW in principle |9| if the damping of the exciton is small, but they can be excited only in particular experimental configurations similar to surface polaritons due to momentum conservation. Second, the electronic states of excitons in QW are between purely 3D and two-dimensional (2D) excitons. The effective mass theory of 3D Wannier excitons |1| can be modified for the ideal 2D case |10| , leading to an increase of the 1s exciton binding energy by a factor of four and a decrease of the Bohr radius to $a_o/2$ (a_o= 3D Bohr radius). The maximum radial density is at $a_{2D}=a_o/4$ instead of a_o for the 3D exciton.

The theoretical treatment of real QW with finite thickness normal to the layers (z - direction) and finite height of the barrier layers taking into account the actual band structure is more complicated |11,12,13|. The most important results are:

1) Two exciton series are obtained in GaAs/AlGaAs QW due to the lifting of the degeneracy of the light and heavy hole bands at the center of the Brillouin zone, refered to as heavy hole (hh) and light hole(lh) excitons. Outside k=0 the different hole subbands couple, resulting in a complicated dispersion and highly nonparabolic effective hole masses |14,15|.

2) The exciton binding energy increases continuously with decreasing QW thickness L_z for $L_z \leq a_o$. For extremely narrow QW, however, the exciton wave

function will extend considerably into the barrier layers due to their finite height and for $L_z \to 0$ the exciton will not correspond to a 2D exciton but to a 3D exciton in the barrier material. In accordance with the increase of the exciton binding energy in QW the separation between the ground state and the excited states increases as well [16,17].

3) The selection rules for optical transitions, which are $\Delta n=0$ (n = subband number for electrons and holes) in the ideal 2D case are weakened in real QW due to the different extension of the electron and hole wave-functions into the barrier material.

Another striking difference between bulk and QW excitons is the minor importance of excitons bound to neutral or charged impurities in the optical spectra of undoped QW opposite to undoped bulk material. Bound excitons dominate only in the case of rather high doping. The neutral acceptor bound exciton could be identified in Be doped GaAs/AlGaAs QW in addition to the free exciton and band to acceptor transition [18]. Instead of "classical" binding, free excitons in QW, however, may get localized within potential fluctuations of interface states. This point will be discussed separately in Sect. II.

The most widely used experimental tool to investigate the optical properties of excitons in QW are "standard" photoluminescence, "optical pumping" using circular polarized excitation and detection [19], reflectively measurements [20,21], excitation spectroscopy and absorption. Dynamical properties of excitons in QW have been studied by time-resolved luminescence [22,23,24], excite and probe experiments [25-28], nonlinear hot luminescence correlation [29] and transient grating experiments [30,31,32].

II. Transition Strength of Excitons in Quantum Wells (An Experimentalist's View)

The dominant recombination channel of optically excited carriers in undoped QW is free exciton recombination [19]. Nevertheless, the emission of the lowest (hh) exciton generally is slightly shifted to lower energies (typically a few meV) with respect to the absorption or excitation spectrum maximum, as shown in Fig. 1 for a double QW with L_z=9nm for example.

This red shift (Stokes-shift) of the emission, as well as the inhomogeneous broadening of the exciton recombination line, is attributed to localization of excitons within potential fluctuations |33,34,35| due to well width fluctuations originating from the two-dimensional-like growth. The low-energy electronic states become localized and a mobility edge occurs if the size of the two-dimensional islands with thickness differences of the order of one monolayer are smaller than the exciton Bohr

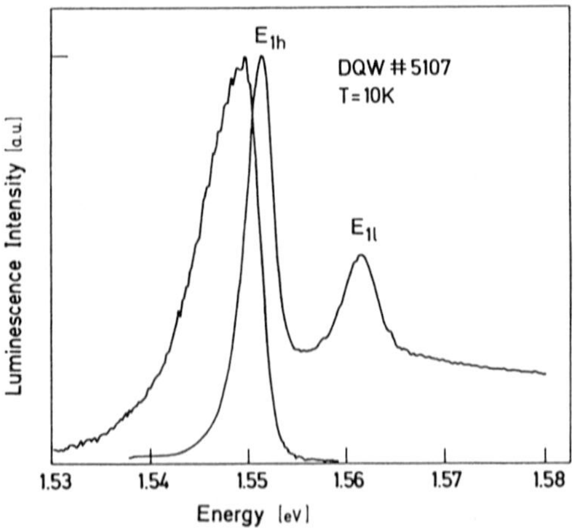

Fig. 1
Photoluminescence and photoluminescence excitation spectrum of a double quantum well with L_z=9nm at T=5K

radius |36,37|, resulting in a pronounced energy dependence of the relaxation rate |38,39|, homogeneous linewidth |35|, and diffusion coefficient |37|. The homogeneous linewidth |35| and diffusion coefficient both decrease for energies below the mobility edge because of reduced exciton scattering and mobility, respectively. The exciton emission lineshape is then inhomogeneously broadened |33,40,41,42| and red shifted with respect to the maximum absorption due to relaxation of the excitons into the lower energy localized states before recombination. Additionally, well width fluctuations may occur on a scale with characteristic dimensions larger than the exciton Bohr radius, which cause an additional, but conceptionally different localization: Localized and delocalized states may coexist energetically (but not spatially) in the latter case, whereas this is not possible in the former case |36,43|. Well width fluctuations characterized by island size larger than the Bohr radius thus cause an additional broadening of the exciton emission characterized by the distribution function of the fluctuations or even splitting of the exciton recombination line into two or more components corresponding to the islands with different width |42,44|. The quality of the GaAs on AlGaAs (normal interface) and the AlGaAs on GaAs (inverted interface) is quite different in MBE growth |45| and the island size in particular of the normal interface can be considerably increased by growth interruption, and therefore the exciton lineshape may be quite different for samples grown under different conditions |42,46|. Actually, homogeneous 2D exciton transitions together with bound exciton recombination |31| have been reported for MBE grown QW with $L_z > 13.5$ nm. Bound exciton recombination in QW samples with localization of the free excitons therefore seems to be supressed because of the reduced mobility of the "free" excitons and hence reduced probability for finding a binding center.

Having identified localization effects of excitons in QW the question comes up, whether the oscillator strength of excitons in QW is affected. The oscillator strength of free carrier or free exciton transitions is referred to a unit cell of the crystal and is determined by the overlap of the electron and hole wave functions |1|. For 3D excitons we obtain for the oscillator strength per unit cell of the 1 s exciton

$$f_{3D} = \Omega \cdot g \, |M_{VC}|^2 \cdot |\phi_{3D}^{1S}(0)|^2 , \tag{1}$$

where Ω is the volume of the unit cell, $|M_{CV}|^2$ is the dipole matrix element connecting Bloch states in the valence and conduction band and ϕ^{1s} is the hydrogen 1s envelope wavefunction of the exciton taking account of the electron hole relative motion. g is a factor inversely proportional to the frequency of the transition. Since $|\phi_{3D}^{1s}(0)|^2 = 1/(\pi a_o^3)$, eq.(1) may be written

$$f_{3D} = \frac{3}{4} \frac{\Omega}{V_X} \cdot g \cdot |M_{CV}|^2 = \frac{\Omega}{V_X} f_{CV} , \tag{2}$$

where V_X is the "volume" of the exciton corresponding to the relative motion of the electron and hole. The number of unit cells which contribute to the transition has to be known in order to determine the real transition strength of a free exciton. The entire crystal contributes for an ideal 1s free exciton created by the absorption of light with wavevector K=0 and the transition strength (this term is used to distinguish it from the oscillator strength) is given by

$$F_{3D}^{1S} = \frac{N \cdot \Omega}{V_X} f_{CV} = \frac{V}{V_X} f_{CV} , \tag{3}$$

where N is the number of unit cells in the crystal and hence V the crystal volume.

The oscillator strength of bound excitons is usually referred to one atom, which is the binding center for the exciton. The electron-hole relative motion and center of gravity motion can be separated for strong electron-hole correlation and the oscillator strength is given by |4|

$$f_{BX} = \frac{N_{BX} \cdot \Omega}{V_X} \cdot f_{CV} = \frac{V_{BX}}{V_X} f_{CV} , \qquad (4)$$

where N_{BX} now is the number of unit cells covered by the center of gravity motion of the bound exciton. Equation (4) is often referred to as the "giant oscillator strength" of bound excitons. The total transition strength for bound excitons in a crystal is obtained by multiplying eq. (4) by the number of binding centers, N_B, in the crystal:

$$F^{BX} = N_B \frac{V_{BX}}{V_X} f_{CV} . \qquad (5)$$

Compared to the free exciton oscillator strength per unit cell (eq. 2) the oscillator strength of a bound exciton (per atom) in fact is giant. However, to compare the transition strength of free and bound excitons eq. (3) and (5) must be considered.

The calculated oscillator strength of a bound exciton can be directly transformed into a radiative lifetime |47| and good agreement with experimental data is achieved |48|. The transition strength for free excitons in 3D, (eq. (3)), however, cannot simply be related to a lifetime because of the polariton nature and due to the fact that not all excitons occupy K=0 states in recombination.

The oscillator strength of a quasi 2D exciton in a QW with infinitely high barriers is given by |6|

$$f_{2D}^{1S} = \frac{\Omega}{\pi a_o^2(L_z) \cdot L_z} \cdot g \, |M_{CV}|^2 = \frac{\Omega}{V_X^{2D}} f_{CV} , \qquad (6)$$

where $a_o^2(L_z)$ is the two-dimensional Bohr radius and V_X^{2D} is the "volume" of the quasi 2D exciton. The transition strength for the 2D free exciton with K=0 again is obtained by multiplying eq. (6) with the number of unit cells in the quantum well which yields |49|

$$F_{2D}^{1S} = \frac{N \cdot \Omega}{\pi \, a_o^2 (L_z) \cdot L_z} \cdot g \cdot |M_{CV}|^2 = \frac{A}{\pi a_o^2 (L_z)} \cdot f_{CV} ; \qquad (7)$$

$N \Omega$ corresponds to the volume of the QW, which is equal to $A L_z$, where A is area of the QW layer. The QW thickness L_z obviously cancels in eq. (7) and the transition strength depends on L_z only via the 2D Bohr radius $a_o(L_z)$, which describes the in-plane relative motion of the electron and hole.

In order to relate the transition strength (eq.(7)) to the radiative lifetime polariton effects can be neglected in QW, however, the finite homogeneous linewidth of the exciton has to be taken into account. This finite homogeneous linewidth corresponds via the dispersion relation to a certain width in K-space, i.e. not only excitons with K=0 take part in the recombination. The effective transition strength for exciton recombination is then obtained by sharing the K=0 transition strength (eq.(7)) equally amongst all states within the exciton linewidth, leading to |49|

$$F_{2D}^{1S} = \frac{E_B}{\Delta(T)} \cdot \frac{\mu}{M} \cdot f_{CV} , \qquad (8)$$

where $M = m_e + m_h$, $\mu^{-1} = (m_e + m_h)^{-1}$, E_B the 2D exciton binding energy $E_B = -4h^2/\mu a_o^2$ and $\Delta(T)$ is the temperature dependent homogeneous linewidth of the exciton.

Finally, we have to take into account the thermal distribution of excitons. Since only excitons within the homogeneous width can contribute to recombination. The fraction of excitons r(T) within $\Delta(T)$ is determined by the Boltzmann factor, which for 2D with an energy-independent density of states leads to

$$r_{(T)} = 1 - e^{-\Delta(T)/kT}$$

and the transition strength then is given by |49|

$$F_{2D}^{1S} = \frac{E_B}{\Delta(T)} \cdot \frac{\mu}{M} f_{CV} (1 - e^{-\Delta(T)/kT}). \qquad (9)$$

Bound excitons in 2D can be treated in analogy to the 3D case, which yields for the oscillator strength per atom

$$f_{2D}^{BX} = \frac{A_{BX}}{\pi a_o^2 (L_z)} f_{CV}, \qquad (10)$$

where A_{BX} is the area corresponding to the center of gravity motion of the bound exciton.

Excitons localized in well width fluctuations with a typical size larger than the Bohr radius can be considered as free within the localization islands, and the transition strength is obtained by replacing A in eq. (7) by the area of the island, which again leads to eq. (9) for the transition strength.

The effect on transition strength of localization of the exciton by well width fluctuations with a typical size much smaller than the Bohr radius has not been investigated. Theoretically, however, one would expect a continuous transition from a quasi-free exciton above the mobility edge to localized excitons at low energies.

In any case, an increase of the oscillator or transition strength and a corresponding decrease of the radiative recombination times according to the decrease of $a_o(L_z)$ with decreasing L_z |11,12| is predicted for free, bound and localized excitons. This, in fact, has been supported qualitatively by absorption |50| and time-resolved photoluminescence measurements |51,52|.

Figure 2 shows the time dependence of the exciton luminescence for two QW with L_z=15 nm and 2.5 nm respectively, for example.

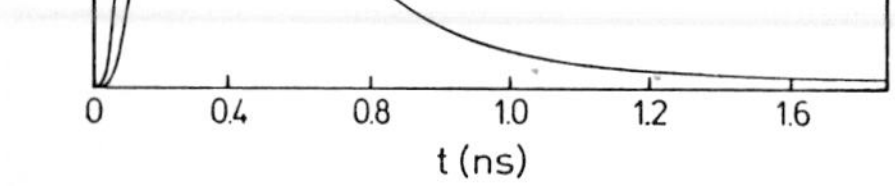

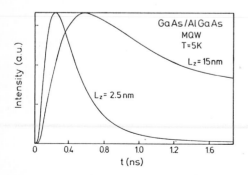

Fig. 2
Temporal variation of the spectrally integrated exciton luminescence for two multi-quantum-well samples with L_z= 15 nm and 2.5 nm at T=5K. The data are obtained by direct excitation of the QW, $\hbar\omega_{exc} < E_g$, GaAlAs by 5 ps pulses of a synchronously pumped dye laser and time-resolved detection of the luminescence with a synchroscan camera. Time resolution amounts to 20ps.

Obviously the decay times decrease with decreasing L_z, however, the relative quantum efficiency remains constant even for QW as thin as 2.5nm, demonstrating that the decrease in decay times cannot be attributed to an increase of nonradiative recombination. Recent data on the decrease of the luminescence decay times with decreasing L_z are summarized in Fig. 3 |49|. The number for the decay times refers to the decay of the spectrally integrated emission intensity of the exciton, which is exponential in all samples. A decrease of τ from 1.7ns to .32ns is found. when L_z decreases from 15nm to 2.5nm, which is even more then the expected maximum decrease by a factor of 4 corresponding to a decrease of $a_0(L_z)$ from the 3D value to $a_0/2$ for an ideal 2D exciton.

Further insight into the recombination mechanism can be obtained by investigating the temperature dependence of the decay kinetics. A temperature dependence according to eq.(9) is expected for free 2D excitons and excitons localized in large-size islands, whereas the decay times should be independent on temperature for bound and strongly localized excitons and only the intensity is expected to decrease with increasing temperature due to dissociation. Figure 4 depicts experimental results for the variation of the decay times with temperature for QW with different L_z |49|. A pronounced increase of τ is found for all samples but the intensity of the emission remains constant, at least up to 50K, which gives further support that radiative recombination prevails in the low-temperature regime. An increase of τ for free exciton recombination is expected on the basis of eq. 9 due to the increase of the homogeneous width of the exciton by acoustic phonon interaction |39,53| and the relative increase of τ roughly agrees with recent data for $\Delta(T)$ obtained by dephasing experiments |31|. We therefore conclude that the radiative recombination in QW at low temperatures is governed by the properties of the free 2D exciton, i.e. the shrinkage of the exciton wavefunction and the interaction of excitons with acoustic phonons determines the exciton lifetime, and localization, though evident in the spectra, does not affect the recombination strength dramatically.

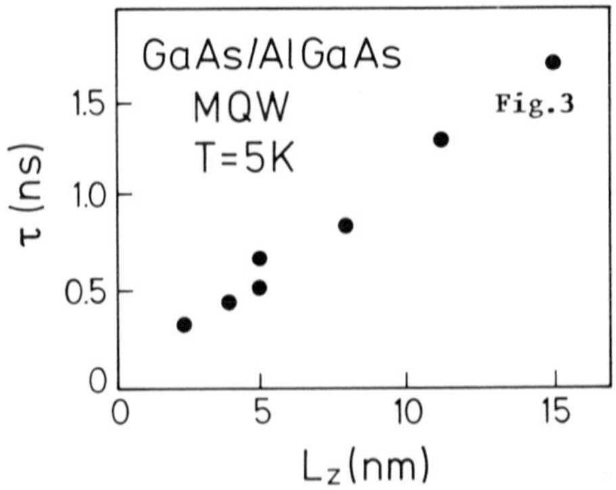

Fig. 3
Photoluminescence decay times versus quantum well thickness L_z at T=5K

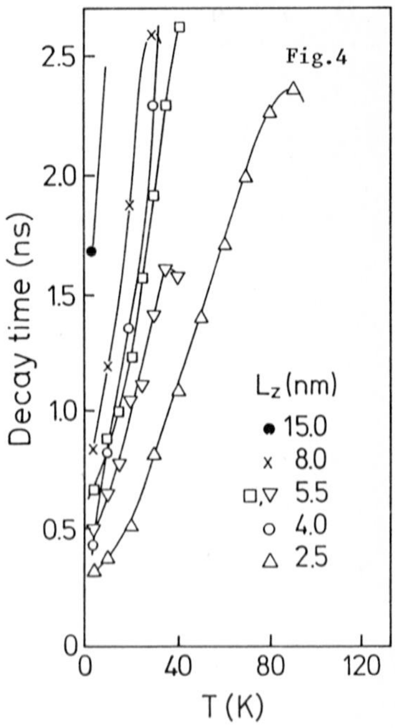

Fig. 4
Temperature dependence of the photoluminescence decay times for multi-quantum-well samples with different L_z

III. Electric Field Dependence of Exciton Lifetimes (The Quantum Confined Stark Effect)

Electric fields applied to bulk semiconductors cause an absorption tail below the fundamental absorption edge as well as characteristic oscillations of the above band gap absorption known as the Franz-Keldysh effect |54,55|. An extension of the "classical" Franz-Keldysh effect, which is attributed to field-induced photon-assisted tunneling and field-induced interference of Bloch states below and above the band gap, respectively, to uncorrelated electron hole transitions in QW has been recently reported by Miller et al. |56|.

A complete description of interband optical transitions under the presence of electric fields, however, has to take into account excitonic effects |57,58|. The excitonic resonance of 3D excitons broadens as an electric field is applied reflecting the lifetime reduction due to field ionization |59| and in addition a small red shift should occur which can be viewed as the exciton analogon to the Stark shift of the energy levels in hydrogen atoms. Large Stark shifts, however, cannot be obtained for 3D excitons due to the easy ionization of excitons at relatively small fields in the order of 10^4 V/cm as illustrated in Fig. 5a).

Qualitatively similar behavior is found for excitons in QW with electric fields applied parallel to the layers |60|. Completely different results are obtained, however, for electric fields applied perpendicular to the layers | 60,61|. Clearly resolved exciton resonances maintain up to electric fields which are higher than the ionization field by orders of magnitude and the red shift of the exciton resonance exceeds the binding energy of the exciton as well by orders of magnitude. The basic difference between the perpendicular and the parallel configuration (or the 3D case) is that field ionization of excitons is prevented in the parallel configuration due to the confinement of electrons and holes by the barrier wells (Fig. 5b). The excitonic interaction of the electron and hole, i.e. the in-plane relative motion, however, is almost unchanged as long as the well width is smaller or comparable to the Bohr radius. This behavior of excitons in QW has been demonstrated originally in absorption experiments by Miller et al. and is referred to as the "Quantum Confined Stark Effect" (QCSE) |61|. A theoretical analysis of the QCSE is found in ref. |60,62 and 63|. A Stark shift has been demonstrated recently also in intersubband optical transitions |64|.

The QCSE was first demonstrated in low temperature luminescence by Polland et al. |65|. A plot of a series of photoluminescence spectra of a GaAs/AlGaAs QW sample with three decoupled QW with nominal thickness L_z=5nm, 10nm, and 20nm respectively and a 100nm thick GaAs buffer layer is depicted in Fig. 6.

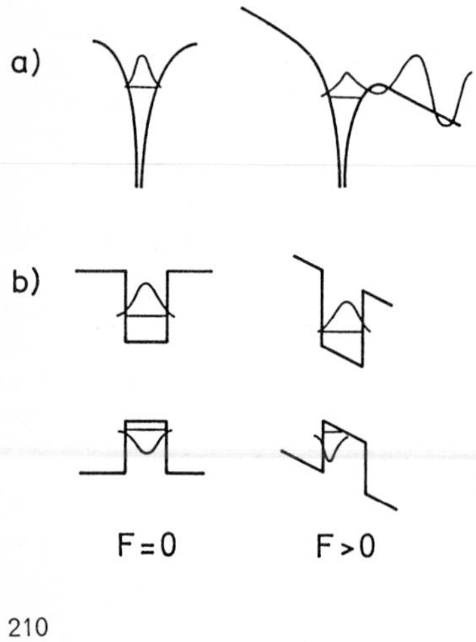

Fig. 5
Schematic representation of the electric field effect on 3D excitons (a) and the basic mechanisms responsible for the QCSE (b). A Coulomb bound state and its field-induced broadening (ionization) is indicated in (a). The change of single particle energies and wave functions for electric fields applied perpendicular to the layers is illustrated in (b)

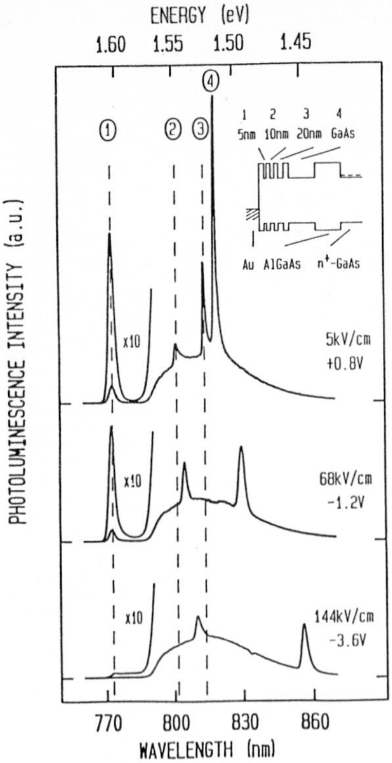

Fig. 6
Photoluminescence spectra (T=7K) of a QW sample containing three QW with different L_z (inset) and a "bulk" GaAs layer at various electric fields

The upper spectrum with an external voltage of 0.8V applied to a semitransparent Au-Schottky contact corresponds roughly to "flat band conditions". The narrow lines labelled 1,2 and 3 are due to exciton recombination in the QW with L_z=5nm, 10nm and 20nm, respectively and line 4 arises from recombination of 3D excitons in the GaAs buffer layer. The broad background emission is due to free carrier recombination in the n^+-GaAs substrate. This substrate luminescence is independent of the external voltage because of the negligible voltage drop across the highly doped substrate layer. The 3D exciton emission disappears rapidly with increasing field (decreasing external voltage) due to field ionization of the excitons as discussed before. Opposite, the intensity of the QW exciton luminescence is hardly affected by the field, clearly demonstrating the stability of the confined exciton. A pronounced red shift is found in addition for the QW with L_z=20nm (almost 100meV!), which exceeds the exciton binding energy by an order of magnitude. This Stark shift is weaker for L_z=10nm and almost absent in the QW with L_z=5nm.

The shift of the exciton resonance by the QCSE reflects both the changes of the confined single particle energies and the changes in the excitonic part, i.e. the binding energy of the exciton | 60,62 |. The latter, however, is small compared to the entire shift found e.g. for the L_z=20nm QW and the QCSE can be basically explained by field-induced changes of the single particle energies and wave functions. This is demonstrated in Fig. 7, where the induced red shift of the QW exciton luminescence is shown for QW with different thickness.

The experimental data (symbols in Fig. 7) are compared to calculations (full lines) taking into account the changes of the single particle energies only. The theoretical curves are obtained by iterative numerical solution of the "particle in a box" Schrödinger equation assuming finite barrier height (ΔE_c=0.57ΔE_g) and electron and heavy hole masses as given in ref. 60. The agreement between theory and experiment

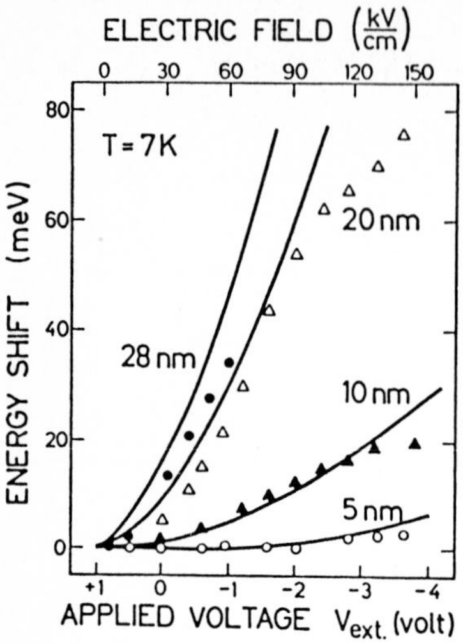

Fig. 7
Energy shift of the recombination with applied voltage for QWs with different thickness L_z

is satisfactory, demonstrating that the QCSE in fact can be attributed basically to the electric field induced spatial separation of the electron and hole in QW. Similar data have been also reported recently by Vina et al. |66|.

An electric field induced decrease of the oscillator and transition strength of the exciton is therefore expected corresponding to the change in $|M_{cv}|^2$ in eq. (1). The radiative exciton lifetime in the QCSE regime thus can be written as

$$\tau = \frac{\tau_o}{|M_{CV}|^2}, \qquad (11)$$

where τ_o is the zero-field lifetime. Early attempts to demonstrate the field-induced increase of the recombination lifetime failed because of either extrinsic recombination |67| or field-induced tunneling of carriers thru the barrier wells |68,69|. Polland et al. |65|, however, finally could demonstrate that the red shift of the exciton resonance by the QCSE is accompanied by a corresponding increase in photoluminescence decay times. Experimental results are depicted in Fig. 8 for the same QW samples as shown in Fig. 7. The change in photoluminescence decay time normalized to the zero-field value is plotted versus the applied voltage. The decay times increase with increasing field except for the QW with L_z=5nm and this increase is more pronounced for thicker wells, where the electrons and holes can be stronger polarized by the field. It is important to note that the intensity of the QW exciton luminescence remains unchanged up to rather high fields in spite of the increase of the decay times with field, which once again demonstrates that radiative recombination dominates at low temperatures. The full lines represent calculations according to eq. (11), i.e. only the change in electron hole overlap in the z-direction is considered, and good agreement with experiment again is achieved.

At very high fields a sudden drop of the photoluminescence intensity and decay times is observed, depending on QW thickness and barrier height and thickness. This is seen for example in Fig. 8 for the QW with L_z=5nm at an applied voltage of about -3V, which corresponds to a field of about 130kV/cm. This decrease of the emission intensity and decay constants at high electric fields is attributed to field-induced tunneling of carriers thru the barrier wells. Tunneling thru the barriers is of course expected to be strongest in thin QW where the confined single particle states are at high energies and in QW samples with thin barrier layers.

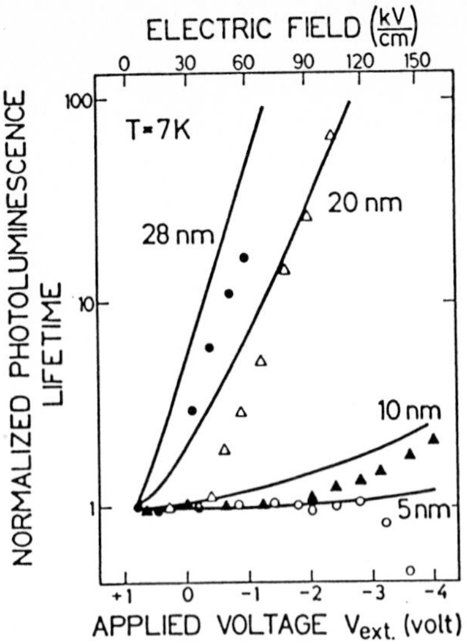

Fig. 8
Photoluminescence lifetimes normalized to the zero-field values versus applied voltage for QW with different thickness L_z

Recent results on the effect of tunneling on the exciton dynamics are summarized in Fig. 9. |70|. The photoluminescence decay times of two QW samples with different well and barrier layer thickness are plotted as a function of the external voltage applied to the Schottky barrier. Sample 1 contained three QW with thickness of 5nm. 10.7nm and 21.4nm seperated by barrier layers with composition x=0.3 and thickness of 24nm. Sample 2 contained two QW with L_z=13.5nm and 27.7nm clad by barriers again with x=0.3 but a thickness of 13nm. The decay times reflect the characteristic QCSE behavior at low fields, showing a more or less pronounced increase depending on QW thickness. This increase turns over into a decrease at an electric field strength (applied voltage) marked by arrows due to tunneling. This onset voltage for single particle tunneling obviously depends on QW and barrier layer thickness as qualitatively expected.

A quantitative analysis of the experimental results including the QCSE and the tunneling of carriers is depicted in Fig. 10 for the two samples shown in Fig. 9. The recombination lifetime τ_r is again calculated according to eq. (11) and the tunneling time τ_t which is determined by the energetic width of the tunneling resonances |60| is estimated within the WKB approximation |70 |. The residual carrier lifetime τ_c is then determined via $1/\tau_c = 1/\tau_r + 1/\tau_t$. The calculations performed this way describe the characteristic features found experimentally quite well and demonstrate that the QCSE regime is limited by intrinsic carrier tunneling thru the wells at high electric fields strength.

The effect of tunneling manifests itself also in an increase of the photocurrent thru the reverse biased Schottky barrier structure. A reduction of the photoluminescence intensity in the field regime where tunneling is important is accompanied by a corresponding increase of the photocurrent |67,71 |, which can be applied to measure photocurrent excitation spectra |71|. This is demonstrated in Fig. 11.

The upper part depicts the photocurrent excitation spectrum and the lower part the corresponding decrease of the photoluminescence intensity by changing the applied voltage from 0.6V (which corresponds to nearly flat band conditions) to the values listed on the right hand side of each spectrum. The heavy and light hole excitons are clearly resolved in both sets of data and the signal-to-noise ratio in the photocurrent excitation spectrum is comparable to or even better than in the photoluminescence excitation spectra at optimized conditions for V_{ext}. This together with the simpler detection scheme suggests photocurrent excitation spectra as a con-

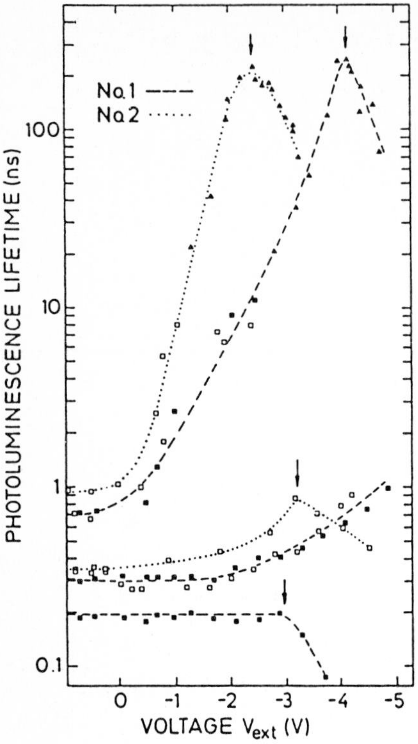

Fig. 9
Photoluminescence lifetimes versus applied voltage for two QW samples. Sample 1 contains three QW with L_z=5nm, 10.7nm, and 21.4nm with $Al_{0.3}Ga_{0.7}As$ barriers of 24nm thickness. Sample 2 contains two QW with L_z=13.5nm and 27.7nm separated by a 14nm barrier with the same composition.

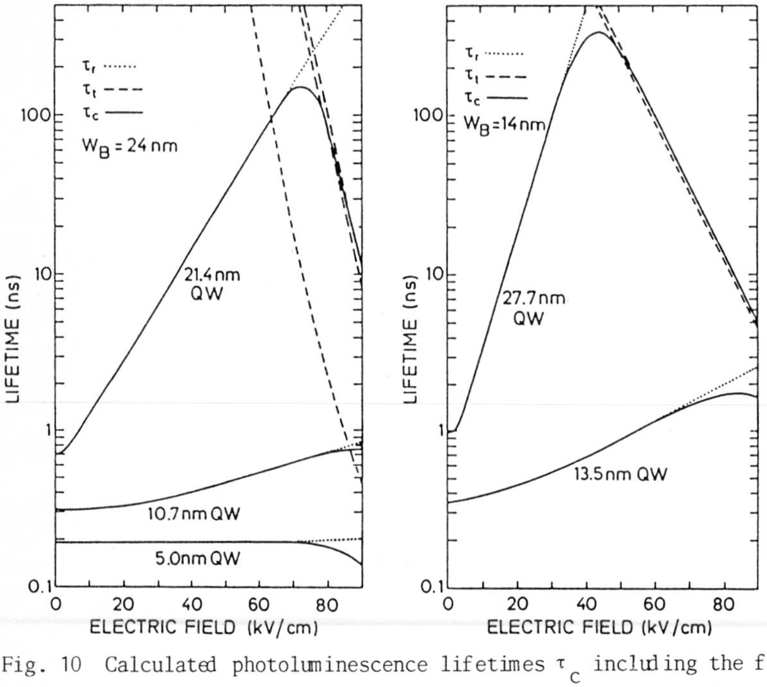

Fig. 10 Calculated photoluminescence lifetimes τ_c including the field-induced changes of the radiative recombination lifetime τ_r and tunneling τ_t

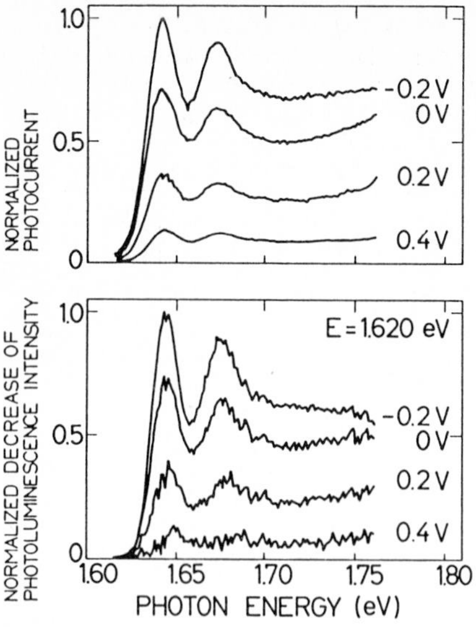

Fig. 11
Photocurrent excitation spectrum (upper part) and corresponding differential decrease of the photoluminescence intensity (lower part) for a single QW with L_z=5nm, x=0.4. The lower spectra are obtained by decreasing the external voltage from 0.6V (which corresponds to about flat band conditions) to the value listed at the right hand side and the corresponding decrease of the luminescence intensity is plotted

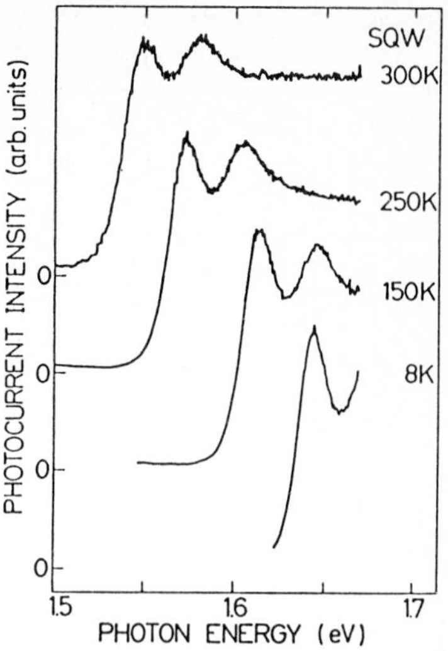

Fig. 12
Photocurrent excitation spectra for the sample shown in Fig. 11 at different temperatures from 8K to 300K. The heavy and light hole exciton are clearly resolved with good signal-to-noise ratio up to room temperature

venient characterization technique, which can be easily applied also at room temperature, as finally demonstrated in Fig. 12.

IV. Summary and Conclusion

We have briefly summarized some recent results on the optical properties of excitons in GaAs/AlGaAs QW in Sect. I. Free exciton emission dominates the radiative recombination in undoped QW at low temperatures, however, lateral localization of excitons due to well width fluctuations is generally observed as manifested e.g. by the Stokes shift between luminescence and absorption. The exciton recombination dynamics is discussed in Sect. II. An expression for the transition strength of free excitons in QW is derived including the effect of finite linewidth and exciton distribution. The exciton radiative decay times depend on the QW thickness L_z only via the dependence of the in-plane Bohr radius on L_z. An increase of the exciton recombination lifetime with temperature according to the increase of homogeneous linewidth is predicted and verified experimentally.

The fundamentals of the quantum confined Stark effect are briefly described in Sect. III and our recent data on time-resolved luminescence of excitons in QW under the presence of electric fields are summarized. The QCSE manifests itself in a red shift of the exciton emission line and an increase of the recombination time constants, which is stronger for thicker wells. Carrier tunneling thru the barrier wells is found at very high fields and this intrinsic tunneling process limits the QCSE.

Altogether, the basic and characteristic features of the optical properties of excitons in QW including electric field effects are understood and experimentally demonstrated, at least at low temperatures. Many details, however, require further consideration, like the influence of the real valence band structure or the effect of exciton localization on optical properties, for example. In addition, different structures like type I/type II quantum wells or superlattices as well as other material systems, e.g. GaInAs/AlInAs have to be explored further. In anv case. optical spectroscopy, including time-resolved studies, will contribute to an improved understanding of the new physics as well as to the development of new devices based on quantum effects in solids.

Acknowledgement: The results reported in this article were obtained in collaboration with the Max-Planck-Institut für Festkörperforschung in Stuttgart (FRG), Philips Research Laboratory in Redhill (England) and AT&T Bell Labs, Murray Hill (USA). I am particularly indebted to J.Kuhl, H.-J.Polland and K.Ploog (MPI), P.Dawson (Philips), C.W.Tu (AT&T) and to J.Feldmann and G.Peter (Marburg) for their various contributions and support. I also acknowledge helpful discussion with R.Elliot concerning the oscillator strength of QW excitons.

References

1. R.J. Elliot, Phys.Rev. 108, 1384 (1957)
2. J.J.Hopfield, Phys.Rev. 112, 1555 (1958)
3. D.G. Thomas, M. Gershenzon, J.J.Hopfield, Phys.Rev. 131, 2397 (1963)
4. E.I. Rashba, G.E. Gurgenishvili, Sov.Phys.-Solid State 4, 759 (1962) (orig.Fiz.Tverd.Tecla 4, 1029 (1962))
5. D.S.Chemla, Helvetia Physica Acta 56, 607 (1983)
6. D.A.B. Miller, D.S. Chemla, S.Schmitt-Rink in "Nonlinear Optical Properties of Semiconductors" (H.Haug ed.) (Academic Press, N. Y. 1987)
7. D.S.Chemla, S.Schmitt-Rink, D.A.B.Miller, ibid
8. G.Bastard, unpublished
9 M.Nakayama, Sol. State Commun. 55, 1053 (1985); M.Nakayama, M. Matsuura, Surf.Science 170, 641 (1966)
10. S.Shinada, S.Sugano, J.Phys.Soc.Jpn. 21, 1936 (1966)
11. G.Bastard, E.E.Mendez, L.L.Chang, L.Esaki, Phys.Rev. B26, 1974 (1982)
12. Y.Shinozuka, M.Matsuura, Phys.Rev. B28, 4878 (1983)
13. R.L.Greene, K.K.Bajaj, D.E.Phelps, Phys.Rev. B29, 1807 (1984)
14. J.C.Chang, J.N.Schulman, Appl.Phys.Lett. 43, 536 (1983)
15. J.C.Maan, G.Belle, A.Fasolino, M.Altarelli, K.Ploog, Phys.Rev. B30, 2253 (1984)
16. R.C.Miller, D.A.Kleinmann, W.T.Tsang, A.C.Gossard, Phys.Rev. B24, 1134 (1981)
17. P.Dawson, K.J.Moore, G.Duggan, H.I.Ralph, C.T.B.Foxon, Phys.Rev. B34, 6007 (1986)
18. R.C.Miller, A.C.Gossard, W.T.Tsang, O.Munteanu, Solid State Commun. 43, 519 (1982)
19. R.C.Miller, D.A.Kleinmann, Journ.Luminesc. 30, 520 (1985)
20. L.Schultheis, K.Ploog, Phys.Rev. B30, 1090 (1984)
21. B.V.Shanabrock, O.J.Glembocki, W.T.Beard, Phys.Rev. B35, 2540 (1987)
22. D.Bimberg, J.Christen, A.Steckenborn, G.Weimann, W.Schlapp, Journ. Luminesc. 30, 562 (1985)
23. J.F.Ryan, Physica 134B, 403 (1985)
24. E.O.Göbel, J.Kuhl, R.Höger, Journ.Luminesc. 30, 541 (1985); E.O.Göbel: In Springer Proc.Phys. 7, 104 (Springer, Berlin, Heidelberg 1986)
25. C.V.Shank, R.L.Fork, B.I.Greene, C.Weisbuch, A.C.Gossard, Surf.Science 113, 108 (1982)
26. W.H.Knox, C.Hirlimann, D.A.B.Miller, J.Shah, D.S.Chemla, C.V.Shank, Phys. Rev.Lett. 56, 1191 (1986)
27. A.Mysioriwicz, D.Hulin, A.Antonetti, A.Migus, W.T.Masselink, H.Morkoc, Phys.Rev.Lett. 56, 2748 (1986)
28. A.von Lehmen, D.S.Chemla, J.E.Zucker, J.P.Heritage, Opt.Lett. 11, 609 (1986)
29. A.J.Taylor, D.J.Erskine, C.L.Tang, J.Opt.Soc. Am. B2, 663 (1985)
30. J.Hegarty, M.D.Sturge, J.Opt.Soc.Am. B2, 1143 (1985)
31. L.Schultheis, A.Honold, J.Kuhl, K.Köhler, C.W.Tu, Phys.Rev. B34, 9027 (1986)

32. L.Schultheis, same issue of this book
33. C.Weisbuch, R.D.Miller, R.Dingle, A.C.Gossard, W.Wiegmann, Solid State Commun. 37, 29 (1981)
34. H.Jung, A.Fischer, K.Ploog, Appl.Phys. A33, 97 (1984)
35. J.Hegarty, M.D.Sturge, C.Weisbuch, A.C.Gossard, W.Wiegmann, Phys.Rev.Lett. 49, 930 (1982)
36. N.F.Mott, E.A.Davis: Electronic Processes in Non-Crystalline Materials, 2nd ed., (Clarendon Press, Oxford 1979)
37. J.Hegarty, L.Goldner, M.D.Sturge, Phys.Rev. B30, 7346 (1984)
38. Y.Masumoto, S.Shionoya, H.Kawaguchi, Phys.Rev. B29, 2324 (1984)
39. H.Stolz, D.Schwarze, W. von der Osten, Proc. 18th Int.Conf.Phys.Semicond., Stockholm, 1986, in print
40. O.Goede, L.John, D.Henning, Phys.Stat.Sol. B89, K183 (1978)
41. J.Singh, K.K.Bajaj, S.Chandhuri, Appl.Phys.Lett. 44, 805 (1985); J.Singh, K.K.Bajaj, J.Appl.Phys. 57, 5433 (1985)
42. D.Bimberg, D.Mars, J.N.Miller, R.Bauer, D.Oertel, J.Vac. Sci.Techn.B4, 1014 (1986); D.Bimberg, D.Mars, J.N.Miller, R.Bauer, D.Oertel, J.Christen, Supperlatt. & Microstructures, 3, 79 (1987)
43. M.H.Cohen, Can.J.Chem. 55, 1906 (1976)
44. D.C.Reynolds, K.K.Bajaj, C.W.Litton, J.Singh, P.W.Yu, P.Pearah, J.Klein, H.Morkoc, Phys. Rev. B33, 5931 (1986)
45. J.H.Neave, B.A.Joice, P.J.Dolson, N.Norton, Appl.Phys. A31, 1 (1983)
46. D.Bimberg, J.Christen, T.Fukunaga, H.Nakashim, D.E.Mars, J.N.Miller, J.Voc.Sci.Technol., in press
47. D.L.Decter, In Solid State Physics, vol. 6 (ed. F.Seitz and D.Turnbull, 1958)
48. see e.g. C.H.Henry, K.Nassau, Phys,Rev. B1, 1628 (1970)
49. J.Feldmann, G.Peter, E.O.Göbel, P.Dawson, K.Moore, C.Foxon, R.Elliot, to be published
50. Y. Masumoto, M.Matsuura, S.Tarucha, H.Okamoto, Phys.Rev. B32, 4275 (1985)
51. E.O.Göbel, H.Jung, J.Kuhl, K.Ploog, Phys.Rev.Lett. 51, 1588 (1983)
52. J.Christen, D.Bimberg, A.Steckenborn, G.Weimann, Appl.Phys.Lett. 44, 84 (1984)
53. J.Lee, E.S.Koteles, M.O.Vassell, Phys.Rev. B33, 8 (1986)
54. W.Franz, Z.Naturforschung 13a, 484 (1958)
55. L.V.Keldysh, Sov.Phys. JETP 7, 788 (1958)
56. D.A.B.Miller, D.S.Chemla, S.Schmitt-Rink, Phys.Rev. B33, 6976 (1986)
57. J.D.Dow, D.Redfield, Phys. Rev. B1, 3358 (1970)
58. I.A.Merkulov, V.I.Perel', Phys.Lett. 45A, 83 (1973)
59. I.A.Merkulov, Sov.Phys.-JETP 39, 1140 (1974) (Zh. Epshk. Teor.Fiz. 66, 2314 (1974)
60. D.A.B.Miller, D.S.Chemla, T.C.Damen, A.C.Gossard, W.Wiegmann, T.H.Wood, C.A.Burrus, Phys. Rev. B32, 1043 (1985)
61. D.A.B.Miller, D.S.Chemla, T.C.Damen, A.C.Gossard, W.Wiegmann, T.H.Wood, C.A.Burrus, Phys.Rev.Lett. 53, 2173 (1984)
62. G.Bastard, E.E.Mendez, L.L.Chang, L.Esaki, Phys.Rev. B28, 3241 (1983)
63. J.A.Brum, G.Bastard, Phys.Rev. B31,3893 (1985)
64. A.Harwit, J.T.Harris, Jr., Appl.Phys.Lett. 55, 685 (1987)
65. H.-J.Polland, L.Schultheis, J.Kuhl, E.O.Göbel, C.W.Tu, Phys.Rev.Lett. 55, 2610 (1985)
66. L.Vina,R.T.Collins, E.E.Mendez, W.I.Wang, L.L.Chang, L.Esaki, Superlattices & Microstructures 39, (1987); L.Vina, R.T.Collins, E.E.Mendez, W.I.Wang, Phys.Rev. B33, 5939 (1986)
67. H.-J.Polland, Y. Horkoshi, R.Höger, E.O.Göbel, J.Kuhl, K.Ploog, Physica 134B, 412 (1985)
68. J.A.Kash, E.E.Mendez, H.Morkoc, Appl.Phys.Lett. 46, 173 (1985)
69. E.J.Austin, M.Jaros, Appl.Phys.Lett. 47, 274 (1985)
70. H.-J.Polland, K.Köhler, L.Schultheis, J.Kuhl, E.O.Göbel, C.W.Tu, Superlattices & Microstructures, 2, 309 (1986)
71. H.-J.Polland, Y. Horikoshi, E.O.Göbel, J.Kuhl, K.Ploog, Surf.Science,174, 278 (1986)

Excitons in II–VI Multiquantum Wells

S. Datta, M. Yamanishi, R.L. Gunshor, and L.A. Kolodziejski*

School of Electrical Engineering, Purdue University,
West Lafayette, IN 47907, USA
*Permanent Address: Department of Physical Electronics,
Hiroshima University, Higashi-hiroshima-shi, Saijo-cho, 724 Japan

1. Introduction

The recent growth of high quality (Cd,Mn)Te and (Zn,Mn)Se multiquantum wells (MQW) by molecular beam epitaxy (MBE) on (100) GaAs substrates has aroused a lot of interest in their optical properties [1,2]. (Cd,Mn)Te superlattices have been grown controllably in both (100) and (111) orientations despite the large lattice mismatch with GaAs [3]. Stimulated emission has been demonstrated in both (Cd,Mn)Te and (Zn,Mn)Se MQW systems [4,5]. The presence of the magnetic ion (Mn) in the barrier layers has led to a variety of new and interesting experiments. Novel interface localization effects have been observed in the luminescence from (111) (Cd,Mn)Te MQW's [2]. The relative strengths of the yellow luminescence from internal Mn d-electron transitions and the blue luminescence from band-to-band transitions in (Zn,Mn)Se MQW's has been used to gain insight into the rate of capture of photoexcited carriers by the wells [6]. The presence of Mn enhances the g-factor in the barriers over that in the wells, so that the shift in the luminescence in a magnetic field indicates the extent of penetration of the exciton wavefunction into the barrier layers. Magneto-optic experiments with (Zn,Mn)Se MQW's have shown a relatively large g-factor indicating a small valence band offset [7]. As a result the large strains in the layers play a significant role in determining the nature of excitonic wavefunctions [8].

This paper is organized as follows. In Section 2 we first discuss the important role played by strains in determining the nature of the excitons in II-VI MQW's. We then discuss our experimental results showing a higher oscillator strength for the TM mode near the bandedge compared to the TE mode in (Zn,Mn)Se MQW's indicating that the light hole band lies above the heavy hole band. The opposite is observed in (Cd,Mn)Te MQW's which is ascribed to the opposite sense of the strain. In Section 3 we discuss an experiment demonstrating the ionization of excitons as the electric field parallel to the layers is increased. This experiment could provide a useful technique for measuring the exciton binding energy in highly resistive materials.

2. Excitons in Strained Layer MQW's

In both (Cd,Mn)Te and (Zn,Mn)Se systems there is a large change in the lattice constant with increasing Mn mole fraction (about 20 to 40 times the change in (Ga,Al)As systems). For example, in a $ZnSe/Zn_{0.7}Mn_{0.3}Se$ superlattice with equal well and barrier widths the strain is $\sim$.7%; assuming a deformation potential $\sim$5 eV, this leads to strain-induced shifts $\sim$35 meV in the band edge energies. This is comparable to or larger than the valence band offsets in these systems so that small differences in the strain can have a profound effect on the hole and hence the exciton wavefunction. The precise values of strain can vary widely depending on the thickness of the buffer layer, the presence of dislocations and the temperature. In (Zn,Mn)Se the presence of Mn *increases* the lattice constant while in (Cd,Mn)Te it *decreases* the lattice constant. Assuming that the composite structure is coherently strained, it will adjust to a single lattice constant a_0' (in the x-y plane, Fig. 1a) intermediate between the lattice constants of the wells and barriers. a_0' can be computed theoretically by minimizing the overall strain energy with respect to a_0'. The result is what one might expect intuitively: a_0' is the average lattice constant of all the layers, each

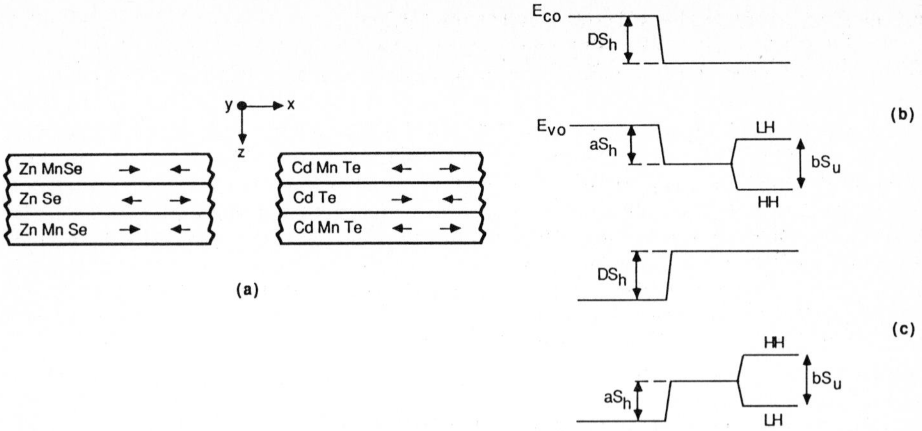

Fig. 1: Strain shifts in (Zn,Mn)Se and (Cd,Mn)Te MQW's
(a) Sense of strain in the x-y plane
(b) Strain shift of the band edge in ZnSe wells and CdMnTe barriers ($S > 0$)
(c) Strain shift of the band edge in CdTe wells and ZnMnSe barriers ($S < 0$)

weighted according to its elastic constant c_{11}. It is believed that the elastic constants do change significantly with the incorporation of Mn [9] so that this factor should be taken into account in calculating a_0'. However, the greatest uncertainty in calculating a_0' arises because it is not clear to what extent the buffer layer (between the substrate and the superlattice) should be considered part of the coherently strained structure; this is determined by the extent and nature of the dislocations.

The strain in the x-y plane in the well (or barrier) is given by

$$S_{xx} = S_{yy} \equiv S = (a_0' - a_0)/a_0 \,, \tag{1a}$$

where a_0 is the natural lattice constant of the well (or barrier). As we might expect, there is a strain in the z-direction as well whose sign is opposite to that in the x-y plane. Near the surface we may assume $T_{zz}=0$ so that

$$S_{zz} = -\frac{2C_{12}}{C_{11}} S \,. \tag{1b}$$

The *hydrostatic* strain S_h is defined as

$$S_h = S_{xx}+S_{yy}+S_{zz}$$

$$= 2(1 - \frac{C_{12}}{C_{11}})S, \tag{2a}$$

while the *uniaxial* strain S_u is defined as

$$S_u = S_{xx}+S_{yy}-2S_{zz}$$

$$= 2(1 + \frac{2C_{12}}{C_{11}})S \,. \tag{2b}$$

In both (Zn,Mn)Se and (Cd,Mn)Te systems C_{12}/C_{11} is close to .5 so that $S_h \sim S$ and $S_u \sim 4S$. In (Zn,Mn)Se, the strains S_h and S_u are positive in the wells and negative in the barriers, while in (Cd,Mn)Te the signs are reversed. The hydrostatic strain causes shifts in the band edge energies while the uniaxial strain causes the 'light hole' and 'heavy hole' band edges to split (Figs. 1b,1c). We thus expect a larger confining potential for 'light holes' relative to 'heavy holes' in (Zn,Mn)Se MQW's; the reverse is true in (Cd,Mn)Te MQW's.

The reason we have written light and heavy holes within quotes is that this nomenclature is accurate only if we neglect motion in the x-y plane, that is, if we set $k_x=k_y=0$. For non-zero k_x and k_y there is significant 'band-mixing' meaning that the electronic wavefunctions are not purely 'light' or 'heavy' hole type but linear combinations thereof. This is important in considering the exciton problem since exciton wavefunctions are constructed out of single particle states with non-zero $\vec{k}$; the smaller the Bohr radius, the larger the range of k-values that are needed. It has been shown that in (Ga,Al)As the band-mixing leads to an increase in the exciton binding energy by ~10-15% [10]. However, band-mixing is neglected in our estimate of the dependence of the binding energy on the valence band offset (Fig. 2).

The effective mass equation for electrons in a MQW is given by

$$\left(E_c(z_e) + DS_h(z_e) + \frac{p_e^2}{2m_e} \right) C(\vec{r}_e) = E_e C(\vec{r}_e), \tag{3}$$

where the subscript 'e' denotes electronic coordinates. D is the conduction band deformation potential [11]. Neglecting the split-off band, the effective mass equation for holes is written as a set of 4 coupled differential equations [12]:

$$\begin{pmatrix} F & O & M & L \\ O & F & -L^* & M^* \\ M^* & -L & G & O \\ L^* & M & O & G \end{pmatrix} \begin{pmatrix} H^+(\vec{r}_h) \\ H^-(\vec{r}_h) \\ L^-(\vec{r}_h) \\ L^+(\vec{r}_h) \end{pmatrix} = E_h \begin{pmatrix} H^+(\vec{r}_h) \\ H^-(\vec{r}_h) \\ L^-(\vec{r}_h) \\ L^+(\vec{r}_h) \end{pmatrix} \tag{4}$$

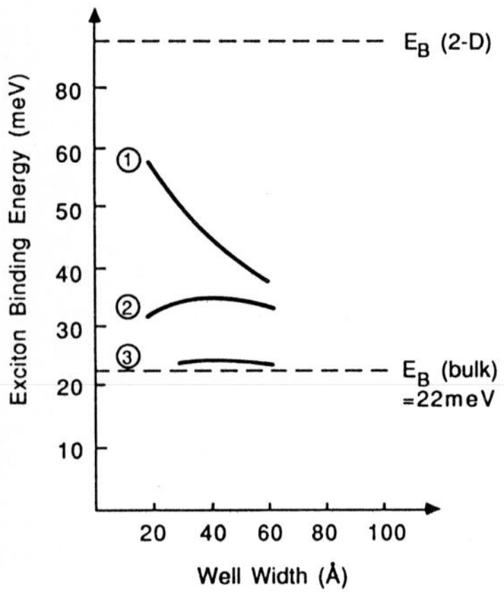

Fig. 2: Binding energy E_B for the LH1-C1 exciton vs. well width W calculated variationally neglecting band-mixing The parameters used correspond to ZnSe: $m_e=.17\ m_0$, $\gamma_1=3.77$, $\gamma_2=1.24$, $\epsilon=8\ \epsilon_0$. (1) ΔE_c, $\Delta E_v \rightarrow \infty$, (2) $\Delta E_c=200$, $\Delta E_v=50$ meV, (3) $\Delta E_c=200$, $\Delta E_v=0$ meV. In (1) an isotropic hole mass of m_0/γ_1 was used

$$\text{where} \quad F = E_{vH}(z_h) - \frac{p_{xh}^2+p_{yh}^2}{2m_0/(\gamma_1+\gamma_2)} - \frac{p_{zh}^2}{2m_0/(\gamma_1-2\gamma_2)} \tag{5a}$$

$$G = E_{vL}(z_h) - \frac{p_{xh}^2+p_{yh}^2}{2m_0/(\gamma_1-\gamma_2)} - \frac{p_{zh}^2}{2m_0/(\gamma_1+2\gamma_2)} \tag{5b}$$

$$E_{vH}(z_h) = E_v(z_h) + a\, S_h(z_h) + \frac{b}{2} S_u(z_h) \tag{5c}$$

$$E_{vL}(z_h) = E_v(z_h) + a\, S_h(z_h) - \frac{b}{2} S_u(z_h) \tag{5d}$$

$$M = \frac{p_{yh}^2-p_{xh}^2}{2m_0/\sqrt{3}\gamma_2} + \frac{ip_{xh}p_{yh}}{2m_0/2\sqrt{3}\,\gamma_3} \tag{5e}$$

$$L = \frac{i(p_{xh}-ip_{yh})p_{zh}}{2m_0/2\sqrt{3}\gamma_3} , \tag{5f}$$

where the subscript 'h' denotes hole coordinates and $\gamma_1,\gamma_2,\gamma_3$ are the Luttinger parameters [13,14]. The deformation potentials a and b for the valence band can be obtained from Ref. 11. It can be seen from (5a) and (5b) that if we neglect 'band-mixing' due to the off-diagonal operators M and L, the 'heavy hole' has a mass $m_0/(\gamma_1-2\gamma_2)$ in the z-direction and a mass $m_0/(\gamma_1+\gamma_2)$ in the x-y plane, while the 'light-hole' has a mass $m_0/(\gamma_1+2\gamma_2)$ in the z-direction and a mass $m_0/(\gamma_1-\gamma_2)$ in the x-y plane. Since the 'light hole' is heavier than the 'heavy hole' for motion in the x-y plane, it has a larger exciton binding energy. Equations (3) and (4) describe the motion of non-interacting electrons and holes. The exciton Hamiltonian H_{ex} for electrons and holes interacting via the Coulomb force can be written as [15]

$$H_{ex} = \left(H_c(\vec{r}_e) - \frac{e^2}{4\pi\epsilon|\vec{r}_e-\vec{r}_h|} \right) I - H_v(\vec{r}_h) , \tag{6}$$

where $H_c(\vec{r}_e)$ is the differential operator on the left hand side of (3), I is the (4×4) identity matrix and $H_v(\vec{r}_h)$ is the (4×4) differential operator on the left hand side of (4).

Neglecting the off-diagonal operators M,L, the wavefunctions and binding energies of the heavy and light hole excitons can be calculated separately from $H_{ex,H}$ and $H_{ex,L}$ [16]:

$$H_{ex,H} = H_c(\vec{r}_e) - F(\vec{r}_h) - \frac{e^2}{4\pi\epsilon|\vec{r}_e-\vec{r}_h|} \tag{7a}$$

$$H_{ex,L} = H_c(\vec{r}_e) - G(\vec{r}_h) - \frac{e^2}{4\pi\epsilon|\vec{r}_e-\vec{r}_h|} \quad . \tag{7b}$$

Fig. 2 shows the results obtained [17] for the light hole exciton in (Zn,Mn)Se using a variational wavefunction of the form

$$\Psi(\vec{r}_e,\vec{r}_h) = \phi_e(\Delta\tilde{E}_c,z_e)\phi_h(\Delta\tilde{E}_v,z_h)\, \exp(-|\vec{\rho}_e-\vec{\rho}_h|/\tilde{a}) , \tag{8}$$

where $\phi_e(z_e)$ and $\phi_h(z_h)$ are the wavefunctions for the lowest subbands in quantum wells with band discontinuities $\Delta\tilde{E}_c$ and $\Delta\tilde{E}_v$ respectively. $\Delta\tilde{E}_c$, $\Delta\tilde{E}_v$, and $\tilde{a}$ are treated as variational parameters. The physical motivation behind this choice of parameters is that the Coulomb attraction between electrons and holes makes the wells look a little deeper than they actually are; thus $\Delta\tilde{E}_c$, $\Delta\tilde{E}_v$ turn out to be a little bigger than the physical band offsets ΔE_c, ΔE_v. Since only the lowest subband is used, the calculated binding energy is appropriate for the LH1-C1 exciton; moreover, the calculation is not accurate for wide wells. For infinitely deep wells the binding energy increases to 4 times the bulk value as the well width is reduced, as we would expect for an isotropic 2D hydrogen atom. But with $\Delta E_v = 0$, there is hardly any change in the exciton binding energy as the well width is decreased indicating that the exciton is hardly perturbed from its 3D wavefunction. When ΔE_v is increased to 50 meV it acquires a quasi-2D character as the well width becomes comparable to the bulk Bohr radius (~40Å for the parameters used). Since the unstrained valence band offset in these materials is small and the strain-shift is large it appears that the nature of the excitons can vary widely from one MQW to another. The theory discussed here does not include magnetic polaron effects [18]. Moreover, even ordinary polaron effects might affect the excitonic wavefunctions significantly since the electron-phonon coupling constant is fairly large in II-VI compounds [19]. For example, there is the possibility of a large 'Lamb shift' in the n=1 and n=2 exciton levels (as well as linewidth broadening) and a modification of the relative oscillator strengths.

The splitting of the heavy hole and light hole exciton levels leads to anisotropic optical properties of the MQW even though the constituents are individually isotropic. For heavy hole to conduction band transitions the oscillator strength f_h is proportional to

$$f_h \propto |\hat{e}_\nu \cdot (\hat{e}_x + i\hat{e}_y)|^2, \tag{9a}$$

where $\hat{e}_\nu$ is a unit vector in the direction of polarization of the light. $\hat{e}_x$, $\hat{e}_y$ and $\hat{e}_z$ are unit vectors along the x,y and z-directions. For light-hole to conduction band transitions the oscillator strength f_ℓ is proportional to

$$f_\ell \propto \frac{1}{3}|\hat{e}_\nu \cdot (\hat{e}_x - i\hat{e}_y)|^2 + \frac{4}{3}|\hat{e}_\nu \cdot \hat{e}_z|^2 . \tag{9b}$$

Neglecting 'band-mixing' effects we thus expect the heavy hole exciton to have oscillator strengths proportional to 1 and 0 for TE (x- or y-polarized) and TM(z-polarized) modes respectively while the light hole exciton should have oscillator strengths proportional to 1/3 and 4/3 [20]. We are also neglecting the difference in the Bohr radii of light and heavy hole excitons.

In bulk cubic materials, heavy and light hole excitons are degenerate so that both TE and TM modes have oscillator strengths proportional to 4/3 ($= 1 + 1/3 = 0 + 4/3$). But in MQW's the degeneracy is lifted due to two independent factors - quantum confinement and strain. The former leads to a blue shift of the light hole relative to the heavy hole while the latter could work either way depending on the sign of the strain. In CdTe wells, strain effects add to those of quantum confinement so that the heavy hole absorption edge is lower than the light hole absorption edge. Consequently below the band edge the TE mode has a higher oscillator strength than the TM mode. In ZnSe wells, the order of the bands is reversed due to strain effects so that the TM mode has a higher oscillator strength than the TE mode. This has been observed experimentally (Figs. 3a, 3b). The absorption coefficient was deduced from the change in the photoluminescence (PL) intensity as the excitation spot is moved away from the edge. It should be noted that the TM mode can only be observed in the emission from the edge and not in the emission from the front surface. It is thus important to have a buffer layer with a lower index of refraction so that the light is guided along the superlattice. The absorption coefficient could only be measured in the bandtail region before it gets too large; it is thus difficult to make quantitative comparisons with theory since the light and heavy hole excitons may have different

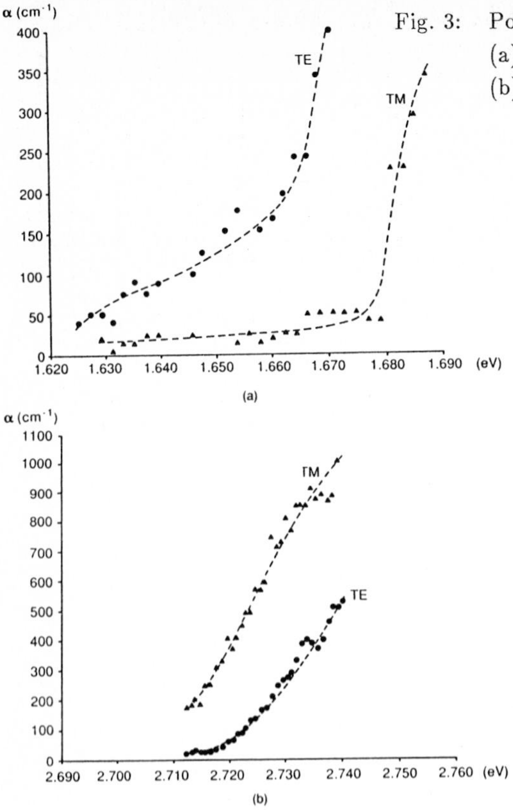

Fig. 3: Polarization-dependent absorption coefficient
(a) (Cd,Mn)Te superlattice
(b) (Zn,Mn)Se superlattice

degrees of bandtailing. The absorption coefficient can be measured deeper into the band by designing the superlattice to occupy a small fraction of the guided wave profile [21]. Stimulated emission has been obtained for the TM mode in the (Zn,Mn)Se superlattice [22]; this is in sharp contrast to (Ga,Al)As systems where lasing is always obtained for the TE mode.

3. Exciton Ionization

The exciton ionization [23] experiments were performed using the superlattice structure shown in Fig. 4a. The photoluminescence from this structure showed two peaks at 2.77 eV and 2.81 eV at 77 K. The lower peak agrees very well with the position of the C1-LH1 transition from the 68Å well. The upper peak could, in principle, be either from the C1-LH1 transition from the 29Å well or from the C1-HH1 transition from the 68Å well. However, we rule out the latter possibility since the two peaks were of comparable magnitude over a wide range of temperatures. If both peaks came from the 68Å well we would expect the ratio of their intensities to be approximately proportional to the Boltzmann factor $\exp(-\Delta E/k_B T)$; but if the two peaks come from different wells there is no reason for them to be in quasi-equilibrium, especially at low temperatures.

An electric field was applied parallel to the layers using indium contacts diffused into the top surface (Fig. 4b). The PL intensity of both peaks was monitored as a function of the applied electric field, together with the current I through the structure. The high energy photoluminescence peak from the narrow wells ($\simeq$29Å) showed now significant reduction up to a field of 13kV/cm. But the low energy peak from the wide well ($\simeq$68Å) is quenched at a field of $\simeq$6.5kV/cm. The increased threshold could be result of the

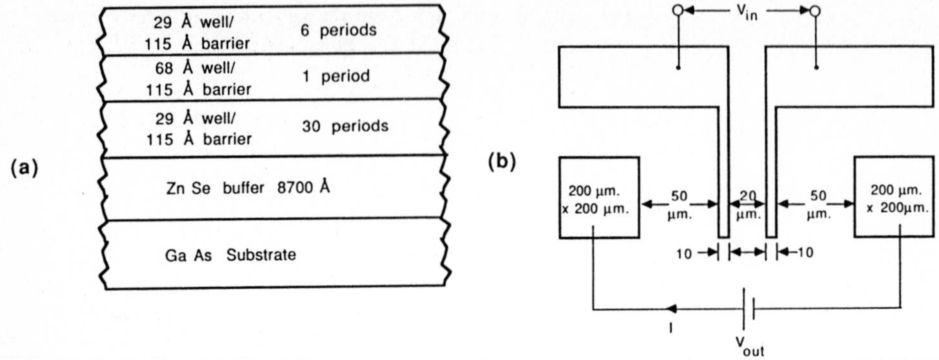

Fig. 4: Exciton ionization experiments
(a) Superlattice structure (side view)
(b) Contact configuration (top view)
The relation between V_{in} and V_{out} is approximately linear

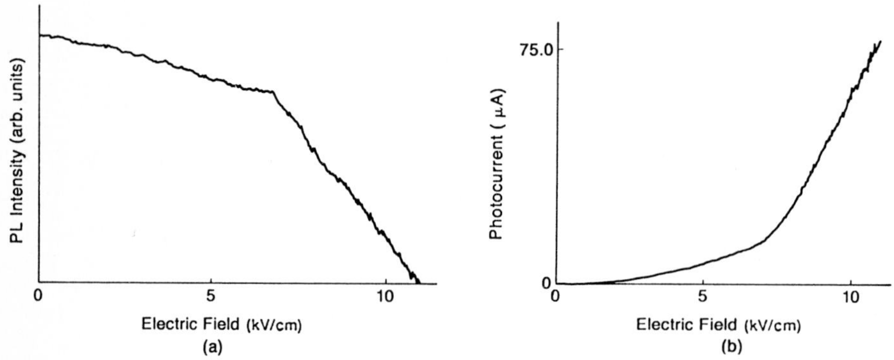

Fig. 5: (a) PL intensity of the lower peak (2.77 eV) as a function of $\mathcal{E}$ at 10K
(b) Current I as a function of $\mathcal{E}$ at 10K

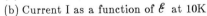

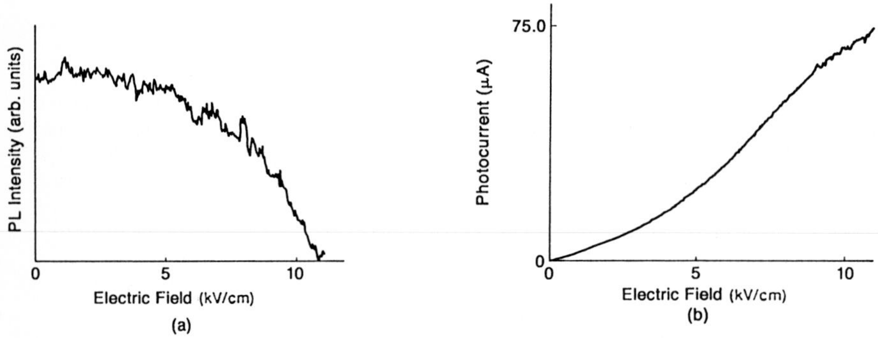

Fig. 6: (a) PL intensity of the lower peak (2.77 eV) as a function of $\mathcal{E}$ at 75K
(b) Current I as a function of $\mathcal{E}$ at 75K

increased binding energy of the quasi-2D exciton or it could arise from the increased localization due to interface roughness. Figs. 5 and 6 show the PL intensity of the lower peak (2.77 eV) and the current I as a function of the parallel electric field $\mathcal{E}$ at 10 K and at 77

K, respectively. At 10 K, both the PL intensity and the current I shows a clear change in slope at 6.5 kV/cm at 10 K; however, at 77K there is no clear breakpoint especially in the current-field characteristic. We interpret this to indicate that at 77K, the excitons are significantly ionized even without any applied field so that the ionization of the excitons has no significant effect on the conductivity; but at 10 K, the thermal ionization of excitons is negligible and the field-induced ionization into free carriers changes the conductivity noticeably.

To explore this point further, let us consider the formation of free carriers from excitons and vice-versa. Fig. 7 shows a simple model depicting the basic processes. An_x is the rate at which excitons ionize into free carriers while Bn_en_h is the rate at which electrons and holes coalesce to form excitons. Electron-hole pairs are generated by the incident light at a rate g. Excitons recombine radiatively with a lifetime τ_x, while electrons and holes have non-radiative lifetimes of τ_e and τ_h respectively. At steady state,

$$\frac{dn_x}{dt} = -\frac{n_x}{\tau_x} + Bn_en_h - An_x \quad = 0 \tag{10a}$$

$$\frac{dn_e}{dt} = -\frac{n_e}{\tau_e} - Bn_en_h + An_x + g = 0 \tag{10b}$$

$$\frac{dn_h}{dt} = -\frac{n_h}{\tau_h} - Bn_en_h + An_x + g = 0 \; . \tag{10c}$$

If we assume that $\tau_e = \tau_h \equiv \tau_c$, then we have $n_e = n_h \equiv n_c$. (10a-c) are then simplified to yield,

$$g = \frac{n_x}{\tau_x} + \frac{n_c}{\tau_c} \tag{11a}$$

$$0 = \frac{n_x}{\tau_x} + An_x - Bn_c^2 \; . \tag{11b}$$

(11a,b) can be solved for n_x and n_c:

$$\frac{n_c}{g\tau_c} = \frac{2}{x}\left(\sqrt{1+x} - 1\right) \tag{12a}$$

$$\frac{n_x}{g\tau_x} = 1 - \frac{n_c}{g\tau_c} \; , \tag{12b}$$

$$\text{where} \quad x \equiv 4\,\frac{Bg\,\tau_c^2}{A\tau_x + 1} \; . \tag{12c}$$

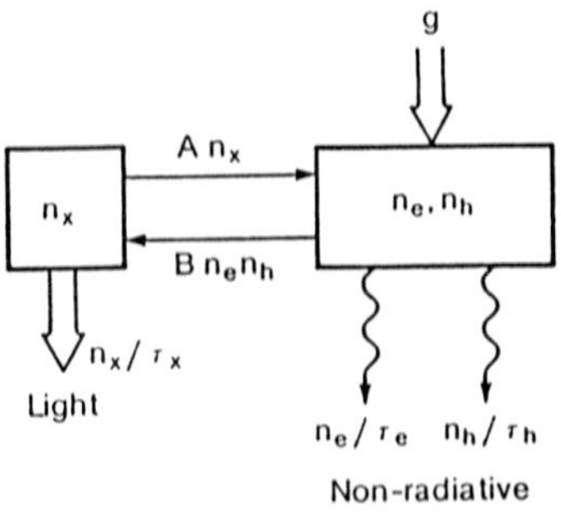

Fig. 7: Simple model for exciton-free carrier system

The parameter B describing the formation of excitons from electrons and holes does not change very much with an applied electric field. However, the rate of ionization of excitons into free carriers is strongly field dependent:

$$A = A_T + A_E + A_I n_c \ ; \tag{13}$$

A_T is the field-independent thermal ionization rate. A_E is the rate at which excitons are directly ionized by the field while $A_I n_c$ is the rate of impact ionization of excitons by free carriers accelerated to high energies by the field. Exciton ionization experiments have been reported in Ge,Si [24,25] GaP, InP, CdS [26] and GaAs [27]. Usually the ionization of excitons is ascribed to the impact ionization mechanism because the ionizing field is much lower than that needed for the direct field ionization rate A_E to become significant. However, in our present experiment the ionizing field agrees very well with the direct field ionization mechanism. In the following discussion we will neglect the impact ionization mechanism, though further theoretical and experimental work is needed to assess the relative importance of this mechanism.

The thermal ionization rate A_T is given by [24]

$$A_T = B \left(\frac{2\pi m_r k_B T}{h^2} \right)^{3/2} \exp\left(- \frac{E_B}{k_B T} \right) , \tag{14a}$$

where E_B is the binding energy of an exciton ($\simeq$20meV for ZnSe) and m_r is the reduced mass ($\simeq$·1m_0). The field ionization rate A_E is given by [28]

$$A_E = 16 \ \frac{E_B}{\hbar} \frac{E_0}{E} \exp\left[- \frac{4E_0}{3E} \right] , \tag{14b}$$

$$\text{where} \quad E_0 = E_B \, / \, e a_B \ , \tag{14c}$$

a_B is the Bohr radius of the exciton ($\simeq$35Å for ZnSe). At 10K, $A_T\,\tau_x$ is very small compared to 1. $A_E\,\tau_x$ is much less than 1 for low fields but rises steeply to a value much greater than 1 around $E = 6.5$ kV/cm, assuming $\tau_x = 1$ ns. As a result, the parameter x (in (12c)) decreases abruptly around electric fields $\simeq$6.5 kV/cm leading to photoluminescence quenching (decrease in n_x, (12b)) and rise in conductivity (increase in n_c, (12a)). Figs. 8a,b show the calculated values of $n_x/g\tau_x$ and photocurrent as a function of the electric field at 10K, assuming the following parameters: $\tau_x = 1$ ns, $\tau_c = 300$ ps, $g = 10^{25}/\text{cm}^3\text{s}$ (corresponding to an excitation power density of 80 W/cm^2), $B = 3.1 \times 10^{-5}$ cm^3/s [24]. The theoretical results are very similar to the experimental observed data. The threshold ionization field is relatively insensitive to our choice of parameters. It depends only on E_B,a_B and τ_x since it is basically the field at which $A_E\tau_x$ increases rapidly from a value much less than 1 to a value much greater than 1. It will be noted that at this value of the field there should be little effect on the absorption spectrum since $A_E \sim 10^9$/s. corresponds to a broadening of the absorption peak by only $\sim$1 μeV. The effect on the PL intensity, however, is drastic as we can see from Fig. 8a.

At 77K, A_T is much larger (14a) so that even without any electric field $A\tau_x$ is larger than 1 and x is small. Thus the excitons are significantly ionized even without any field and the electric field induced ionization does not change the conductivity very much. Figs. 9a,b show the calculated values of $n_x/g\tau_x$ and photocurrent as a function of the electric field at 77K assuming the same parameters as before. The results are in good agreement

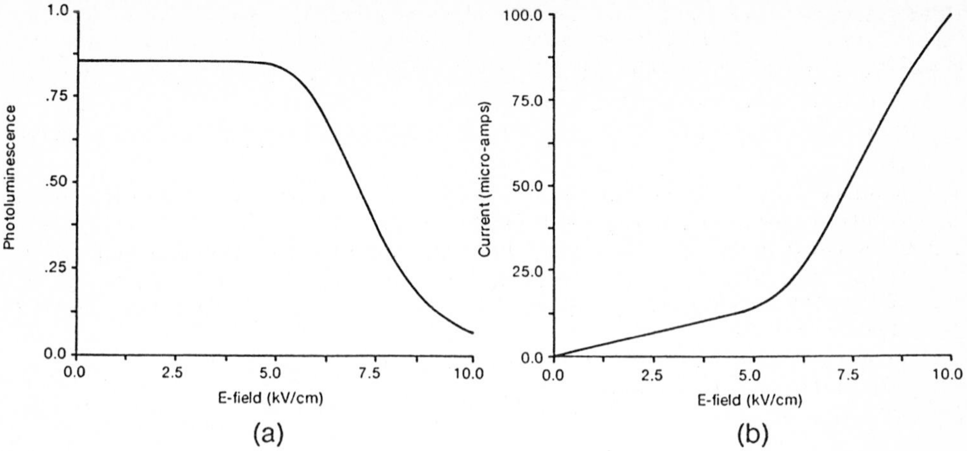

Fig. 8: Theoretical results at 10K (a) PL intensity ($n_x/g\tau_x$) vs. $\mathcal{E}$ (b) Current I vs. $\mathcal{E}$

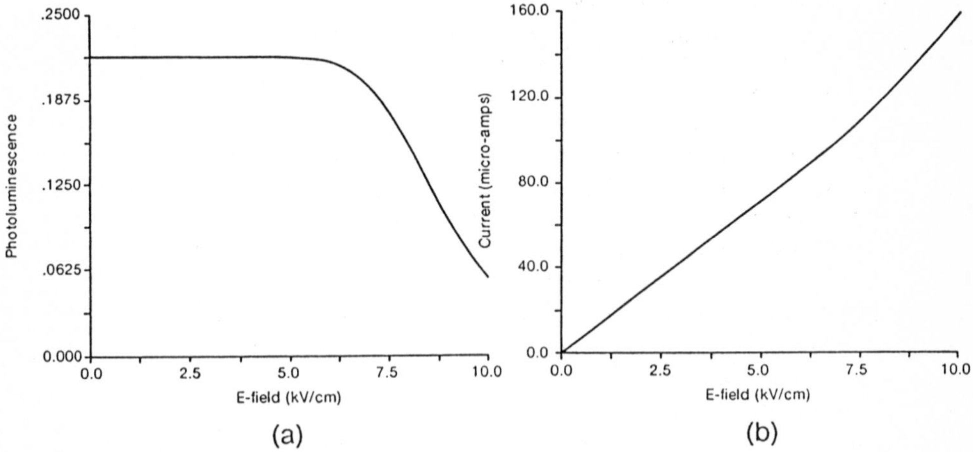

Fig. 9: Theoretical results at 77K (a) PL intensity ($n_x/g\tau_x$) vs. $\mathcal{E}$ (b) Current I vs. $\mathcal{E}$

with experiment. Thus, this simple model neglecting impact ionization of excitons appears to explain the experiment quite well; however, further experimental work is needed investigating the effect of varying temperature and excitation intensity. The effect of impact ionization by free carriers could be minimized by resonant generation of excitons.

ACKNOWLEDGEMENTS

The authors are grateful to T. C. Bonsett for the polarization-dependent absorption coefficient data, to B. Das and D. R. Andersen for the exciton ionization data, to B. Das for Figs. 8 and 9, to Y. Lee for calculating the exciton binding energy versus well width and to S. Bandyopadhyay for his help. It is a pleasure to acknowledge our collaborator A. V. Nurmikko for many valuable discussions and for his numerous contributions to the project. This work was supported by the Office of Naval Research under contract no. N00014-82-K-0563, the Air Force Office of Scientific Research under contract no. 85-0185, the NSF/MRG under grant no. 85-20866 and the DARPA/URI under contract no. N00014-86-K-0760.

REFERENCES

1. L. A. Kolodziejski, R. L. Gunshor, N. Otsuka, S. Datta, W. M. Becker and A. V. Nurmikko, *IEEE Trans. QE-22,* 1666 (1986).
2. A. V. Nurmikko, R. L. Gunshor and L. A. Kolodziejski, *IEEE Trans. QE-22,* 1785 (1986).
3. N. Otsuka, L. A. Kolodziejski, R. L. Gunshor, S. Datta, R. N. Bicknell and J. F. Schetzina, *Appl. Phys. Lett. 46,* 860 (1985).
4. R. N. Bicknell, N. C. Giles-Taylor, J. F. Schetzina, N. G. Anderson and W. D. Laidig, *Appl. Phys. Lett. 46,* 238 (1985); R.N. Bicknell, N. C. Giles-Taylor, D. K. Blanks, J. F. Schetzina, N. G. Anderson and W. D. Laidig, *Appl. Phys. Lett. 46,* 1122 (1985).
5. R. B. Bylsma, W. M. Becker, T. C. Bonsett, L. A. Kolodziejski, R. L. Gunshor, M. Yamanishi and S. Datta, *Appl. Phys. Lett. 47,* 1039 (1985).
6. Y. Hefetz, W. C. Goltsos, A. V. Nurmikko, L. A. Kolodziejski and R. L. Gunshor, *Appl. Phys. Lett. 48,* 372 (1986).
7. Y. Hefetz, J. Nakahara, A. V. Nurmikko, L. A. Kolodziejski, R. L. Gunshor and S. Datta, *Appl. Phys. Lett. 47, 989 (1985).*
8. R. B. Bylsma, J. Kossuth, W. M. Becker, R. L. Gunshor, L. A. Kolodziejski and R. Frohne, *J. Appl. Phys.,* to be published.
9. R. J. Sladek, private communication.
10. G. D. Sanders and Y. C. Chang, *Phys. Rev. B32,* 5517 (1985) and references therein.
11. A. Blacha, H. Presting, and M. Cardona, *Phys. Status Solidi* (b) *126,* 11 (1984).
12. G. L. Bir and G. E. Pikus, *Symmetry and Strain-induced Effects in Semiconductors,* Keter Publishing House Jerusalem Ltd., 1974, Chapter 5.
13. J. M. Luttinger, *Phys. Rev. 102,* 1030 (1956).
14. P. Lawaetz, *Phys. Rev. B4,* 3460 (1971).
15. A. Baldereschi and N. O. Lipari, *Phys. Rev. B3,* 439 (1971).
16. R. L. Greene and K. K. Bajaj, *Solid State Commun. 45,* 831 (1983).
17. Y. Lee, private communication.
18. J. W. Wu, A. V. Nurmikko, and J. J. Quinn, *Phys. Rev. B,* (1986).
19. M. Matsuura and H. Büttner, *Phys. Rev. B21,* 679 (1980).
20. M. Yamanishi and I. Suemune, *Jap. J. Appl. Phys. 23,* L35 (1984).
21. J. S. Weiner, D. S. Chemla, D. A. B. Miller, H. A. Haus, A. C. Gossard, W. Wiegmann and C. A. Burrus, *Appl. Phys. Lett. 47,* 664 (1985).
22. T. C. Bonsett, M. Yamanishi, L. A. Kolodziejski, R. L. Gunshor and S. Data, to be presented at the International Quantum Electronics Conference, Baltimore, 1987.
23. B. Das, D. R. Andersen, M. Yamanishi, T. C. Bonsett, R. L. Gunshor, L. A. Kolodziejski, S. Datta, SPIE Conference on quantum Well and Superlattice Physics, Bay Point, Florida, 1987.
24. Ch. Nöldeke, W. Metzger, R. P. Huebener and H. Schneider, *Phys. Status Solidi (b) 129,* 687 (1985).
25. a. V. M. Asnin, A. A. Rogachev, and S. M. Ryvkin, *Sov. Phys.-Semicond. 1,* 1445 (1968).
    b. T. Yao, K. Inagaki, and S. Maekawa, *Solid State Commun. 13,* 533 (1973).
    c. D. L. Smith, D. S. Pan, and T. c. McGill, *Phys. Rev. B12,* 4360 (1975).
26. a. M. V. Lebedev and V. G. Lysenko, *Sov. Phys.-Solid State 24,* 1721 (1982).
    b. B. M. Ashkinadze, S. M. Ryvkin, and I. D. Yaroshetskii, *Sov. Phys.-Semicond. (3,* 455 (1969).

c. B. J. Skromme and G. E. Stillmann, *Phys. Rev. B28,* 4602 (1983).

27. W. Bludau and E. Wagner, *Phys. Rev. B13,* 5410 (1976).

28. D. A. B. Miller, D. S. Chemla, T. C. Damen, A. C. Gossard, W. Wiegmann, T. H. Wood and C. A. Burrus, *Phys. Rev. B32,* 1043 (1985).

Optical Spectroscopy of Excitons in Quantum Wells Under an Electric Field

L. Viña+, R.T. Collins, E.E. Mendez, W.I. Wang, L.L. Chang, and L. Esaki

IBM T.J. Watson Research Center, P.O. Box 218,
Yorktown Heights, NY 10598, USA
+Permanent address: Instituto de Ciencia de Materiales, Universidad de Zaragoza, Consejo Superior de Investigaciones Cientificas, E-50009 Zaragoza, Spain

The effects of an external electric field on the excitonic spectrum of GaAs/$Ga_{1-x}Al_xAs$ quantum wells have been studied by means of low-temperature photoluminescence, photoluminescence-excitation, and photocurrent spectroscopies. A strong dependence of the field-induced shift of the excitonic recombination on well thickness has been observed. A corresponding quenching of the luminescence intensity has also been measured. A comparison of the effects of the field on the photocurrent and the photoluminescence clarifies the role that the impurities play in the Stoke shifts observed between luminescence and absorption measurements. For well thickness larger than ~120Å, the spectra exhibit two excitonic peaks in an energy range where only a single excitonic feature is expected to occur. Uniaxial-stress and polarization-dependent measurements have been used to identify these peaks as arising from a mixing between the first light-hole and the second heavy-hole valence subbands. In addition, the excited states of the excitons have been resolved. Coupling between the excited states of the heavy-hole and the ground state of the light-hole excitons has been observed. The coupling introduces fine structure in the spectra, which is attributed to the 2p state of the heavy-hole exciton.

1.- Introduction

The concept of superlattices and quantum wells (QWs), consisting of periodic structure of alternating ultrathin layers, was introduced by ESAKI and TSU in 1970 [1]. Shortly thereafter, the predominant role of excitonic effects in the optical properties of these structures was demonstrated by DINGLE et al., who measured the optical absorption of isolated QWs [2] and superlattices [3]. Although the use of electric fields has been proven to be a very powerful method to understand the optical spectra in atomic physics, it took almost one decade before the first study of the effects of an electric field on the optical properties of GaAs/$Ga_{1-x}Al_xAs$ QWs was performed [4]. Since then, the interest in this field has grown considerably and different techniques such as photoluminescence (PL) and photoluminescence excitation (PLE) [5-9], optical absorption [10-13], photocurrent (PC) [7,8,14-16], and electroreflectance [17] have been employed. The fields have been applied mainly perpendicularly to the layers (longitudinal fields), but also in a parallel configuration [11-13]. The main effects in the former case are a Stark shift of the excitons in the wells together with a decrease of the oscillator strength, whereas in the latter case Stark shifts are hardly observable, similarly to the situation in bulk GaAs [18]. Time-resolved PL measurements have clarified the mechanisms of the observed quenching [15,19,20]. For thin wells [19,20] (≤100Å) or thin barriers [15], the decrease of the lifetime is consistent with the Fowler-Nordheim tunneling model. The increase in lifetime observed for wider wells [20] can be explained by the field-induced polarization of the electron and hole wavefunctions [21].

In this paper we review three main aspects of the electric-field effects on the optical properties of $GaAs/Ga_{1-x}Al_xAs$ QWs, limiting our discussion to longitudinal fields, for which field ionization of the excitons is prevented by the lateral confinement. After a short description of the experimental techniques, we will focus on the well-thickness dependence of the Stark shifts of the ground state excitons [9] and the role of impurities in the Stokes shifts between PL and absorption measurements [5]. Strong evidence of valence-band mixing, obtained from PL-excitation, polarization and uniaxial stress dependent PC measurements [16], will be presented in Section 4. Finally, the observation of fine structure in the excited states of the heavy-hole exciton and their coupling with the ground state of the light-hole exciton [22] will be discussed.

2.- Experiments

The samples used in our studies were grown by molecular beam epitaxy on (100)-oriented n+-GaAs substrates. One or several GaAs QWs were grown on a n+-GaAs buffer, followed by a $Ga_{0.65}Al_{0.35}As$ layer. The wells, separated by ~250Å $Ga_{0.65}Al_{0.35}As$ barriers, were capped either with a thin Ni film to form a Schottky contact, or with a p+-GaAs layer to result in a p+-i-n+ configuration. The region between the top and the n+-GaAs layers was nominally undoped, except in some cases where a few hundred angstroms of the $Ga_{0.65}Al_{0.35}As$ adjacent to the buffer layer were slightly Si doped. The magnitude of the electric fields, perpendicular to the layers, was estimated from growth parameters and from the bias corresponding to flat band condition, and it is believed to be accurate to about 10% [9]. Flat band corresponds to a certain positive bias, therefore a decrease of the external voltage implies an increase in the applied electric field. For the polarization-dependent PC measurements, the thickness of the $Ga_{0.65}Al_{0.35}As$ layers and the number of QWs were chosen to form a leaky waveguide structure [23], allowing illumination from the edge of the samples.

The samples were mounted in a variable temperature cryostat. The QW luminescence was excited either indirectly, with the 5145-Å line of an Ar+-laser, above the $Ga_{0.65}Al_{0.35}As$ band gap, or selectively with the 6471-Å line of a Kr+-laser. The measurements were performed at 5K with low excitation power densities (≤0.5Wcm^{-2}) to avoid excessive carrier generation. The PL signal was dispersed by a 3/4-m double monochromator and detected with a cooled GaAs photomultiplier. Excitation spectra were recorded at 5K with a resolution of 0.2meV. An LD700 dye-laser, pumped by the Kr+-laser, was used to excite the sample, while the spectrometer was set at the heavy-hole emission wavelehgth, which increases with increasing applied field. For the PC spectra the sample was illuminated with the light from a tungsten lamp dispersed from a grating monochromator, and the photocurrent was detected as a function of incident wavelength using standard lock-in techniques.

3.- Thickness Dependence of Stark Shifts

Photoluminescence spectra of a sample with 230Å QWs are shown in Fig.1. These spectra were measured with the Ar+-laser, using a power density of 80$mWcm^{-2}$. We assign the structures at the two largest voltages to the free

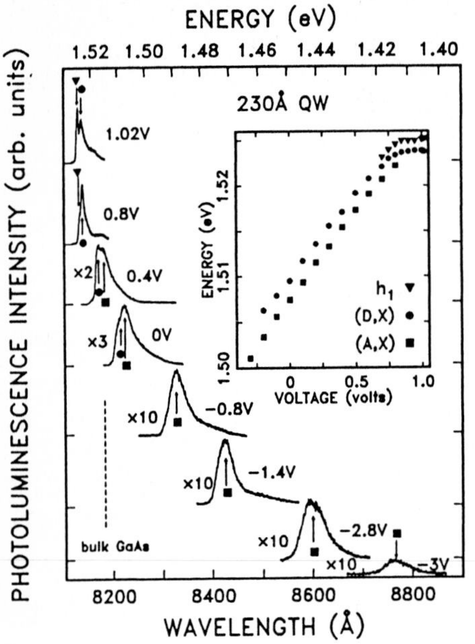

Fig. 1: Photoluminescence spectra of a sample with 230Å quantum wells. Free heavy-hole, and donor- and acceptor-bound excitons are marked with triangles, dots and squares, respectively. Note the large shift (~100meV) of the PL below the band gap of bulk GaAs. The inset shows the peak positions as a function of applied voltage, in the range of coexistence of all structures

heavy-hole exciton (▼, h_1) and to a donor-bound exciton [•, (D,X)]. The former assignment is based on a comparison of the observed energy with envelope-function calculations [24], using the parameters given in Ref.9. The latter assignment is based on the temperature and excitation density dependence of this structure, and on its energy difference with the free exciton (~1.2meV), close to the corresponding difference in bulk GaAs.

As the electric field is increased, a strong quenching of the free exciton peak is observed, which is seen only as a shoulder at 0.8V. A further increase of the field quenches the donor-bound exciton, but simultaneously makes resolvable a third structure [■ ,(A,X)], which we assign to an acceptor-bound exciton. We should mention that a complicated behavior of free and bound excitons in Be-doped $GaAs/Ga_{1-x}Al_xAs$ superlattices has been reported before, although no significant Stark shifts were observed in this case [5]. The inset in the figure shows the excitonic energies as a function of external bias. The energy difference between the structures remains constant, indicating that the changes of binding energies of free and bound excitons with electric fields are the same.

Figure 2 shows the Stark shifts versus electric field of the dominant peak in the PL spectra for four samples. An increasing shift with increasing well thickness, for the same electric field, as expected from perturbation-theory arguments [25], is clearly observed. The Stark effect results in a shift of the excitonic

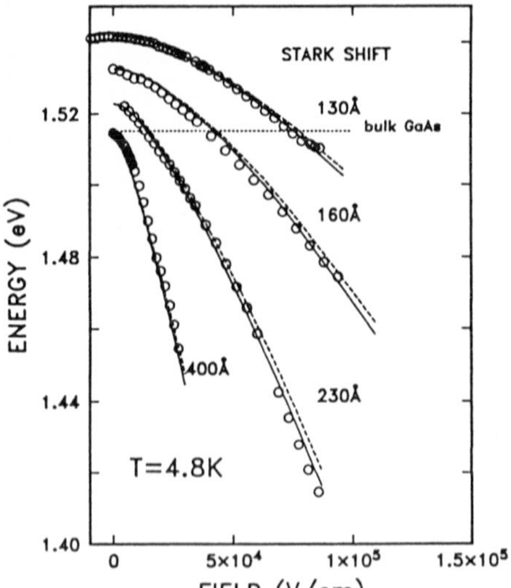

Fig. 2: Experimental heavy-hole exciton energies as a function of electric field for different well thickness (points). Theoretical calculations (see text) are shown by dashed and solid lines. The energy of the free exciton in bulk GaAs is indicated by the horizontal line

recombination below the bulk GaAs exciton, whose energy is indicated by a dotted line in the figure. The results of two different calculations [9] are also depicted in this figure. The dashed lines correspond to variational calculations [21], while the solid lines represent the exact numerical solution using a matrix formalism [9]. It is also remarkable that for wide wells, larger than the three-dimensional exciton diameter (~300Å), the electron-hole interaction remains strong enough so that excitons still exist when the Stark shift considerably exceeds the binding energy. This demonstrates that, even for wide wells, impact ionization is not important for the quenching of the PL. We have also shown that for very wide wells (~1000Å) the Stark shifts are extremally small, and the behavior of bulk GaAs is recovered [26].
The variational calculations provide wavefunctions that can be used to compute the overlap integral between electrons and holes, which can be compared with the integrated PL efficiency. A comparison between our experimental data and the calculations [9] indicates that the decrease in the wavefunction overlap is the main mechanism for the PL quenching in thick QWs (≥100Å), in agreement with time-resolved measurements [20].

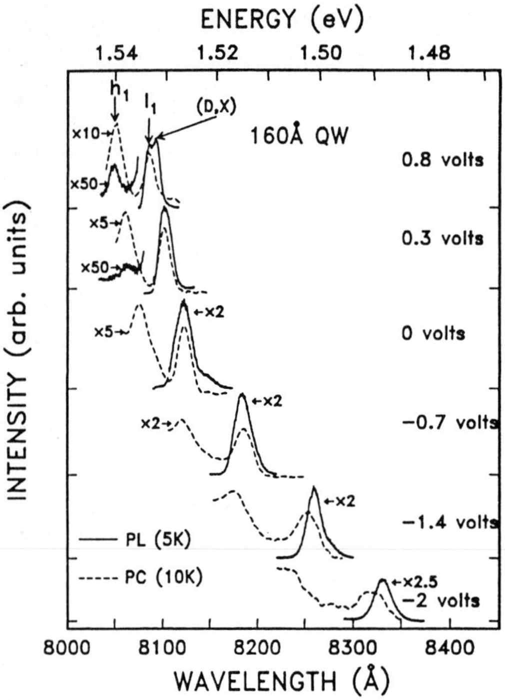

Fig. 3: Photoluminescence and photocurrent spectra at different bias for a sample with 160Å QWs. The total applied voltage to the sample is shown on the right side of the spectra. l_1, h_1, and (D,X) indicate free light-hole, heavy-hole and donor-bound excitons, respectively

Photoluminescence and photocurrent spectra of a sample with 160Å-wide wells for six selected voltages (fields between ~0V/cm and ~77kV/cm) are shown in Fig.3. The ground state of the light- and heavy-hole excitons, as well as a donor-bound exciton, are seen at the largest voltage. The overall behavior of the PL is similar to that of the sample of Fig.1. The energies of the free excitons found in PL and PC agree to within 0.3meV up to fields of ~50kV/cm. The remarkably good agreement between the linewidths observed in both spectra indicates that both techniques are sampling the same electronic states. An increasing Stokes shift between the PL and PC, amounting to 2.5meV at -2V, is observed at higher fields. This effect can be explained as the result of enhanced bound-exciton luminescence as the carriers are skewed closer to the interfaces, where more impurities and defects exist. This also illustrates the important role that impurities can play in the Stokes shifts, between emission and absorption measurements, reported in the literature [27].

4.- Valence-Band Mixing

Figure 4 shows several PLE (solid lines) and PC (dashed lines) spectra of a p+-i-n+ sample with 160Å QWs [28]. At low fields, the spectra are dominated by the ground states of the heavy- (h_1) and light-hole (l_1) excitons (h_1 is not shown in the PLE spectra because the spectrometer was set at the wavelength of h_1, and the sample presented a negligible Stokes shift). In the presence of an electric field, the QWs become asymmetric and forbidden transitions ($\Delta n \neq 0$) become allowed [14]. According to envelope-function calculations [21,28], the ground state of the exciton corresponding to the n=1 conduction subband and n=2 heavy-hole subband (h_{12}), and only this exciton, should appear in the energy range where the two transitions labelled h_{12a} and h_{12b} are observed. These two structures have been observed in all the samples studied with well widths larger than ~120Å. The excited states of the heavy-hole exciton ($h_1(2x)$, where x stands for both the s and p states) are also observed at low fields in this figure, their coupling with the l_1 exciton will be discussed in the next section.

The energies of some of the excitons observed in Fig.4 are plotted in Fig.5 as a function of the electric field. The solid lines represent envelope function calculations [28] of the energies of h_1, l_1 and h_{12}. At low fields, the excited states of the light-hole exciton ($l_1(2x)$) are close to the h_{12b} exciton. With increasing field, $l_1(2x)$ shifts closer to h_{12a} until they have almost the same energy. This behavior suggests a possible contribution of the light-hole subband to the presence of the two excitons in the region where only one is predicted. For fields larger than ~50kV/cm, h_{12b} dominates h_{12a}, and its shift agrees with the results of the calculation for h_{12}, suggesting that its character should be predominantly heavy at high fields.

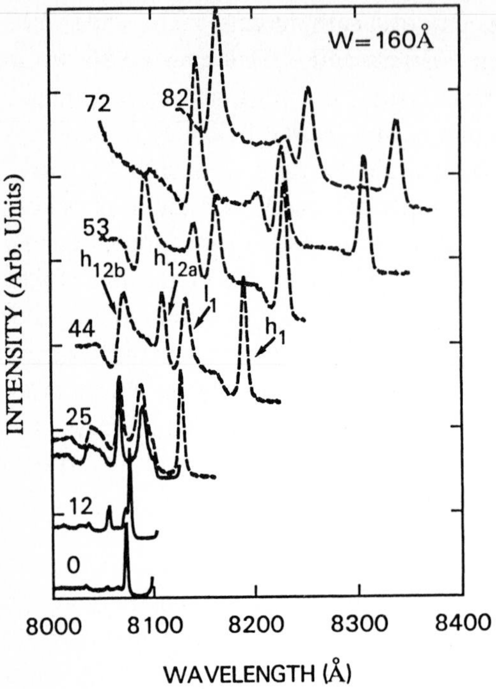

Fig. 4: Photoluminescence excitation (solid) and photocurrent (dashed) spectra of a p+-i-n+ sample with 160Å QWs. The electric fields are shown on the left side of the spectra. h_1 and l_1 are the free heavy- and light-hole excitons, respectively (h_1 is not shown in the PLE spectra). h_{12a} and h_{12b} are excitons arising from mixing between the n=1 light-hole and the n=2 heavy-hole valence subbands

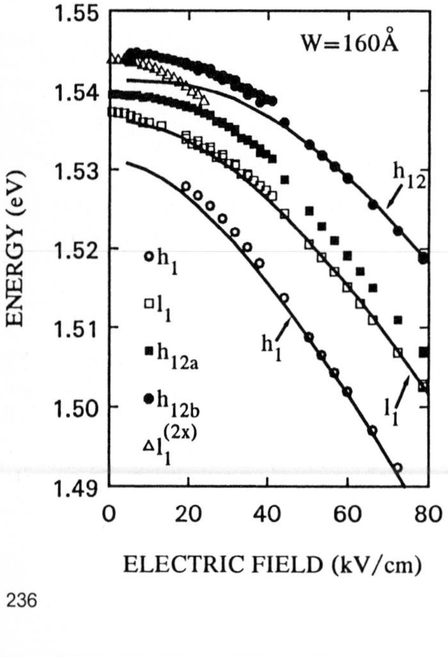

Fig. 5: Stark shifts of the excitons shown in Fig.4. The solid lines represent calculations in the envelope-function approximation

Polarization and uniaxial-stress dependent PC measurements were performed to clarify the origin of h_{12a} and h_{12b}. Figure 6a presents polarization-dependent PC measurements for a sample with 130Å wells imbedded in a $Ga_{1-x}Al_xAs$ waveguide [23]. For XY-polarized light (in the plane of the layers) both heavy- and light-hole excitons are allowed. However, for Z-polarized light only the light-hole excitons are allowed [23]. Only a large l_1 peak is observed in the Z polarization in Fig.6a. Therefore, h_{12a} and h_{12b} have *oscillator strengths* arising primarily from the second heavy-hole subband.

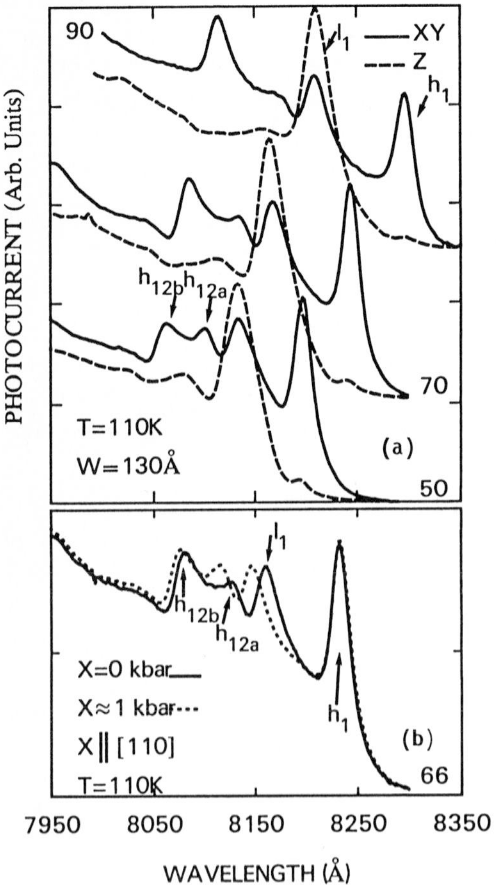

Fig. 6: a). Polarization dependence of photocurrent measurements for two 130Å QWs imbedded in a $Ga_{1-x}Al_xAs$ waveguide. Notice that the temperature is higher than in Fig.4. XY represents light polarized in the plane of the layers, while Z is perpendicular to the layers
b).Uniaxial stress dependence of photocurrent measurements for the same sample as a). The stress is applied perpendicularly to the growth direction. The h_1 peaks have been aligned

The results of the uniaxial-stress experiments are shown in Fig.6b. According to calculations for compressive stress along a [110] axis [29], with increasing stress, the light-hole excitons should show a larger blue shift than the heavy-hole excitons. This has been borne out from the experimental results in Fig.6b, where the h_1 peaks are aligned by neglecting a rather small shift. An examination of the spectra shows that h_{12a} and h_{12b} have a predominant light- and heavy-hole character, respectively, for the electric field of the figure. The seemingly contradictory conclusions drawn from the polarization and uniaxial-stress measurements can be easily understood by realizing that the uniaxial-stress measurements provide information on the entire wavefunction, while the polarization-dependent measurements involve only the part of the wavefunction which contributes to the absorption. Strong mixing between the n=1 light-hole and the n=2 heavy-hole subbands has been predicted for well widths in the range used in our studies [30]. Based on recent theoretical results by CHAN [31], we believe that the presence of the two peaks h_{12a} and h_{12b} arises from an interaction between the excited states of l_1 and the ground state of h_{12}.

5.- Fine Structure in the Excited States of the Heavy-Hole Exciton

The presence of excited states in the optical spectra of GaAs/$Ga_{1-x}Al_xAs$ QWs was first reported by MILLER et al. [32]. Recently, DAWSON et al. have observed well-resolved excited-state peaks in PLE [33], as well as in PL spectra [34]. We present in Fig.7 a detailed low-field PLE spectrum of the sample used in Fig.4. Also depicted in the figure is the heavy-hole PL, which shows a rising background from the p+-GaAs PL on its low energy side. The peaks in this spectrum are extremely sharp, with a full width at half maximum of 0.5meV, comparable to the thermal broadening. This reflects the exceptional quality of the sample, which enables us to resolve the excited states of excitons as clearly defined peaks. Based on agreements with previously published experimental [32,33] and theoretical [32-35] data, and with recent PLE experiments in the presence of a magnetic field [36], we assign the shaded structures to the excited states of the heavy- and light-hole excitons, $h_1(2x)$ and $l_1(2x)$, respectively. Calculations of the oscillator strength give a ratio of the excited to the ground states of 0.11 [35,37], in reasonable agreement with the experimental value of 0.07.

PLE spectra for different voltages, in the spectral range covering the l_1 and the $h_1(2x)$ excitons, are shown in Fig.8. At zero field, $h_1(2x)$ lies 1.5meV higher than l_1,and the ratio of their oscillator strengths is ~1/4. With increasing electric field, $h_1(2x)$ shifts more than l_1, because of its heavy-hole character; their relative energy spacing decreases, and their intensities become comparable. At

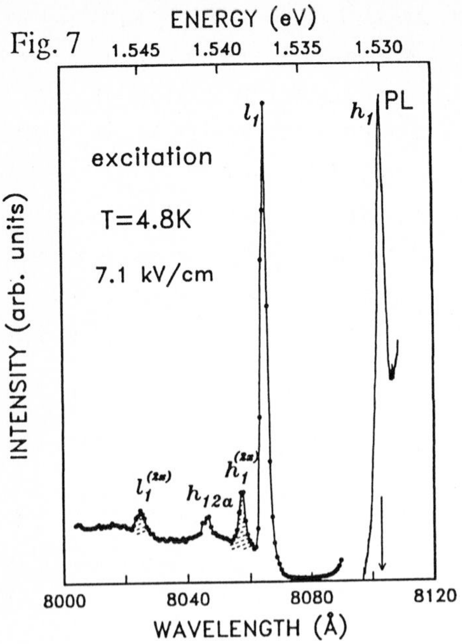

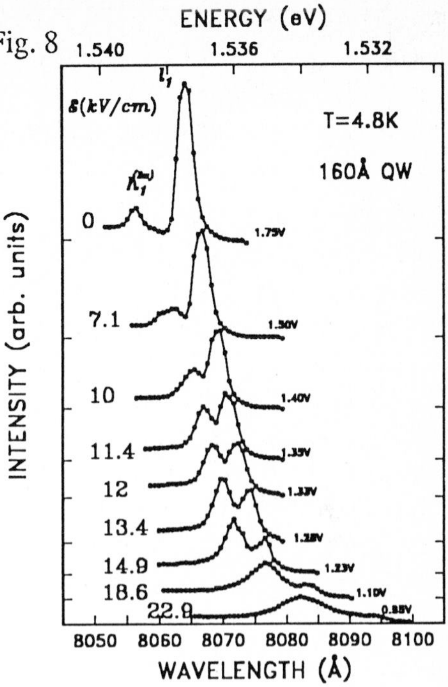

Fig. 7: Excitation spectrum of the same sample as in Fig. 4. The shaded structures $h_1(2x)$ and $l_1(2x)$, correspond to excited states of the heavy- and light-hole excitons, respectively. The photoluminescence (PL) spectrum of the heavy-hole exciton is also shown. The detection energy for the PLE spectrum is indicated by the arrow

Fig. 8: Excitation spectra for several fields in the spectral range of the light-hole exciton (l_1) and the excited states of the heavy-hole exciton ($h_1(2x)$). The sample is the same as in Fig.7

still higher fields the excitons separate; $h_1(2x)$ moves to the low-energy side of l_1, reversing the original order, and its intensity decreases steadily.
Figure 9, where the energies of $h_1(2x)$ and l_1 are plotted as a function of field, shows clearly the coupling between these excitons, which manifests itself as an anticrossing. The dashed and dotted lines represent calculations of the Stark shifts of uncoupled heavy- and light-hole excitons, respectively [22]. As seen in the figure, the excitons follow closely the theoretical predictions outside of the range of strong coupling. The inset in the figure shows the integrated intensity of $h_1(2x)$, normalized to that of l_1, as a function of electric field. A strong correlation exists between the maximum in the intensity and the minimum in the

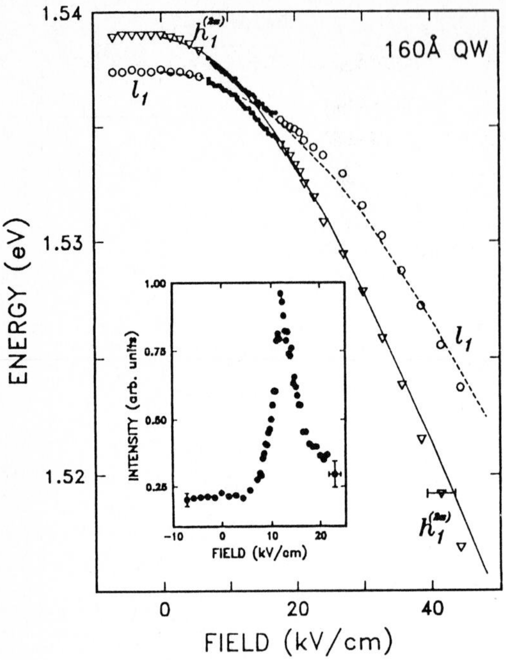

Fig. 9: Stark shifts of the light-hole exciton (open circles) and excited states of the heavy-hole exciton (open triangles). The data are shown as solid circles in the region of strong coupling. The lines correspond to calculations in the envelope-function approximation, neglecting changes in binding energy and coupling between excitons. The integrated intensity of $h_1(2x)$, normalized to that of l_1, is shown in the inset

energy separation of the excitonic structures. The minimum separation between l_1 and $h_1(2x)$, 0.7±0.2meV, indicates an enhancement of the exchange interaction between these excitons of ~2 with respect to bulk GaAs [38].
As a consequence of the interaction, fine structure is resolved in the excited states of the heavy-hole exciton. Figure 10 shows spectra of the excited states of h_1 from zero field up to 9.1kV/cm. A shoulder, which can be guessed already at flat band, is resolved with increasing field. At 6kV/cm, it becomes comparable in amplitude to the original peak and is the dominant feature at higher fields. We attribute the high- (low-) energy component of the doublet to the 2s (2p) state of the h_1 exciton. Their energy separation, ~0.45meV, agrees with the calculated difference of their binding energies [34,35].

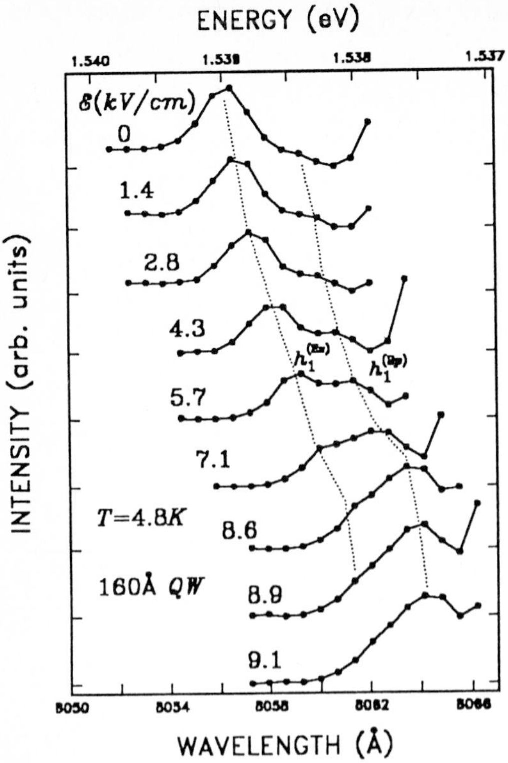

Fig. 10: Excitation spectra of the excited states of the heavy-hole exciton ($h_1(2x)$) for different fields where fine structure is resolved. The lines are a visual aid to follow the structures

6.- Summary

We have demonstrated the strong dependence of Stark shifts in GaAs/ $Ga_{1-x}Al_xAs$ QWs on well thickness. Shifts as large as 100meV for fields of ~10^5V/cm have been observed. For the range of well widths used in our experiments, a satisfactory agreement has been obtained between the PL quenching and the calculated decrease of the oscillator strenght, due to the polarization of electron and hole wavefunctions. The combination of PL and PC techniques allowed us to establish the role that impurities play in the Stokes shift between absorption and emission measurements. For quantum wells wider than ~120Å we have observed two excitonic peaks in the energy range of the h_{12} exciton, and correlated this observation with mixing between the second heavy- and the first light-hole valence subbands. An anticrossing behavior, in the presence of an electric field, between the excited states of the heavy-hole and the ground state of the light-hole excitons has been illustrated. Fine structure in the excited states of the heavy-hole exciton has been resolved and assigned to the 2s and 2p states of the h_1 exciton.

Acknowledgments
We thank L. Alexander for his assistance in sample preparation, and F. Stern, C. Mailhiot and D.L. Smith for many valuable discussions.

REFERENCES

1. L. Esaki and R. Tsu, IBM J. Res. Develop. **14**, 61 (1970).
2. R. Dingle, W. Wiegmann, and C.H. Henry, Phys. Rev. Lett. **33**, 827 (1974).
3. R. Dingle, A.C. Gossard, and W. Wiegmann, Phys. Rev. Lett. **34**, 1327 (1975).
4. E.E. Mendez, G. Bastard, L.L. Chang, L. Esaki, H. Morkoc, and R. Fischer, Phys. Rev B **26**, 7101 (1982); Physica **117&118B**, 711 (1983).
5. R.C. Miller and A.C. Gossard, Appl. Phys. Lett. **43**, 954 (1983).
6. Y. Horikoshi, A. Fischer, and K. Ploog, Phys. Rev. B **31**, 7859 (1985).
7. H.J. Polland, Y. Horikoshi, R. Höger, E.O. Göbel, J. Kuhl, and K.Ploog, Physica **134B**, 412 (1985).
8. L. Viña, R.T. Collins, E.E. Mendez, and W.I. Wang, Phys. Rev. B **33**, 5939 (1986).
9. L. Viña, E.E. Mendez, W.I. Wang, L.L. Chang, and L. Esaki, J. Phys, C: Solid State Phys. **20**, 2803 (1987).
10. D.S. Chemla, T.C. Damen, D.A.B. Miller, A.C. Gossard, and W. Wiegmann, Appl. Phys. Lett. **42**, 864 (1983); T.H. Wood, C.A. Burrus, D.A.B. Miller, D.S. Chemla, T.C. Damen, A.C. Gossard, and W. Wiegmann, *ibid* **44**, 16 (1984).
11. D.A.B. Miller, D.S. Chemla, T.C. Damen, A.C. Gossard, W. Wiegmann, T.H. Wood, and C.A. Burrus, Phys. Rev. Lett. **53**, 2173 (1984).
12. H. Iwamura, T. Saku, and H. Okamoto, Jpn. J. Appl. Phys. **24**, 104 (1985).
13. D.A.B. Miller, D.S. Chemla, T.C. Damen, A.C. Gossard, W. Wiegmann, T.H. Wood, and C.A. Burrus, Phys. Rev. B **32**, 1043 (1985).
14. R.T. Collins, K.v. Klitzing, and K. Ploog, Phys. Rev. B **33**, 4378 (1986).
15. Y. Masumoto, S. Tarucha, and H. Okamoto, Phys. Rev. B **33**, 5961 (1986).
16. R.T. Collins, L. Viña, W.I. Wang, L.L. Chang, L. Esaki, K.v. Klitzing, and K. Ploog, in *Proceedings of the 18th International Conference on the Physics of Semiconductors*, edited by O. Engström (World Scientific, Singapore, 1987), p. 521.

17. C. Alibert, S. Gaillard, J.A. Brum, G. Bastard, P. Frijlink, and M. Erman, Solid State Commun. **53**, 457 (1985).
18. W. Bludau and E, Wagner, Phys. Rev. B **13**, 5410 (1976).
19. J.A. Kash, E.E. Mendez, and H. Morkoc, Appl. Phys. Lett. **46**, 173 (1985).
20. H.J. Polland, L. Schultheis, J. Kuhl, E.O. Göbel, and C.W. Tu, Phys. Rev. Lett. **55**, 2610 (1985).
21. G. Bastard, E.E. Mendez, L.L. Chang, and L. Esaki, Phys. Rev. B **28**, 3241 (1983).
22. L. Viña, R.T. Collins, E.E. Mendez, L.L. Chang, and L. Esaki, Phys. Rev. Lett. **58**, 832 (1987).
23. J.S. Weiner, D.S. Chemla, D.A.B. Miller, H.A. Haus, A.C. Gossard, W. Wiegmann, and C.A. Burrus, Appl. Phys. Lett. **47**, 664 (1985).
24. G. Bastard, Phys. Rev. B **24**, 5693 (1981).
25. F.M. Fernandez and E.A. Castro, Physica **11A**, 334 (1982).
26. L. Viña, R.T. Collins, E.E. Mendez, W.I. Wang, L.L. Chang, and L. Esaki, Superlattices and Microstructures **3**, 9 (1987).
27. G. Bastard, C. Delalande, M.H. Meynadier, P.M. Frijlink, and M. Voos, Phys. Rev. B **29**, 7042 (1984).
28. R.T. Collins, L. Viña, W.I. Wang, L.L. Chang, L. Esaki, K.v. Klitzing, and K. Ploog, Phys. Rev. B **36**, 1531 (1987).
29. C. Mailhiot and D.L. Smith, private communication.
30. J.N. Schulman and Y.C. Chang, Phys. Rev. B **31**, 2056 (1985).
31. K.S. Chan, J. Phys. C: Solid State Phys., **19**, L125 (1986).
32. R.C. Miller, D.A. Kleinman, W.T. Tsang, and A.C. Gossard, Phys. Rev. B **24**, 1134 (1981).
33. P. Dawson, K.J. Moore, G. Duggan, H.I. Ralph, and C.T.B.Foxon, Phys. Rev. B **34**, 6007 (1986).
34. K.J. Moore, P. Dawson, and C.T.B. Foxon, Phys. Rev. B **34**, 6022 (1986).
35. Y. Shinozuka and M. Matsuura, Phys. Rev. B **28**, 4878 (1983).
36. L. Viña, E.E. Mendez, R.T. Collins, W.I. Wang, E. Isaacs, and D. Heiman (unpublished).
37. R.L. Greene, K.K. Bajaj, and D.E. Phelps, Phys. Rev. B **29**, 1807 (1984).
38. C. Jagannath and E.S. Koteles, Solid State Commun. **58**, 417 (1986).

Observation of Forbidden Transitions in Coupled Quantum Wells

E.S. Koteles, Y.J. Chen, and B.S. Elman

GTE Laboratories Inc., 40 Sylvan Road, Waltham, MA 02254, USA

Introduction

Recently, coupled double quantum well (CDQW) structures have attracted increasing attention since the coupling of electronic levels in the two wells, which can be finely tuned, provides an additional parameter for modifying the properties of bandgap-engineered structures and devices.[1-9] In an isolated, single quantum well (SQW), transitions between the ground state of the electron in the conduction band and the ground states of the heavy and light holes in the valence band give rise to a strong doublet in the absorption [and photoluminescence excitation (PLE)] spectra related to ground state heavy and light-hole excitons. When two SQWs are brought into close proximity, wavefunctions of the confined electrons and holes can overlap, leading to mixing and split levels in the conduction and valence wells in a manner analogous to the splitting observed in atomic levels of a diatomic molecule.[2] Since each of the original levels is doubled, a maximum of eight transitions is possible, as illustrated in Fig. 1. If the two SQWs are identical and unperturbed, the split levels have well-defined symmetries and selection rules dictate that transitions are allowed only between levels with the same symmetry. This restriction results in four allowed transitions (labelled 1, 3, 6, and 8 in Fig. 1) and these have been observed.[2,6]

To completely characterize these transitions, five band structure parameters are required (as shown in Fig. 1, these are the modified energy of the heavy-hole exciton, **a**, the energy splitting between the highest heavy-hole and lowest light-hole levels, **b**, and the splittings of the conduction, Δ_e, and two valence bands, Δ_h and Δ_l, levels in the coupled case). Thus, if only the four allowed transitions are observed, it is not possible to determine these parameters without recourse to a theoretical model. However, if the energies of five or more transitions were measured, modeling would not be necessary in order to determine all the level splittings in the CDQW. We report here the observation

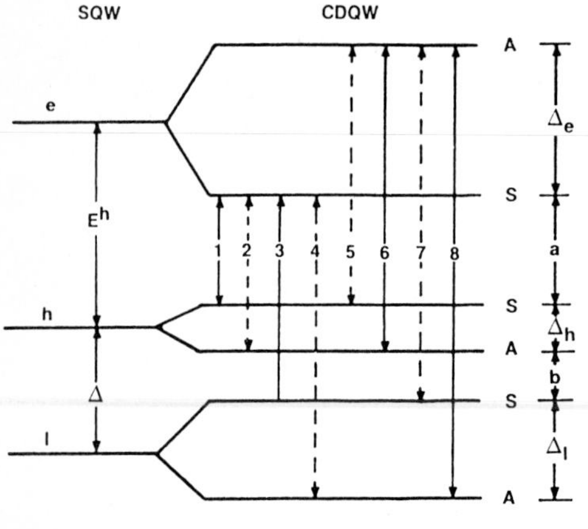

Fig. 1. Schematic diagram illustrating the splitting of the conduction and valence band levels of a single quantum well when two identical wells are brought into close proximity. **e** (Δ_e), **h** (Δ_h), and **l** (Δ_l) refer to the electron, heavy-hole and light-hole levels (and their splittings in the CDQW case). $\mathbf{E^h}$ (**a**) and Δ (**b**) label the energies of the heavy-hole exciton and heavy-hole-light-hole splitting (and their modified values in the CDQW system). **S** and **A** label symmetric and antisymmetric levels in the unperturbed (electric field free) case.

of forbidden optical transitions between split levels in the conduction and valence bands of a coupled pair of identical quantum wells separated by a narrow barrier. This is possible because of the slight alteration of the symmetry of the CDQW due to fringing electric fields. This enables us to derive, directly from experiment, splittings of quantum confined levels in a CDQW structure.

Experiment and Results

Low temperature PLE spectroscopy was employed to study optical transitions in a CDQW structure grown on an n-doped GaAs substrate using molecular beam epitaxy. The samples were mounted, strain-free, in an exchange gas LHe cryostat and excited with a tunable dye laser pumped with a Kr-ion pump laser. The photoluminescence intensity was monitored with a double grating spectrometer coupled to a cooled GaAs photomultiplier.

The CDQW sample consisted of a 500 nm silicon doped GaAs buffer, followed by 100 nm of undoped $Al_{0.27}Ga_{0.73}As$, on which was grown a pair of 7.5 nm GaAs quantum wells separated by a 1.8 nm barrier which, in turn, was topped with a 100 nm undoped $Al_{0.27}Ga_{0.73}As$ layer. Another sample, consisting of a single 7.5 nm quantum well embedded in similar layers, exhibited the standard heavy-hole light-hole split exciton doublet in its PLE spectrum (in Fig. 2a, $\mathbf{E^h}$ is the energy of the heavy-hole exciton and Δ is the difference between the energies of the ground state heavy- and light-hole excitons). On the other hand, the PLE spectrum of the CDQW displayed seven peaks (Fig. 2b). If the CDQW was simultaneously illuminated with above-bandgap light of low intensity (~ $100 mW/cm^2$ from a HeNe laser) while the PLE spectrum was measured, the number of peaks decreased to four (Fig. 2c).

Discussion

These results can be understood by invoking the effect of fringing electric fields from the doped buffer layer on the electronic levels of the CDQW. These fields, although small, distort the potential profile of the coupled wells enough that wavefunctions are modified, selection rules are relaxed and forbidden transitions (i.e. transitions between states of opposite symmetry, labelled 2, 4, 5, and 7 in Fig. 1) become possible. These forbidden transitions, in fact, give rise to the extra peaks seen in Fig. 2b.

From another viewpoint, the effect of electric fields is to transform a strong allowed transition into a weak one. This is due to the fact that, in an electric field, the symmetric and antisymmetric wavefunctions of electrons and holes have a tendency to concentrate in different wells.[5] Thus, for example, the symmetric level of the ground state of the conduction band is concentrated in a different well than the symmetric levels of the hole bands. Therefore, for example, transition 1, which is symmetry-allowed in zero electric field, becomes a transition between electrons primarily in one well and holes in the adjacent well when an electric field is applied to the CDQW. Since this transition is essentially spatially indirect in real space (i.e., the wavefunction overlap of these two states is significantly reduced), its oscillator strength declines rapidly with increasing field. On the other hand, transition 2, symmetry-forbidden in zero field, remains spatially direct in an electric field and thus does not lose oscillator strength.[5] In fact, its intensity appears to increase as the electric field is increased.

When these fringing electric fields were reduced by the presence of photo-generated carriers (using above-bandgap light from the HeNe laser), the electronic wavefunction symmetries, and consequently the selection rules, were restored and the

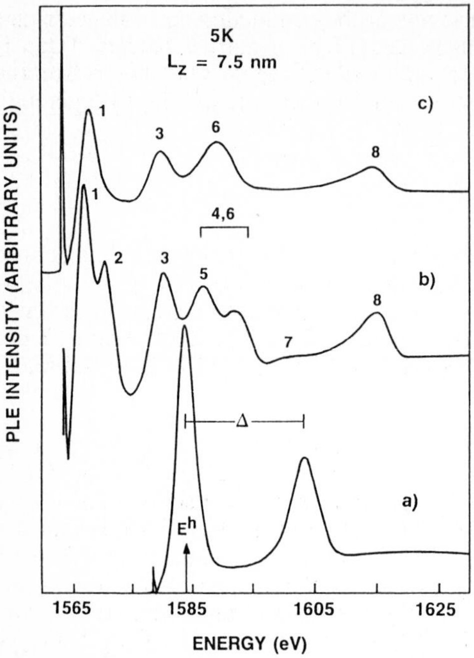

Fig. 2. a) 5K PLE spectrum of a single 7.5 nm wide quantum well. Δ is the splitting of the heavy and light-hole excitons and $\mathbf{E}^h$ is the energy of the ground state heavy-hole exciton. b) 5K PLE spectrum of a CDQW structure. The peaks are labelled in accordance with the transitions shown in Fig. 1. c) Same as (b) except that the sample is simultaneously illuminated with 100 mW/cm^2 of above-bandgap light.

number of transitions returned to the standard four allowed transitions (Fig. 2c). Consistent with this model, intensities of the peaks attributed to forbidden transitions were observed to decrease monotonically as the charge concentration in the sample was increased (by intensifying the above-bandgap light). Further, in an identical structure, grown for comparison purposes on an undoped buffer (and which, therefore, lacked fringing electric fields), only the four allowed transitions were observed and there was no change in the PLE spectrum when the sample was simultaneously excited with a HeNe laser. These observations were replicated in another sample, grown in a pin configuration, to which an external electric field was applied.[10]

Using the measured energies of the allowed and forbidden transitions, we were able to directly determine the level splittings caused by the mixing of wavefunctions between the coupled wells. We found that the ground state levels of the electron, heavy-hole (hh) and light-hole (lh) in our structure were split 22.1, 3.9 and 15.5 meV respectively. Further, with these values and the modified hh exciton energy (**a** in Fig. 1) and the modified hh-lh splitting, (**b** in Fig. 1), (both of which were also determined from the experimental spectra), we were able to derive the hh exciton energy $\mathbf{E}^h$ (1582.2 meV) and hh-lh exciton splitting, Δ, (18.8 meV) in the wells in the absence of coupling by employing simple arithmetic manipulation. Agreement with values measured in the SQW sample (1584 and 19 meV, respectively) was very good, especially since it was assumed that the splittings due to coupling were isotropic and that the small field-induced addition to the splitting could be ignored. These are reasonable assumptions for small electric fields.[5]

Conclusions

In conclusion, in the presence of a small perturbation (i.e., fringing electric fields from a doped GaAs buffer), we were able to observe optical transitions between split levels in a coupled double well system which are normally forbidden due to symmetry considerations. By measuring the experimental energies of all the possible transitions, we were able to uniquely determine level splittings due to well coupling without recourse to a theoretical model.

Acknowledgements

We would like to acknowledge stimulating discussions with Drs. Paul Melman and C. Jagannath and the expert technical assistance of Doug Owens and Joseph Powers.

References

1. R. Dingle, A.C. Gossard, and W. Wiegmann: *Phys. Rev. Lett.* **34**, 1327 (1975)
2. C. Delalande, U.O. Ziemelis, G. Bastard, M. Voos, A.C. Gossard, and W. Wiegmann: *Surf. Sci.* **142**, 498 (1984)
3. H. Kawai, J. Kaneko, and N. Watanabe: *J. Appl. Phys.* **58**, 1263 (1985)
4. A. Yariv, C. Lindsey, and U. Sivan: *J. Appl. Phys.* **58**, 3669 (1985)
5. E.J. Austin and M. Jaros: *J. Phys. C.* **19**, 533 (1986)
6. H.Q. Le, J.J. Zayhowski, W.D. Goodhue, and J. Bales: to be published in the **Proceedings of the 1986 Symposium on GaAs and Related Compounds.**
7. Y.R. Lee, A.K. Ramdas, F.A. Chambers, J.M. Meese, and L.R. Ram Mohan: *Appl. Phys. Lett.* **50**, 600 (1987)
8. D.A.B. Miller: to be published in the **Proceedings of the SPIE Conference on Quantum Well and Superlattice Physics, 1987**
9. A. Torabi, K.F. Brennan, and C.J. Summers: to be published in the **Proceedings of the SPIE Conference on Quantum Well and Superlattice Physics, 1987**
10. Y.J. Chen, Emil S. Koteles, B.S. Elman, and C. Armiento: to be published

Excitons and Biexcitons in Semiconductor Quantum Wires

L. Banyai, I. Galbraith, C. Ell, and H. Haug

University of Frankfurt, Robert-Mayer-Str. 8–10,
D-6000 Frankfurt a.M. 1, Fed. Rep. of Germany

Following the success of semiconductor quantum well structures in permitting the study of quasi-two-dimensional phenomena [1] there is a growing experimental [2] and theoretical [3,4] interest in similar quasi-one-dimensional structures. The Coulomb potential in one dimension has some pathological features [5] (e.g. the ground state energy becomes infinite) so it is interesting to see how this affects the spectra of states in very slender semiconductor wires - so-called 'quantum-well wires' (QWWs).

In this paper we study the exciton and biexciton ground state binding energies in such QWWs. We shall consider here the ideal quantum confinement of electrons and holes within an infinite cylindrical potential well of radius R.

If the cladding medium outside the wire has a different dielectric constant ϵ_2 than that within the wire ϵ_1, the Coulomb interaction between the electrons and holes will be distorted. We have computed this dielectric polarisation effect via the associated variation in the electrostatic energy. By restricting the motion in the transverse plane via the quantum confinement, the motion along the z axis will be governed by an effective interaction energy[6]

$$U(z_e-z_h)=e\int_0^R \rho_e d\rho_e \int_0^R \rho_h d\rho_h \int_0^{2\pi} d\theta_e \int_0^{2\pi} d\theta_h V\left(\vec{r}_e,\vec{r}_h\right) |\phi_0(\rho_e)|^2|\phi_0(\rho_h)|^2.$$

The interaction potential is

$$V(\vec{r},\vec{r}')= -\frac{e}{\epsilon_1}\frac{1}{|\vec{r}-\vec{r}'|}$$

$$-\frac{2e}{\pi}\left(\frac{1}{\epsilon_2}-\frac{1}{\epsilon_1}\right)\sum_{m=-\infty}^{\infty}\int_0^{\infty} dk\cos(k(z-z'))e^{im(\theta-\theta')}C_m(kR)I_m(k\rho)I_m(k\rho'),$$

where

$$C_m(kR) = \frac{K_m(kR)\, K'_m(kR)}{I_m(kR)K'_m(kR) - \frac{\epsilon_1}{\epsilon_2} I'_m(kR)K_m(kR)} \; .$$

Here I_m, K_m are the modified Bessel functions and $\phi_0(\rho)$ is the ground state wave function for the transverse motion in the cylindrical well.

We have solved numerically the corresponding Schrödinger equation for the one dimensional electron hole relative motion. The binding energies for $\epsilon_1=\epsilon_2$ and $\epsilon_1=1.3\epsilon_2$ are shown in Fig. 1 (with R scaled to the bulk excitonic Bohr radius $a_0=\hbar^2\epsilon_1/me^2$ and the binding energy ($-\epsilon$) scaled to the bulk excitonic Rydberg $E_R=\hbar^2/2ma_0^2$). It can be seen that the binding energy blows up as $R\rightarrow 0$, exceeding the limiting two-dimensional ground state binding energy of $4E_R$ for $R < a_0/2$ which corresponds to the two-dimensional exciton Bohr radius.

In quantum wires it can be seen that the effect of the dielectric polarisation on the binding energy is small in comparison to the corresponding two-dimensional case, where the ratio of the binding energies should be $(\epsilon_1/\epsilon_2)^2$, a factor of 1.69 compared with our enhancement of $\simeq 1.2$. This is attributable to the differing behaviour in the limit of small thicknesses. For two-dimensional systems the wave function width tends to a constant $a_0/2$ while in the QWW case the wave function width collapses to zero [5]. Since the materials currently being proposed for the manufacture of QWWs have dielectric constant ratios less than the 1.3 case considered here we shall concentrate now exclusively on the case $\epsilon_1=\epsilon_2$.

For very small radii one must return to a bulk problem in the cladding material, a limit not present within our model as we assumed an infinite cylindrical potential for the wire.

Consideration of the biexciton problem in QWWs leads one to a one-dimensional effective Hamiltonian with pair-wise interaction ± U. We wish

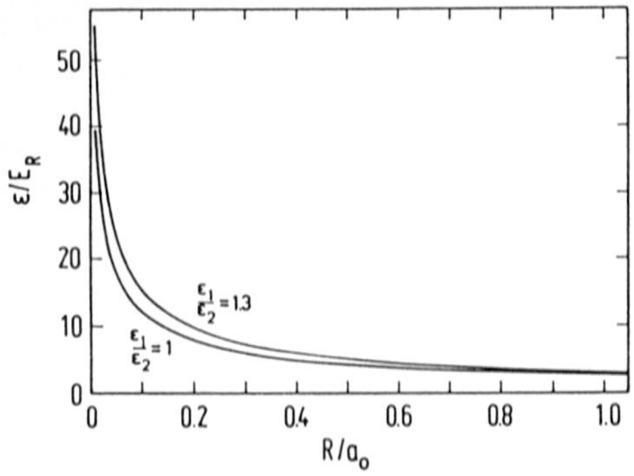

Fig. 1 Excitonic binding energy as a function of wire radius for two different dielectric constant ratios.

to solve for the ground state biexcitionic binding energy, and we look for a solution of the form [7] (in the coordinates $y_1=z_{e_1}-z_{h_1}$, $y_2=z_{e_2}-z_{h_2}$, $x=z_{h_1}-z_{h_2}$).

$$\Psi(y_1,y_2,x)=\psi_b(|x|)\Phi(y_1,y_2,x)$$

with

$$\Phi(y_1,y_2,x)=\frac{1}{S(x)}\phi(y_1,y_2,x)$$

and the normalisation function

$$S^2(x)=\int_{-\infty}^{\infty}\int_{-\infty}^{\infty} dy_1 dy_2\ \phi^2(y_1,y_2,x).$$

For ϕ we use the Heitler-London [8] approximation, i.e.

$$\phi(y_1,y_2,x)=\psi_x(y_1)\psi_x(y_2) + \psi_x(y_1+x)\psi_x(y_2-x).$$

One may then write [7] a Schrödinger equation for the relative hole-hole wave function, ψ_b,with the effective hole-hole potential

In this paper we study the exciton and biexciton ground state binding energies in such QWWs. We shall consider here the ideal quantum confinement of electrons and holes within an infinite cylindrical potential well of radius R.

If the cladding medium outside the wire has a different dielectric constant ϵ_2 than that within the wire ϵ_1, the Coulomb interaction between the electrons and holes will be distorted. We have computed this dielectric polarisation effect via the associated variation in the electrostatic energy. By restricting the motion in the transverse plane via the quantum confinement, the motion along the z axis will be governed by an effective interaction energy[6]

$$U(z_e-z_h)=e\int_0^R \rho_e d\rho_e \int_0^R \rho_h d\rho_h \int_0^{2\pi} d\theta_e \int_0^{2\pi} d\theta_h V\left(\vec{r}_e,\vec{r}_h\right)\ |\phi_0(\rho_e)|^2|\phi_0(\rho_h)|^2.$$

The interaction potential is

$$V(\vec{r},\vec{r}\,')= -\frac{e}{\epsilon_1}\frac{1}{|\vec{r}-\vec{r}\,'|}$$

$$v(x) = -2\epsilon + \int_{-\infty}^{\infty}\int_{-\infty}^{\infty} \Phi(y_1,y_2,x)\, H\, \Phi(y_1,y_2,x)\, dy_1 dy_2.$$

In Fig. 2 we show this effective hole-hole potential with an electron-hole mass ratio of 0.1 for various wire radii. We see that the potential becomes deeper and narrower for smaller radii until a maximum depth is reached at $R \simeq 0.083$ after which the depth decreases.

The corresponding Schrödinger equation for these potentials was solved numerically and the molecular binding energy is given as a function of wire radius in Fig. 3. One may see that the binding energy has a maximum at $R=0.025a_0$ and vanishes for $R=0$ as the potential is zero everywhere except $x=0$. Notwithstanding this down turn for small radii, the actual value of the exciton binding energies are relatively large - at $R < 0.5a_0$ they exceed $0.5E_R$ and are more than a factor 5 greater than the bulk molecular binding energy [7] of $\simeq 0.12E_R$.

An interesting connection to the bulk is the electron-hole mass ratio dependence of the molecular binding energy when scaled to the corresponding exciton binding energy. The calculated dependence is qualitatively the same as for the bulk [7,8], indeed for $R=0.53$ it coincides almost exactly with the bulk values.

Our results for the excitonic and biexcitonic spectra in QWWs show that for small wire radii one may obtain a strong enhancement of the binding energies. The optical properties of QWWs will therefore exhibit large exci-

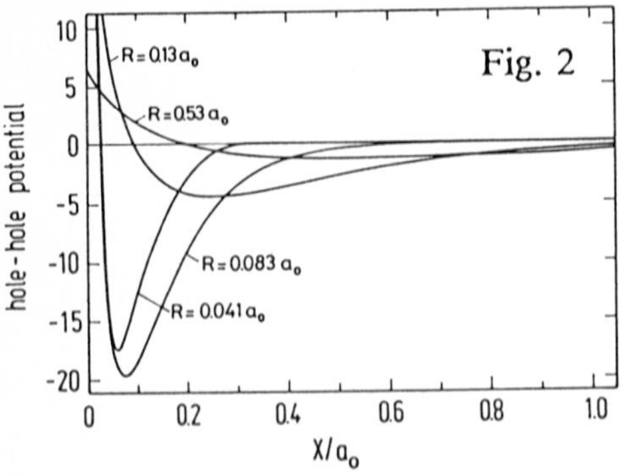

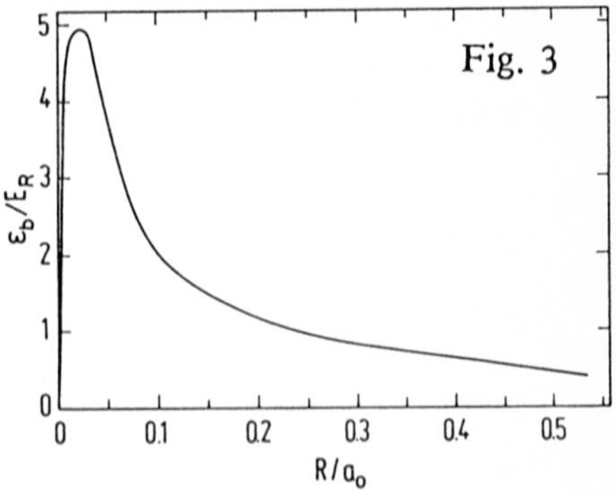

Fig. 2 Hole-hole potentials for various wire radii for $\frac{m_e}{m_h}=0.1$.

Fig. 3 Biexcitonic binding energy as a function of wire radius for $\frac{m_e}{m_h}=0.1$.

tonic and biexcitonic nonlinearities even at room temperature. The question of whether or not condensation into plasma clusters will occur remains an open one.

I.G. acknowledges receipt of the European Community grant STI-0168-D(CD). This work was supported by the Deutsche Forschungsgemeinschaft through the Sonderforschungsbereich 185 Frankfurt-Darmstadt

1. R. Dingle, Festkörperprobleme (Advances in Solid State Physics), edited by H.J. Queisser (Pergamon, Oxford, 1975), **15**, p21.
2. P.M.Petroff, A.C. Gossard, R.A. Logan and W. Wiegmann, Appl. Phys. Lett., **41**, 635 (1982).
A.P. Fowler, A. Hartstein and R.A. Webbl, Phys. Rev. Lett. **48**, 196 (1982).
Y.Arakwa, K. Vahala, A. Yariv and K. Lau, Appl. Phys. Lett., **47**, 1142 (1985).
Yia-Chung Chang, L.L. Chang and L. Esaki, Appl. Phys. Lett., **47**, 1324 (1985).
J. Cibert, P.M. Petroff, G.J. Dolan, S.J. Pearton, A.C. Gossard and J.H. English, Appl. Phys. Lett. **49**, 1275 (1986).
3. H. Sakaki, J. Journ. Appl. Phys.,**19**, L735 (1980)
H. Adamska, H.N. Spector, J. Appl. Phys., **59**, 619 (1986).
4. G.W. Bryant, Phys. Rev., **B29**, 6632 (1984).
5. R. Loudon, Am. J. Phys., **44**, 1064 (1976).
6. L. Banyai,I. Galbraith, C. Ell and H Haug, to appear Phys. Rev. **B** October (1987).
7. W.F. Brinkman, T.M. Rice, B. Bell, Phys. Rev., **B8**, 1570 (1973).
8. E. Hanamura, H. Haug, Physics Rep., **33**, 209 (1977).

Index of Contributors